Introduction to Multivariate Analysis for the Social Sciences

9.50

Introduction to Multivariate Analysis for the Social Sciences

John P. Van de Geer
UNIVERSITY OF LEIDEN, THE NETHERLANDS

W. H. Freeman and Company
SAN FRANCISCO

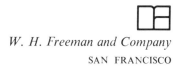

Printed in the United States of America
Library of Congress Catalog Card Number 71-156044
International Standard Book Number 0-7167-0932-5

1 2 3 4 5 6 7 8 9

This text was written while I was a Fellow of the Center for Advanced Study in the Behavioral Sciences, Stanford

Contents

Preface

This text has been developed from courses in multivariate analysis for postgraduate students at the University of Leiden, The Netherlands. My book in Dutch on the same subject appeared in 1967, but it covers much less territory. This book is written for postgraduate students in behavioral and social sciences, but it may also be of interest to researchers who want a fairly systematic, but not overly difficult, introduction to the subject matter.

With this audience in mind, I have had to make several compromises. First, many advanced textbooks on multivariate analysis are unrewarding reading material for the average student or researcher, because he lacks the mathematical background needed to enjoy such texts fully. In my opinion, statistical theory is a substantially more advanced subject than is required to understand what multivariate techniques really do with data. I have therefore decided not to treat statistical matters at all (distribution theory, significance tests, etc.), yet I am somewhat doubtful about this decision, since statistical considerations are becoming increasingly crucial in judging the merits of advanced techniques and all their varieties.

Second, I have not attempted to achieve a high level of mathematical precision. Again, the reason is that mathematical precision requires much more detail than is relevant for practical applications, and would ask for much more background and motivation than the average researcher can be expected to have. Moreover, I am a psychologist by training. (Nevertheless, the first reason is not merely a rationalization for the second.)

Happily, there are also some positive reasons for the approach I have chosen. The first one is that fast computers make advanced techniques available to everybody. In days past, when desk calculators were the last word, application of multivariate techniques required much technical skill: one had to know all the tricks of the trade in order to survive and not be trapped in a hopelessly time-consuming affair. Fast computers are a blessing, of course, but they also remove the natural barriers against a mindless use of canned programs. Now that any type of analysis can simply be ordered, the new generation of consumers of multivariate techniques should concentrate on basic understanding. The time spent on acquiring such understanding is more than repaid by the time gained in using the techniques, since one no longer need be concerned with computational niceties.

A second major reason for my approach is that I emphasize the basic similarities between multivariate techniques. Perhaps this emphasis is due to one of my own personality traits. Personality research has discovered a dimension that differentiates between individuals who emphasize perceived differences, on the one hand, and persons who emphasize perceived similarities, on the other. For instance, if persons are presented with a large number of photographs of facial expressions and are requested to classify them into as many categories as they wish, the "narrow categorizer" will create many groups witth only a few pictures in each, whereas the "broad categorizer" forms only a few categories with many pictures in each. I am apparently a broad categorizer. Faced with the large variety of multivariate techniques currently available, I do not feel that they fall apart into a great many special categories; rather, I perceive them as variations on, at most, three themes. Many researchers do not seem to agree with such a view. Perhaps they are narrow categorizers.

In my opinion, however, all this is not just a matter of personality traits. It is unfortunate, but true, that formal methods tend to become identified with specific applications of them, and that problems in different substantive areas (psychology, sociology, economics) tend to be perceived as so unique and so special that the tools used to deal with them must also be unique and special. These tendencies surely do not create the proper climate for judging methods on their intrinsic merits. What is worse, they lead to a distinct lack of cross-fertilization. Researchers in one field are very often completely unaware of what is going on in a neighboring field, even if what is going on there is highly relevant to their own progress. What we are facing in such a situation is not a question of emphasizing perceived differences, but of not being able to perceive *real* similarities, which is a defense mechanism rather than a personality trait.

But let us come to business. Multivariate techniques are based on matrix algebra; so the first part of the book (chapters 1 to 8) centers on matrix algebra. The exposition is restricted to topics that are immediately relevant to the second part of the book. Given the special nature of the introductory chapters, I have included exercises for them. The second part of the book (chapters 9 to 18) develops multivariate analysis as a straightforward application of the matrix-algebra machinery. The tenor of part II is that, once we have formulated and specified the theoretical model that can be assumed to hold for the data, all that remains is to apply standard tricks.

My coverage is not complete. In particular, it does not extend to analysis of variance, and it is restricted to metric analysis, for two reasons. First, many excellent textbooks on the design of experiments and analysis of variance are available, and I have nothing to add to them. Second, I believe that, for the education of the modern social scientist, an understanding of metric analysis has a clear priority.

Acknowledgements are due to John Wiley & Sons, New York, for permission to reproduce table 12.1, and to The University of Chicago Press, publishers of *The American Journal of Sociology*, for permission to reproduce table 14.3. I feel especially indebted to the Center for Advanced Study in the Behavioral Sciences, Stanford, where I received both hospitality and generous support during the preparation of the first draft of this book.

September 1970 *John P. Van de Geer*

Introduction to Multivariate Analysis for the Social Sciences

PART I

Introduction to Matrix Algebra

1

An Overview of Matrix Algebra

Matrix algebra is, of course, a branch of mathematics, but in this book we will look on it simply as a tool that enables us to work with matrices. Matrices, for our purposes, are nothing but systematic collections of observed data. Generally, in multivariate analysis, we have observations on a certain number of variables. Let us consider two examples.

The first example is from sociology, where one might be interested in factors related to absenteeism in private, middle-large industries. For such a study, data will be collected on such variables as: average number of days a worker is absent because of illness during a given period, ratio of skilled to unskilled laborers, union membership, turnover, job satisfaction, and personnel policy. For each industrial firm included in the study, each variable is "measured"; i.e., we obtain some number that characterizes each firm in terms of each variable. A matrix, then, is a systematic arrangement of these numbers in rows and columns. In general, we shall adopt the convention that the columns refer to the variables, and the rows to the individual observation units, the firms, in this example.

For a second example, the variables could be the results of tests taken by two groups of psychiatric patients. In the matrix, columns will refer to the tests, rows to the patients. This matrix can be logically divided into two parts: in one part the rows correspond to the patients in the first group, in the second to the patients in the other group.

Our first concern is to have a language that makes it possible to express operations on matrices in a simple notation. This language is matrix algebra. The operations we need are often not very complicated by themselves, but if a matrix is large, in that it has many rows and columns, operations on it become extensive. They then require careful bookkeeping. Very often matrix algebra is nothing but a kind of shorthand to describe these extensive procedures on many sets of numbers simultaneously.

Part I of this book presents enough matrix algebra to facilitate the description of the operations we need in the second part of the book, where matrix algebra is applied to problems of multivariate analysis. All we need is a few basic tricks, at least at first. Let us summarize them, so that you will have some idea of what is coming and, more importantly, of why these topics must be mastered.

1. Our point of departure is always a multivariate *data matrix* with a certain number, n, of rows for the individual observation units, and a certain number, m, of columns for the variables.

2. In most applications of multivariate analysis, we shall not be interested in variable means. They have their interest, of course, in each study, but multivariate analysis instead focuses on variances and covariances. Therefore, the data matrix will in general be transformed into a matrix where columns have zero means and where the numbers in the column represent *deviations from the mean*.

3. Such a matrix is the basis for the *variance-covariance matrix* with m rows and m columns. For a variable i, the variance is defined as $\sum x_i^2/n$, whereas for two variables i and j the covariance is defined as $\sum x_i x_j/n$, x_i and x_j being taken as deviations from the mean. Variances and covariances can be collected in the variance-covariance matrix, in that the number in row i, column i (on the diagonal), gives the variance of variable i, while the number in row i, column j $(i \neq j)$, gives the covariance between the pair of variables i and j, and is the same number as in row j, column i.

4. An often useful transformation is to *standardize* the data matrix: we first take deviations from the mean for each column, then divide the

deviation from the mean by the standard deviation for the same column. The result is that values in a column will have zero mean and unit variance.

5. The standardized data matrix is then the basis for calculating a *correlation matrix*, which is nothing but a variance-covariance matrix for standardized variables. In the diagonal of this matrix, we therefore find values equal to unity. In the other cells we find correlations: in row i, column j, we shall find the correlation coefficient $r_{ij} = \sum x_i x_j / n\sigma_i\sigma_j$.

6. Very often we shall need a variable that is a *linear compound* of the initial variables. The linear compound is simply a variable whose values are obtained by a weighted addition of values of the original variables. For example, with two initial variables x_1 and x_2, values of the compound are defined as $y = w_1x_1 + w_2x_2$, where w_1 and w_2 are weights. A linear compound could also be called a weighted sum.

7. For some techniques of multivariate analysis, we need to be able to solve simultaneous equations. Doing so usually requires a computational routine called *matrix inversion*.

8. Multivariate analysis nearly always comes down to finding a *minimum* or a *maximum* of some sort. A typical example is to find a linear compound of some variables that has maximum correlation with some other variable (multiple correlation), or to find a linear compound of the observed scores that has maximum variance (factor analysis). Therefore, among our stock of basic tricks, we need to include procedures for finding extreme values of functions.

9. In addition, we shall often need to find maxima (or minima) of functions where the procedure is limited by certain *side-conditions*. For instance, we are given two sets of variables, and are required to find a linear compound from the first set, and another from the second set, such that the value of the correlation between these two compounds is maximum. This task can be reformulated as follows: find the two compounds in such a way that the covariance between them is maximum, given that the compounds both have unit variance.

10. Very often in multivariate analysis, a maximization procedure under certain side-conditions takes on a very specific and recognizable form, namely, finding *eigenvectors* and *eigenvalues* of a given matrix.

2

Basic Concepts of Matrix Algebra

2.1. Introduction

This chapter is a first introduction to matrix algebra. It defines the basic concepts and describes the most elementary operations on matrices.

2.2. Definition of a Matrix

A *matrix* is a collection of numbers ordered in an array of rows and columns. The numbers are called *elements* of the matrix. If the matrix has n rows and m columns, it is called a matrix of *order* $n \times m$. This is different from a matrix of order $m \times n$, which has m rows and n columns.

A matrix is described by giving the elements and adding rounded brackets. For instance, a matrix of order 2×3 is written as

$$\begin{pmatrix} 3 & 8 & 6 \\ 2 & 1 & 5 \end{pmatrix}.$$

As a symbol for a matrix, we shall use boldface capital letters, **A**, **X**, etc. An element of the matrix is then symbolized by the corresponding small letter with subscripts giving the row and column numbers of the element. The element in row i and column j of matrix **A** is given as a_{ij}.

Example: Let **X** be the matrix

$$\mathbf{X} = \begin{pmatrix} 8 & 6 & 5 & 6 \\ 1 & -4 & 3 & 0 \\ 12 & 2 & 7 & -5 \end{pmatrix};$$

then $x_{22} = -4$, $x_{13} = 5$, $x_{32} = 2$.

A matrix of order $n \times 1$ is called a *vector*, specifically, a column vector. A row vector is a matrix of order $1 \times m$. Vectors will be indicated by small boldface letters, with a prime added if it is a row vector. If the vector is taken from a matrix **X**, we shall use $\mathbf{x}_i'$ to indicate a row of this matrix and $\mathbf{x}_j$ for a column, with appropriate subscript. For instance, for the matrix **X** given above, we have

$$\mathbf{x}_2 = \begin{pmatrix} 6 \\ -4 \\ 2 \end{pmatrix}$$

and $\mathbf{x}_3' = (12 \quad 2 \quad 7 \quad -5)$.

2.3. Basic Operations

The basic operations with matrices are addition, subtraction, multiplication, and division. Many of the operations that are done on matrices are essentially analogous to operations on simple numbers. In fact, a simple number is a matrix of order 1×1. However, there *are* operations on matrices that have no equivalents in simple arithmetic.

2.3.1. Addition. If **A** and **B** are matrices of the same order, they can be added to form a sum matrix $\mathbf{C} = \mathbf{A} + \mathbf{B}$, for which $c_{ij} = a_{ij} + b_{ij}$. In simple words, we add corresponding elements of **A** and **B** to obtain the elements of **C**:

$$\begin{pmatrix} 2 & 4 \\ 1 & -2 \\ 6 & 3 \end{pmatrix} + \begin{pmatrix} 3 & -3 \\ 3 & -6 \\ 15 & 9 \end{pmatrix} = \begin{pmatrix} 5 & 1 \\ 4 & -8 \\ 21 & 12 \end{pmatrix}.$$

Subtraction is defined in a similar way. If $\mathbf{C} = \mathbf{A} - \mathbf{B}$, then $c_{ij} = a_{ij} - b_{ij}$.

2.3.2. Scalar Multiplication. Let **A** be a matrix and p a single number. The operation $p \cdot \mathbf{A}$ then means that each element of **A** is multiplied by p. This is called scalar multiplication, and p is the *scalar* (or scalar factor):

$$3\begin{pmatrix} 2 & 4 \\ -1 & 5 \\ 0 & 6 \end{pmatrix} = \begin{pmatrix} 6 & 12 \\ -3 & 15 \\ 0 & 18 \end{pmatrix}.$$

2.3.3. Transposition. If **A** is a matrix of order $n \times m$, then $\mathbf{A}'$ is called the *transpose* of **A** if $a_{ji}' = a_{ij}$. In other words, $\mathbf{A}'$ is the $m \times n$ matrix that is obtained if we interchange rows with columns. Rows of **A** become columns of $\mathbf{A}'$, and columns of **A** become rows of $\mathbf{A}'$:

$$\mathbf{A} = \begin{pmatrix} 1 & 8 \\ 3 & 5 \\ 2 & 7 \end{pmatrix}; \qquad \mathbf{A}' = \begin{pmatrix} 1 & 3 & 2 \\ 8 & 5 & 7 \end{pmatrix}.$$

2.3.4. Matrix Multiplication. Let **A** be a matrix of order $n \times m$, and **B** a matrix of order $m \times k$. We shall then define a matrix **C**, called the product of **A** and **B** ($\mathbf{C} = \mathbf{A} \cdot \mathbf{B}$, or $\mathbf{C} = \mathbf{AB}$), where **C** is of order $n \times k$, and where

$$c_{ij} = a_{i1} \cdot b_{1j} + a_{i2} \cdot b_{2j} + a_{i3} \cdot b_{3j} + \cdots + a_{im} \cdot b_{mj} = \sum_{g=1}^{m} a_{ig} \cdot b_{gj}.$$

In words: if we take row $\mathbf{a}_i'$ of matrix **A** and column $\mathbf{b}_j$ of matrix **B**, the element c_{ij} is obtained by multiplying the m elements of row $\mathbf{a}_i'$ in succession by the corresponding elements of column $\mathbf{b}_j$, and adding these individual products. It follows that **A** must have as many columns as **B** has rows, otherwise multiplication remains undefined and the product matrix $\mathbf{A} \cdot \mathbf{B}$ has no meaning.

Example:

$$\begin{pmatrix} 8 & 1 & 3 \\ 2 & -6 & 4 \end{pmatrix} \cdot \begin{pmatrix} 5 & 2 & 9 & 1 \\ 1 & 7 & 4 & 3 \\ 8 & 3 & 2 & 6 \end{pmatrix} = \begin{pmatrix} 65 & 32 & 82 & 29 \\ 36 & -26 & 2 & 8 \end{pmatrix}.$$

It follows, also, that if **A** is of order $n \times m$, and **B** of order $m \times n$, both **A** · **B** and **B** · **A** have meaning. However, **A** · **B** will be different from **B** · **A** (except in very special cases); i.e., matrix multiplication is non-commutative. We therefore distinguish between *premultiplication* of **A** by **B**, resulting in **BA**, and *postmultiplication* of **A** by **B**, which produces **AB**.

Example:

$$\mathbf{A} = \begin{pmatrix} 2 & 5 \\ 0 & 7 \\ 4 & 3 \end{pmatrix}; \qquad \mathbf{B} = \begin{pmatrix} 4 & 2 & 1 \\ 6 & 3 & 2 \end{pmatrix};$$

$$\mathbf{AB} = \begin{pmatrix} 38 & 19 & 12 \\ 42 & 21 & 14 \\ 34 & 17 & 10 \end{pmatrix}; \qquad \mathbf{BA} = \begin{pmatrix} 12 & 37 \\ 20 & 57 \end{pmatrix}.$$

Note that the sum of the elements in the diagonal of **AB** (the main diagonal, which runs from upper left to lower right) is equal to the sum of the elements in the diagonal of **BA**. This is always true; the proof is left to the reader as an exercise. Note also that a matrix can always be multiplied by its transpose, since **A**′ has as many rows as **A** has columns, and **A** as many rows as **A**′ has columns. However, **A**′**A** is different from **AA**′; e.g., if **A** is of order $n \times m$, then **A**′**A** is of order $m \times m$ and **AA**′ of order $n \times n$.

2.3.5. Matrix Division. For simple numbers, division can be reduced to multiplication by the reciprocal of the divider: 32 divided by 4 is the same as 32 multiplied by 1/4, or 32 multiplied by 4^{-1}, where 4^{-1} is defined by the general equality $a^{-1} \cdot a = 1$.

In working with matrices, we shall adopt the latter idea, and therefore not use the term division at all; instead we take multiplication by an inverse matrix as the equivalent of division. However, I mention this here only in passing, since the complete treatment is rather involved and has to be deferred to a later section (4.5).

2.3.6. Partitioning of Matrices. A matrix can be divided into submatrices, for example,

$$\left(\begin{array}{cc:cc} 3 & 6 & 1 & 8 \\ \hdashline 2 & 4 & 6 & 3 \\ 8 & 5 & 1 & 2 \end{array}\right).$$

This is called *partitioning* of the matrix, and the subdivision are *partitions*. The two obvious possibilities are partitioning by rows, and partitioning by columns. The example shows a combination of the two.

Partitioning can be combined with multiplicative operations. Suppose **X** is of order $n \times m$, and **Y** of order $m \times k$. **X** is partitioned by rows into $\mathbf{X}_1$ and $\mathbf{X}_2$, and **Y** is partitioned by columns into $\mathbf{Y}_1$, $\mathbf{Y}_2$, and $\mathbf{Y}_3$. Then the product matrix **XY** can be partitioned in a corresponding way:

$$\mathbf{XY} = \begin{pmatrix} \mathbf{X}_1 \\ \hline \mathbf{X}_2 \end{pmatrix} (\mathbf{Y}_1 \mid \mathbf{Y}_2 \mid \mathbf{Y}_3) = \left(\begin{array}{c|c|c} \mathbf{X}_1\mathbf{Y}_1 & \mathbf{X}_1\mathbf{Y}_2 & \mathbf{X}_1\mathbf{Y}_3 \\ \hline \mathbf{X}_2\mathbf{Y}_1 & \mathbf{X}_2\mathbf{Y}_2 & \mathbf{X}_2\mathbf{Y}_3 \end{array}\right).$$

The above statement follows from the definition of matrix multiplication. The obvious generalization is that if we partition **X** into p submatrices by rows, and **Y** into q submatrices by columns, then the product **XY** is partitioned into $p \times q$ submatrices.

It is interesting to look at the alternative, where **X** is partitioned by columns and **Y** by rows. We then have, for example,

$$\mathbf{XY} = (\mathbf{X}_1 \mid \mathbf{X}_2) \begin{pmatrix} \mathbf{Y}_1 \\ \hline \mathbf{Y}_2 \end{pmatrix} = \mathbf{X}_1\mathbf{Y}_1 + \mathbf{X}_2\mathbf{Y}_2.$$

A condition here is that the partitioning of **X** conform to that of **Y**; i.e., if $\mathbf{X}_1$ has m_1 columns, then $\mathbf{Y}_1$ must have m_1 rows, otherwise the product $\mathbf{X}_1\mathbf{Y}_1$ would be meaningless, and similarly for $\mathbf{X}_2$ and $\mathbf{Y}_2$.

A combination of the principles in the last two paragraphs shows that if

$$\mathbf{X} = \begin{pmatrix} \mathbf{X}_{11} & \mathbf{X}_{12} \\ \mathbf{X}_{21} & \mathbf{X}_{22} \end{pmatrix}$$

and

$$\mathbf{Y} = \begin{pmatrix} \mathbf{Y}_{11} & \mathbf{Y}_{12} & \mathbf{Y}_{13} \\ \mathbf{Y}_{21} & \mathbf{Y}_{22} & \mathbf{Y}_{23} \end{pmatrix},$$

then

$$\mathbf{XY} = \left(\begin{array}{c|c|c} \mathbf{X}_{11}\mathbf{Y}_{11} + \mathbf{X}_{12}\mathbf{Y}_{21} & \mathbf{X}_{11}\mathbf{Y}_{12} + \mathbf{X}_{12}\mathbf{Y}_{22} & \mathbf{X}_{11}\mathbf{Y}_{13} + \mathbf{X}_{12}\mathbf{Y}_{23} \\ \hline \mathbf{X}_{21}\mathbf{Y}_{11} + \mathbf{X}_{22}\mathbf{Y}_{21} & \mathbf{X}_{21}\mathbf{Y}_{12} + \mathbf{X}_{22}\mathbf{Y}_{22} & \mathbf{X}_{21}\mathbf{Y}_{13} + \mathbf{X}_{22}\mathbf{Y}_{23} \end{array}\right),$$

where it is assumed that the partition by columns of **X** conforms to the partition by rows of **Y**. A generalization is straightforward.

2.4. Special Names

A matrix of order $n \times n$ is called a *square* matrix. The elements on the diagonal from upper left to lower right are called the *diagonal* elements of the square matrix. The sum of these elements is the *trace* or *spur* of the matrix.

A *symmetric* matrix is a square matrix $\mathbf{A}$ for which $\mathbf{A} = \mathbf{A}'$: the matrix is equal to its transpose, which it can be only if its elements are mirrored along the diagonal. A *diagonal* matrix is a square matrix whose elements are zero everywhere except along the diagonal; it is a special case of a symmetric matrix. An even more special case is the *unit* matrix, which is a diagonal matrix with unity elements on the diagonal. Such a matrix will be symbolized $\mathbf{I}$. If the elements on the diagonal of a diagonal matrix are all equal to k, say, then the matrix can be written as $k \cdot \mathbf{I}$ (k being a scalar). A matrix is *skew symmetric* if it is square and if $a_{ij} = -a_{ji}$. A *triangular* matrix has all elements on one side of its diagonal equal to zero.

2.5. Exercises

1. Prove the following theorem. A matrix does not change if it is multiplied by the unit matrix; i.e., $\mathbf{A} \cdot \mathbf{I} = \mathbf{I} \cdot \mathbf{A} = \mathbf{A}$.
2. Prove the theorem $(\mathbf{AB})' = \mathbf{B}'\mathbf{A}'$; that is, the transpose of a product is equal to the product of the transposes in reversed order.
3. Generalize the theorem in exercise 2 to $(\mathbf{ABC})' = \mathbf{C}'\mathbf{B}'\mathbf{A}'$.
4. Prove for any matrix $\mathbf{X}$ that the product matrices $\mathbf{XX}'$ and $\mathbf{X}'\mathbf{X}$ are symmetric.
5. Multiply

$$\begin{pmatrix} 2 & 1 & 4 \\ 3 & 2 & 5 \\ 4 & 3 & 6 \\ 8 & 1 & 9 \end{pmatrix} \cdot \begin{pmatrix} 2 & 0 & 0 \\ 0 & 3 & 0 \\ 0 & 0 & 1 \end{pmatrix},$$

and prove the corresponding theorem: For $\mathbf{A} \cdot \mathbf{D}$, where $\mathbf{D}$ is a diagonal matrix, the ith column of the product matrix is equal to $d_{ii} \cdot \mathbf{a}_i$.

6. Suppose we want to multiply the first row of a matrix **A** by a scalar d_1, the second row by a scalar d_2, etc. How can this be written in matrix notation?

7. Suppose we want to multiply the elements in the second column of a matrix **A** by a factor *d*, leaving all other elements unchanged. Write this in matrix notation.

8. Suppose we want to interchange the first column of a matrix **A** with the second. Can this be written in matrix notation?

9. Suppose **D** is a diagonal matrix. Proof that $\mathbf{D}^2 = \mathbf{D} \cdot \mathbf{D}$ is also a diagonal matrix with elements d_{kk}^2.

10. Let **D** be a diagonal matrix. Find a matrix **E** so that $\mathbf{D} \cdot \mathbf{E} = I$.

11. The following exercise (from Pike and Lowe 1968) shows how matrix operations can handle a situation that at first sight seems rather baffling. Imagine three persons, *A*, *B*, and *C*. The exercise focuses on the following sentence. "*A* says to *C*: I said to him: you said to me: he said to you: I saw him." Question: who saw whom?

For a treatment in terms of matrices, make a distinction between:

the *cast* vector, *A*, *B*, and *C*;
the *case* vector, *S* (subject), *R* (recipient), and *X* (third party);
the *person* vector, *I* (I or me), *U* (you), and *H* (he or him).

The sentence of interest can be divided into five phrases (separated by colons). The first phrase "*A* says to *C*" identifies cast in terms of case: *A* is subject, *C* is recipient, and *B* is third party. In matrix notation,

$$\mathbf{X}_1 = \begin{pmatrix} 1 & 0 & 0 \\ 0 & 0 & 1 \\ 0 & 1 & 0 \end{pmatrix},$$

where rows refer to the cast, and columns to the case.

Any subsequent phrase (take the fourth, "he said to you") can be translated into a matrix with persons for rows, and case for columns. The fourth phrase becomes:

$$\mathbf{X}_4 = \begin{pmatrix} 0 & 0 & 1 \\ 0 & 1 & 0 \\ 1 & 0 & 0 \end{pmatrix}.$$

To identify cast and persons in the final phrase, we just multiply all matrices: $\mathbf{X}_1 \cdot \mathbf{X}_2 \cdot \mathbf{X}_3 \cdot \mathbf{X}_4 \cdot \mathbf{X}_5$. The product matrix has cast for rows, case for columns. In the example, the result is: *C* saw *B*. Prove the validity of this approach.

12. In Potawatomi language (Pike and Erickson 1964) the subject and object of a transitive verb are revealed by pronomial prefixes. For instance, where in our language we use the pronouns I and you to say that "I see you," in Potawatomi this is expressed by a prefix *k* to the verb for "see." In contrast, the prefix *n* means that "He sees me," and a prefix *w* refers to a "fourth party seeing him." Analysis of the language shows that the prefixes can indicate (for subject or object) eight different grammatical persons: I, you, he, fourth person, I and you, we, you (plural), they. The prefixes appear to relate subject and object as in table 2.1. This matrix can be made to look much simpler by appropriate

Table 2.1. The Potawatomi Pronomial Prefixes.

	1	2	3	4	12	1*p*	2*p*	3*p*
1		*k*	*n*	*n*			*k*	*n*
2	*k*		*k*	*k*		*k*		*k*
3	*n*	*k*		*w*	*k*	*n*	*k*	
4	*n*	*k*	*w*		*k*	*n*	*k*	*w*
12			*k*	*k*				*k*
1*p*		*k*	*n*	*n*			*k*	*n*
2*p*	*k*		*k*	*k*		*k*		*k*
3*p*	*n*	*k*		*w*	*k*	*n*	*k*	

interchanges of rows and columns. Try to find suitable interchanges, and express these operations in matrix notation.

13. In a sociometric experiment, members of a group are asked which other members they like. Suppose the data are collected in a choice diagram as given in figure 2.1, where an arrow going from i to j means that i likes j.
    (a) Convert the diagram to a matrix **C**, where $c_{ij} = 1$ if i likes j, and 0 otherwise (diagonal elements of **C** are zero).
    (b) Let **u** be an $n \times 1$ vector with unit elements. Note that **u**′**C** gives scores for "popularity."
    (c) Verify that **Cu** gives scores for "generosity."

    Suppose **B** is a similar matrix of choices, where $b_{ij} = 1$ if i likes j, and $b_{ij} = -1$ if i dislikes j. Imagine a similar matrix **P**, where $p_{ij} = 1$ if j believes that i likes him and -1 if j believes that i dislikes him. Diagonal elements of **B** and **P** are equal to zero.

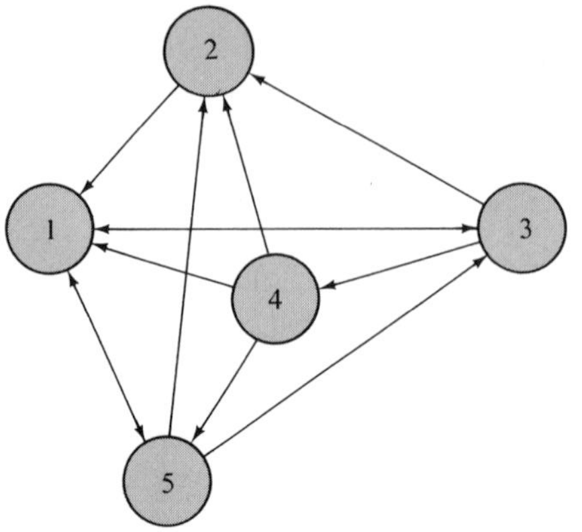

Figure 2.1
Sociometric choice diagram.

(d) If $\mathbf{N} = \mathbf{P}'\mathbf{B}$, check whether we can use n_{kh} as a measure for "identification" (k believes he is liked by the people who, in fact, like h, and believes he is disliked by those who, in fact, dislike h).
(e) Check whether diagonal elements of $\mathbf{N}$ can be interpreted as a measure for "realism" (k is realistic if his beliefs agree with reality).
(f) Check whether diagonal elements of $\mathbf{BP}'$ can be interpreted as a measure for "overtness" (k is overt if beliefs about k conform to k's likes and dislikes).
(g) Compute the matrix $\mathbf{C}^2 = \mathbf{C} \cdot \mathbf{C}$ and interpret the elements of this matrix.
(h) Try also to find an interpretation for the elements of $\mathbf{C}^3$.
(i) Is there any simple interpretation for the elements of $\mathbf{CC}'$ or $\mathbf{C}'\mathbf{C}$?
(j) Since $\mathbf{u}'\mathbf{C}$ simply tells how often a person is chosen, in this sense it is a measure of popularity. Suppose two persons are chosen equally often, but the first one is chosen by popular persons and the second by unpopular persons; then it is only reasonable to say that the first person is more popular than the second, since the first person has more indirect choices. Suppose now we want to use this for a measure of popularity. The measure is defined as follows. Take $0 < p < 1$, and assume that each direct choice contributes an amount p to a person's popularity, an indirect choice with one intermediate person an amount p^2, an indirect choice with two intermediates an amount p^3, etc. Prove that this measure can be found from the vector $\mathbf{u}'\mathbf{T}$, where $\mathbf{T} = p\mathbf{C} + p^2\mathbf{C}^2 + p^3\mathbf{C}^3 + \cdots$. Is it your guess that the elements of $\mathbf{T}$ will converge to some finite value? (Compare exercise 9 in section 4.8.)

14. In a paired comparison experiment, a subject is presented with all pairs of stimuli that can be formed out of a set of stimuli (the stimuli are political candidates, say), and is asked to tell for each pair which stimulus he prefers. The data can be collected in a square matrix $\mathbf{X}$ of order $n \times n$ (n = number of stimuli), where $x_{ij} = 1$ if j is preferred to i and is 0 otherwise (diagonal elements are zero).

(a) Verify that if $x_{ij} = 1$, then $x_{ji} = 0$, where $i \neq j$.
(b) Figure 2.2 gives a diagram of choices, where an arrow going from i to j indicates that i is preferred to j. Convert the diagram to a choice matrix **X**.

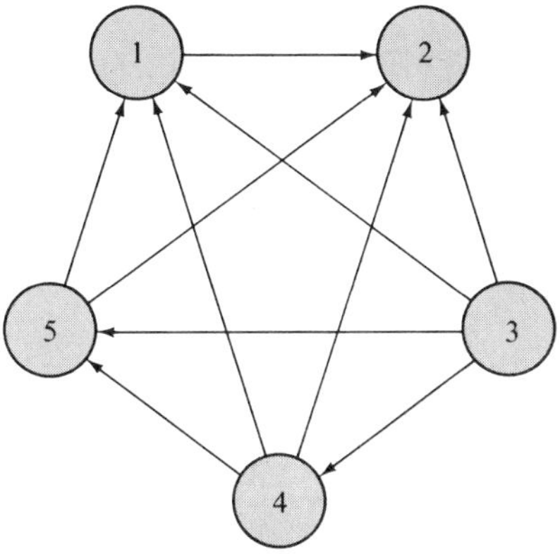

Figure 2.2
Preferential choice diagram.

(c) Calculate $\mathbf{u}'\mathbf{X}$ ($\mathbf{u}$ is a vector with unit elements) and give an interpretation to the resulting vector.
(d) Rearrange rows and columns of **X** in such a way that all unit elements are above the diagonal (the result is a triangular matrix). What is the advantage of this transformed matrix?
(e) Inspection of the diagram reveals that all choices are transitive; i.e., where i is preferred to j and j to k, then k is never preferred to i. If k were preferred, we would have a "circular triad." Draw a diagram in which circular triads do occur, and convert it into the corresponding choice matrix. Calculate $\mathbf{u}'\mathbf{X}$ for this matrix, and compare with the result for a transitive matrix.

15. *Statistical application.* Let **T** be an $n \times m$ matrix in which t_{ij} is the raw score of person i in test j.
(a) Find a matrix expression for the means of the tests.
(b) Suppose that in each column of **T**, we want to subtract the mean in order to obtain a matrix **A** that gives deviations from the mean. How can this be expressed in matrix notation?
(c) Verify that for matrix **A** we have $\mathbf{u}'\mathbf{A} = 0$.
(d) Calculate **A** for the following matrix **T**:

$$\mathbf{T} = \begin{pmatrix} 10 & 22 \\ 9 & 14 \\ 7 & 16 \\ 11 & 20 \\ 13 & 28 \end{pmatrix}.$$

(e) Calculate $\mathbf{A}'\mathbf{A}/n$ and interpret the elements.

(f) Suppose we want to transform $\mathbf{A}$ into a matrix of standard scores, $\mathbf{X}$ (a standard score is defined as the deviation from the mean, divided by the standard deviation). Suppose $\mathbf{S}$ is a diagonal matrix with the values of standard deviations in the diagonal; find a matrix notation for $\mathbf{X}$ in terms of $\mathbf{A}$ and $\mathbf{S}$ (compare exercise 10).

(g) Compute $\mathbf{R} = \mathbf{X}'\mathbf{X}/n$ and interpret the elements of $\mathbf{R}$.

3

Geometric Representation of Matrices

In this chapter we shall see how a matrix can be pictured in geometric space. Such a picture will be equivalent to the matrix itself, and therefore will not contain anything that is new or different. So, to paraphrase *Alice in Wonderland*, one wonders what the use of such pictures is? In fact, the advantage is psychological (being a psychologist, I hesitate to say that it is merely psychological): a picture often enables us to grasp relations intuitively where otherwise we would have to rely on a chain of algebraic reasoning that might be difficult to follow. Therefore, we shall very often use pictures or geometric ideas as a guide.

3.1. Vectors

Given a matrix $\mathbf{X}$ of order $n \times m$, each row of the matrix can be taken as the rectangular coordinates of a point in an m-dimensional space (an

m-space). From now on, we shall use the abbreviation S_m for such a space. The line that joins the origin of the space to a point is called a vector; it is a geometric vector that corresponds to a row vector of the matrix. For an example, let us take the matrix **X**,

$$\begin{pmatrix} 2 & 3 \\ 1 & 4 \\ 6 & 2 \end{pmatrix},$$

which is graphed in figure 3.1.

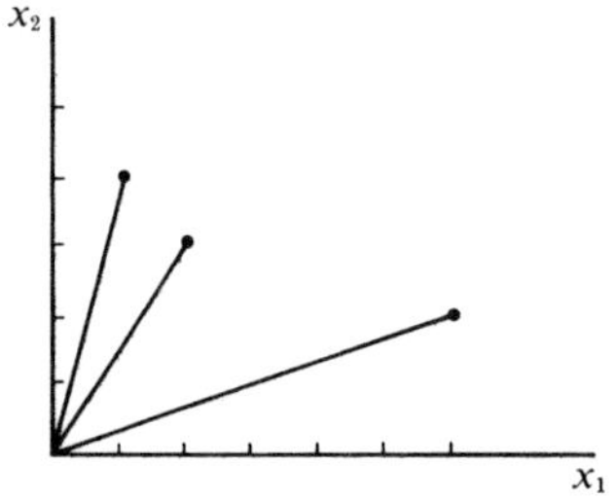

Figure 3.1
Graph of matrix **X** in a 2-space.

Now calculate $\mathbf{XX}' = \mathbf{A}$, say; it is easily verified that diagonal elements of **A** have numerical values equal to the squared length of a vector. This can be directly seen from the calculation, since we have $a_{11} = x_{11}^2 + x_{12}^2$; in figure 3.1, x_{11} and x_{12} can be identified with the right sides of a rectangular triangle of which the vector is the hypothenuse. It follows from Pythagoras's theorem that a_{11} gives the squared length of the vector to point (x_{11}, x_{12}). From now on we shall write ρ_i for a vector length; so we have $a_{ii} = \rho_i^2$.

In the example, we find for **A**:

$$\mathbf{A} = \mathbf{XX}' = \begin{pmatrix} 13 & 14 & 18 \\ 14 & 17 & 14 \\ 18 & 14 & 40 \end{pmatrix},$$

and from diagonal elements we see that $\rho_1 = \sqrt{13}$, $\rho_2 = \sqrt{17}$, and $\rho_3 = \sqrt{40}$.

The obvious question is what the other elements of **A** represent. For instance, take $a_{12} = 14$. We shall now show that this value is equal to $\rho_1\rho_2 \cos(1, 2)$, where $\cos(1, 2)$ is the cosine of the angle between the vectors $\mathbf{x}_1'$ and $\mathbf{x}_2'$ (row vectors of $\mathbf{X}$). The proof takes the general case: $a_{ij} = \rho_i\rho_j \cos(i, j)$.

Let $\mathbf{x}_i'$ be the ith row of $\mathbf{X}$, and $\mathbf{x}_j$ the transpose of $\mathbf{x}_j'$ (i.e., the jth row of $\mathbf{X}$, written as a column). Then we have $a_{ij} = \mathbf{x}_i'\mathbf{x}_j$. The two rows $\mathbf{x}_i'$ and $\mathbf{x}_j'$ represent two points in the space, and their distance will be symbolized as d_{ij}. It follows from the generalized theorem of Pythagoras that d_{ij}^2 is equal to the sum of the squared differences between corresponding coordinate values:

$$d_{ij}^2 = (x_{i1} - x_{j1})^2 + (x_{i2} - x_{j2})^2 + \cdots + (x_{im} - x_{jm})^2. \tag{3.1}$$

This can be more compactly written as

$$d_{ij}^2 = (\mathbf{x}_i - \mathbf{x}_j)'(\mathbf{x}_i - \mathbf{x}_j) = \mathbf{x}_i'\mathbf{x}_i + \mathbf{x}_j'\mathbf{x}_j - \mathbf{x}_i'\mathbf{x}_j - \mathbf{x}_j'\mathbf{x}_i. \tag{3.2}$$

However, the vector product $\mathbf{x}_i'\mathbf{x}_j$ is a single number, and therefore $\mathbf{x}_i'\mathbf{x}_j = \mathbf{x}_j'\mathbf{x}_i$. Also, we know that $\mathbf{x}_i'\mathbf{x}_i = \rho_i^2$, and $\mathbf{x}_j'\mathbf{x}_j = \rho_j^2$. Expression (3.2) therefore can be rewritten as

$$d_{ij}^2 = \rho_i^2 + \rho_j^2 - 2\mathbf{x}_i'\mathbf{x}_j. \tag{3.3}$$

On the other hand, the two points defined by $\mathbf{x}_i'$ and $\mathbf{x}_j'$ form, together with the origin of the space, a flat triangle (since we can always produce a flat plane through three given points, even if these points are located in a higher-dimensional space). For any triangle, the cosine rule says:

$$d_{ij}^2 = \rho_i^2 + \rho_j^2 - 2\rho_i\rho_j \cos(i, j). \tag{3.4}$$

Equating (3.3) and (3.4), we obtain the result

$$\mathbf{x}_i'\mathbf{x}_j = \rho_i\rho_j \cos(i, j), \tag{3.5}$$

which completes the proof. A product of vectors, like $\mathbf{x}_i'\mathbf{x}_j$, is often called a *scalar product* (or an *inner product*). Matrix $\mathbf{XX}'$ therefore can be looked on as a matrix of scalar products. Note that for diagonal elements of $\mathbf{XX}'$, expression (3.5) remains true, since for them $\cos(i, i) = 1$ (the cosine of a zero angle), and therefore $d_{ii}^2 = \rho_i^2 + \rho_i^2 - 2\rho_i^2 = 0$.

For figure 3.1, matrix $\mathbf{A}$ gives the inner products. Vector lengths ρ_i were already identified. Values of the cosines are readily computed to be $\cos(1, 2) = .94$, $\cos(1, 3) = .79$, and $\cos(2, 3) = .54$. These cosines might be collected in a cosine matrix,

$$\begin{pmatrix} 1 & .94 & .79 \\ .94 & 1 & .54 \\ .79 & .54 & 1 \end{pmatrix},$$

and the angles themselves, in arc degrees, are

$$\begin{pmatrix} 0 & 19 & 38 \\ 19 & 0 & 57 \\ 38 & 57 & 0 \end{pmatrix},$$

which may be verified in figure 3.1.

Figure 3.1 shows three points in a 2-space S_2. Matrix **X**, however can be pictured in a different way, by taking each column (instead of each row)

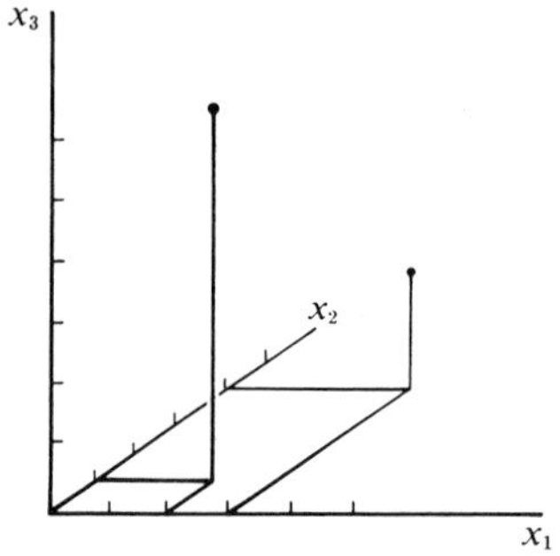

Figure 3.2
Graph of matrix **X** in a 3-space.

as a specification of rectangular coodinates of a point (m points in an S_n, which here is two points in a 3-space). The corresponding picture is given in figure 3.2. The two pictures, 3.1 and 3.2, although they look different, are completely equivalent; if we take the term "picture" in a wider sense than it has in common language, we might say that figure 3.1 is a picture of figure 3.2, and vice versa.

Now let us compare the relative merits of two such pictures in a fresh example. The matrix **X** in table 3.1 can be thought of as a matrix of test scores, for nine persons on two tests, where the test scores are given as deviations from their means (i.e., columns in matrix **X** add up to zero). Figure 3.3 gives two pictures of this matrix. The first picture is obvious

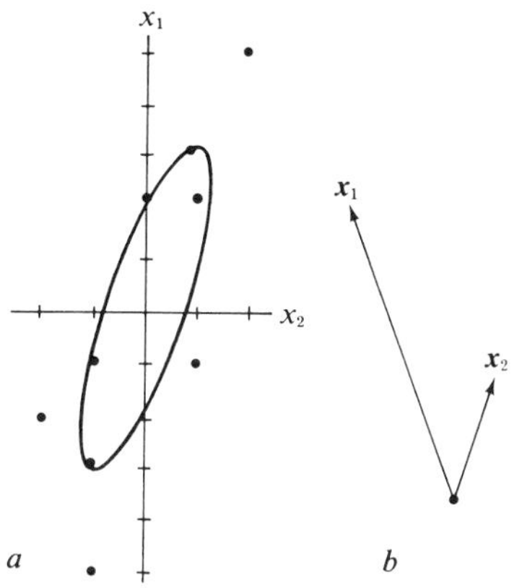

Figure 3.3
Two different graphs of the matrix in table 3.1.

and does not need explanation. The second picture, however, is less direct: here we should actually picture two vectors in a nine-dimensional space. But two vectors in an S_9 can be contained in a flat, two-dimensional subspace, since we can always produce a plane through two intersecting lines; so figure 3.3b actually gives a picture of that plane.

To see how figure 3.3b was arrived at, remember that a matrix $\mathbf{X}'\mathbf{X} = \mathbf{B}$ again gives inner products ($\mathbf{XX}'$ gave scalar products for row vectors, and $\mathbf{B}$ is the corresponding matrix for column vectors). Here we have

$$\mathbf{B} = \begin{pmatrix} 82 & 27 \\ 27 & 14 \end{pmatrix},$$

Table. 3.1 Matrix of test scores for nine persons on two tests, pictured in figures 3.3 and 3.4.

5	2
3	1
2	1
2	0
−1	1
−1	−1
−3	−1
−2	−2
−5	−1

which shows that the first vector has a squared length of 82, the second a squared length of 14. From $b_{12} = \rho_1\rho_2 \cos(1, 2)$, we can then calculate the cosine of the angle between the vectors, which gives us enough information to draw figure 3.3b.

However, the sum of squared deviations divided by n gives the variance σ^2 of a variable. It follows that $b_{11} = n\sigma_1^2$ and $b_{22} = n\sigma_2^2$. But since $b_{11} = \rho_1^2$, we derive immediately that the vector length is equal to $\rho_1 = \sqrt{n\sigma_1^2} = \sigma_1\sqrt{n}$, and similarly for the second vector. In general, diagonal elements of $\mathbf{B} = \mathbf{X}'\mathbf{X}$ relate to standard deviations of the variables multiplied by $\sqrt{n}$, and these values are pictured as the vector lengths.

Element b_{12} gives n times the covariance between the two variables. Remembering that the covariance is equal to $\sigma_1\sigma_2 r_{12}$, where r_{12} is the correlation coefficient, we have

$$b_{12} = \rho_1\rho_2 \cos(1, 2) = n\sigma_1\sigma_2 r_{12}, \tag{3.6}$$

and therefore

$$\cos(1, 2) = r_{12}, \tag{3.7}$$

with the important result that, in pictures like figure 3.3b, angles between vectors have cosines equal to correlations. So figure 3.3b is highly informative: it pictures standard deviations and correlations.

Figure 3.3b could be made somewhat more elegant by multiplying its scale by a factor $1/\sqrt{n}$ (that is, merely changing the scale of the paper on which the figure is drawn), which comes to the same thing as picturing the matrix $\mathbf{X}/\sqrt{n}$. If we make this change of scale, vectors will have a length equal to the standard deviation, whereas cosines of angles will remain equal to the correlation coefficients, since angles do not change if we change the scale.

It is easy to show that the preceding reasoning remains valid for the general case where the matrix is of order $n \times m$. We then have m vectors in an S_n, and $\mathbf{B} = \mathbf{X}'\mathbf{X}$ will be an $m \times m$ matrix of inner products. The matrix $\mathbf{X}'\mathbf{X}/n$ is the variance-covariance matrix, and the picture of $\mathbf{X}/\sqrt{n}$ will show m vectors in S_n with lengths equal to the standard deviations of the variables they stand for, and angles between vectors equal to $\cos^{-1}(r)$.

The next question is: What happens if we graph a matrix of standard scores? In the example in table 3.1, standard scores are obtained by dividing the first column by $\sqrt{82/9}$, and the second by $\sqrt{14/9}$ (compare exercises 2.5 and 2.15f). The two pictures of the standardized matrix are given in figure 3.4. The difference between figure 3.4b and 3.3b is that in figure 3.4b the vectors have unit length (since standard scores have unit

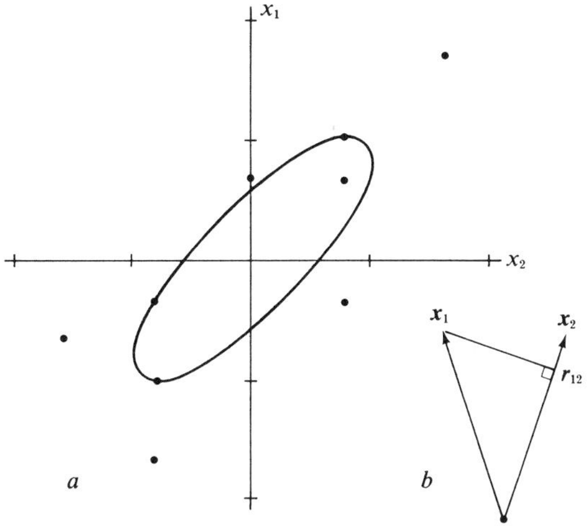

Figure 3.4 Two different graphs of the matrix in table 3.1, after standardization of variables.

variance). The angles between the vectors remain unchanged, but since the vectors now have unit length, a correlation coefficient r_{ij} can also be represented as the projection of vector **i** on vector **j**, since for unit vectors this projection will be equal to the cosine of the angle between them.

The difference between figures 3.3a and 3.4a is not so easily visualized. Remember that in figure 3.3a standard deviations are suggested by the spread of the points parallel to the axes. Since there is much more spread parallel to x_1 than to x_2, the cloud of points has an orientation from SSW to NNE, so to speak. In figure 3.4a, however, the variances have been made equal, so that the cloud of points is oriented at 45° to the axes.

3.2. Direction Cosines

Let **x** be a vector in an n-dimensional space, with certain angles to the n axes. The cosines of these angles are called *direction cosines*. They specify the direction of the vector, irrespective of its length. Figure 3.5 gives an illustration for a 2-space. Vector **x** here is defined by its two coordinates (x_1, x_2), and we easily derive for the direction cosines k_1 and k_2:

$$k_1 = x_1 \rho;$$
$$k_2 = x_2 \rho. \tag{3.8}$$

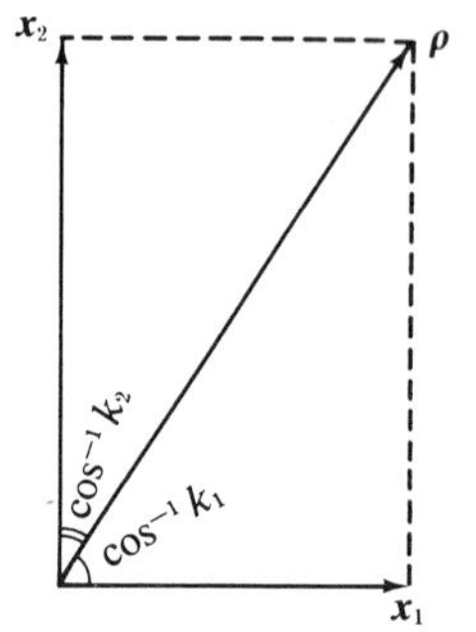

Figure 3.5
Relation between projections of a vector and direction cosines.

It is also easily verified that the sum of the squared direction cosines is equal to unity. In other words, if $\mathbf{k}$ is the 2×1 column vector of direction cosines, then $\mathbf{k}'\mathbf{k} = 1$. That this remains true for higher dimensionality can be intuitively grasped if one realizes that a direction cosine can be interpreted as the projection on the coordinate axes of a vector with unit length. The result $\mathbf{k}'\mathbf{k} = 1$ then follows immediately from the higher-dimensional Pythagorean theorem. But it is also useful to think of the relation $\mathbf{k}'\mathbf{k} = 1$ in another way: it spells out the condition that must be satisfied in order to fit a vector into a space. Obviously, in a plane, we cannot draw a vector that makes an angle of 45° with the abscissa and of 30° with the ordinate; the vector fits only if the squared cosines add up to unity. Similar restrictions apply to a space of higher dimensionality, and they are compactly summarized in the expression $\mathbf{k}'\mathbf{k} = 1$.

3.3. Projections of Points on Vectors

Let $\mathbf{x}_i$ (a vector of order $n \times 1$) specify a point P_i in an n-dimensional space. Suppose we introduce into this space a new vector whose direction cosines are given in a vector $\mathbf{k}$, also of order $n \times 1$, that satisfies $\mathbf{k}'\mathbf{k} = 1$. Then the projection $\mathbf{y}_i$ of point P_i onto this new vector is given by $y_i = \mathbf{x}_i'\mathbf{k}$ (the proof is left to the reader as an exercise; first try a proof for the two-dimensional plane, and then generalize).

If we have a collection of points $P_1, P_2, \ldots, P_m$, the coordinates of which are given in a matrix $\mathbf{X}$ of order $m \times n$, then all projections onto a vector with direction cosines $\mathbf{k}$ are obtained from $\mathbf{y} = \mathbf{Xk}$. This is just a straightforward generalization of the earlier result.

3.4. Angles between Directions

Let a vector $\mathbf{k}_1$ give the direction cosines of a vector in an n-dimensional space, and $\mathbf{k}_2$ those of another vector; $\mathbf{k}_1'\mathbf{k}_1 = 1$, and $\mathbf{k}_2'\mathbf{k}_2 = 1$. Then $\mathbf{k}_1'\mathbf{k}_2$ gives the cosine of the angle between the vectors. Again, the proof is left as an exercise (hint: take $\mathbf{k}_1$ and $\mathbf{k}_2$ as coordinates of a vector with unit length and apply equation 3.5).

It follows that two vectors are perpendicular if $\mathbf{k}_1'\mathbf{k}_2 = 0$. In general, if $\mathbf{K}$ (of order $n \times m$) gives direction cosines of m vectors in an S_n, then these vectors form an orthogonal set if $\mathbf{K}'\mathbf{K} = \mathbf{I}$. Suppose now we have p points in the space, with coordinates given in $\mathbf{X}$ of order $p \times n$. Then their projections on the m new vectors specified by the matrix $\mathbf{K}$ are given by $\mathbf{XK} = \mathbf{Y}$, where $\mathbf{Y}$ is of order $p \times m$.

The latter also gives the solution for the problem of how to find transformed coordinates if we rotate the coordinate system. A rotation of coordinates is identical to the introduction of a new set of n orthogonal vectors. So a matrix $\mathbf{K}$ of order $n \times n$ would specify the rotation, with $\mathbf{K}'\mathbf{K} = \mathbf{I}$. The transformed coordinates then are $\mathbf{Y} = \mathbf{XK}$, where $\mathbf{Y}$ has the same order as $\mathbf{X}$.

3.5. Exercises

1. Draw the two kinds of pictures for the matrices $\mathbf{T}$, $\mathbf{A}$, and $\mathbf{X}$ from exercise 15 in section 2.5. For the first type of picture, check that the transformation from $\mathbf{T}$ to $\mathbf{A}$ is just a translation of the axes, and that the transformation from $\mathbf{A}$ to $\mathbf{X}$ is a change in the scale of the axes.

2. Let $\mathbf{K}$ be a matrix of direction cosines for n vectors in an n-dimensional space (i.e., $\mathbf{K}$ is of order $n \times n$), and let $\mathbf{K}'\mathbf{K} = \mathbf{I}$. Prove that $\mathbf{KK}' = \mathbf{I}$.

3. Using the result of exercise 2, prove that, for an arbitrary point in the n-dimensional space, the distance to the origin is not changed by a transformation with the rotation matrix $\mathbf{K}$, and also that for two points the angle between the vectors to these points remains invariant.

4. Demonstrate whether the same results as in exercise 3 will be obtained if $\mathbf{K}'\mathbf{K} \neq \mathbf{I}$.

4

Determinants and Matrix Inversion

4.1. Introduction

This chapter is rather tedious, but cannot be avoided if we want to understand how we can find the inverse $\mathbf{A}^{-1}$ of a matrix $\mathbf{A}$ (so that $\mathbf{A}^{-1}\mathbf{A} = \mathbf{I}$). Some readers may be satisfied to know that such inverse matrices sometimes exist and can be calculated by a computer. However, not knowing how this calculation is done may leave a hazy region in one's mind; and if you dislike such vagueness, you should work through this chapter.

To begin, we shall need the concept of a determinant in order to compute inverse matrices (and for some other purposes as well). So we shall first explain what a determinant is, and then use this knowledge to see how an inverse matrix is found.

4.2. Definition of a Determinant

A *determinant* will be symbolized as $|\mathbf{A}|$, where $\mathbf{A}$ is a square matrix; $|\mathbf{A}|$ is said to be the determinant of the matrix $\mathbf{A}$. Notation:

$$\begin{pmatrix} 4 & 7 & 3 \\ 2 & 0 & 4 \\ 1 & 4 & 6 \end{pmatrix} \quad \text{is a square matrix, and}$$

$$\begin{vmatrix} 4 & 7 & 3 \\ 2 & 0 & 4 \\ 1 & 4 & 6 \end{vmatrix} \quad \text{its determinant.}$$

If $\mathbf{A}$ is of order $n \times n$, then the determinant is said to be of order n.

The crucial idea is that a determinant stands for a single number, which is called the *value* of the determinant. It requires some steps to show how the value is found.

Given a determinant $|\mathbf{A}|$ of order n, we can form products of n elements in such a way that from each row and column of $|\mathbf{A}|$ one and only one element is selected as a factor for the product. This is more easily seen in an example. Suppose $|\mathbf{A}|$ is of third order:

$$\begin{vmatrix} a_{11} & a_{12} & a_{13} \\ a_{21} & a_{22} & a_{23} \\ a_{31} & a_{32} & a_{33} \end{vmatrix}.$$

Then the product $a_{12} \cdot a_{23} \cdot a_{31}$ would satisfy the requirement: we have a_{12} as the only element from the first row (second column), a_{23} as the only element from the second row (third column), and a_{31} as the only element from the third row (first column). Such a product is called a *term* of the determinant.

However, we can form many terms like this. Other examples are $a_{12} \cdot a_{21} \cdot a_{33}$, or $a_{13} \cdot a_{22} \cdot a_{31}$. The complete set appears to be:

$$\begin{array}{lll} a_{11} \cdot a_{22} \cdot a_{33}, & a_{11} \cdot a_{23} \cdot a_{32}, & a_{12} \cdot a_{21} \cdot a_{33}, \\ a_{12} \cdot a_{23} \cdot a_{31}, & a_{13} \cdot a_{21} \cdot a_{32}, & a_{13} \cdot a_{22} \cdot a_{31}. \end{array} \tag{4.1}$$

There is a systematic basis for this arrangement. First, since we have one element out of each row, we can always rank the elements in each term in such a way that the row indices follow the natural numbers. All that is left is to determine an order for the second subscripts; clearly they can be taken in as many ways as there are permutations of the three natural numbers. Since there are 3! such permutations, we can form $3! = 6$ terms for a determinant of order 3. In general, a determinant of order n will have $n!$ terms.

The next step is to assign a plus or minus sign to each term. Assuming again that the row subscripts are in natural order, the sign depends on the column subscripts only. First, we shall agree that every time a higher subscript precedes a lower, we have an inversion. Looking back at the six terms in (4.1), we find (following reading order) the number of inversions in each term to be 0, 1, 1, 2, 2, and 3, respectively.

If the number of inversions is even, then the permutation is said to be even (or of even class); otherwise the permutation is odd. We shall now assign a plus sign to terms with an even permutation of the column subscripts, and a minus sign to those with an odd permutation. The value of the determinant is then defined, finally, as the algebraic sum of its terms.

For a determinant of third order, we have a value equal to:

$$|\mathbf{A}| = a_{11}a_{22}a_{33} - a_{11}a_{23}a_{32} - a_{12}a_{21}a_{33} + a_{12}a_{23}a_{31} + a_{13}a_{21}a_{32} - a_{13}a_{22}a_{31}.$$

Example:

$$\begin{vmatrix} 4 & 7 & 3 \\ 2 & 0 & 4 \\ 1 & 4 & 6 \end{vmatrix} = 4 \cdot 0 \cdot 6 - 4 \cdot 4 \cdot 4 - 7 \cdot 2 \cdot 6 + 7 \cdot 4 \cdot 1 + 3 \cdot 2 \cdot 4 - 3 \cdot 0 \cdot 1 = -96.$$

4.3. Some Rules

The following rules are sometimes helpful. They are given without proof.

1. The determinant of $\mathbf{A}$ has the same value as the determinant of $\mathbf{A}'$.
2. The value of the determinant changes sign if one row (column) is interchanged with another row (column).
3. If a determinant has two equal rows (columns), its value is zero.
4. If a determinant has two rows (columns) with proportional elements, its value is zero.

5. If all elements in a row (column) are multiplied by a constant, the value of the determinant is multiplied by that constant.

6. If a determinant has a row (column) in which all elements are zero, the value of the determinant is zero.

7. The value of the determinant remains unchanged if one row (column) is added to or subtracted from another row (column). Moreover, if a row (column) is multiplied by a constant and then added to or subtracted from another row (column) the value remains unchanged.

Examples where the rules can be used with profit are given as exercises.

4.4. Expansion of a Determinant

Let **A** be a matrix of order $n \times m$. If we omit one or more rows or columns from **A**, we obtain a matrix of smaller order, called a *minor* of the matrix. Similarly, we have minors of a determinant, and in particular, if we omit from the determinant the ith row and the jth column, the resulting minor will be square and its determinant will be symbolized $|\mathbf{M}_{ij}|$. This determinant is called a *cofactor* (c_{ij}) if we give it a sign equal to $(-1)^{i+j}$, so that $c_{ij} = (-1)^{i+j} |\mathbf{M}_{ij}|$.

Using this notation, we can write a formula for the expansion of a determinant of order n:

$$|\mathbf{A}| = a_{i1}c_{i1} + a_{i2}c_{i2} + \cdots + a_{in}c_{in} = \sum_{g=1}^{n} a_{ig}c_{ig}. \tag{4.2}$$

In this version, the determinant is expanded according to its ith row. A similar formula can be written for expansion according to the jth column:

$$|\mathbf{A}| = a_{1j}c_{1j} + a_{2j}c_{2j} + \cdots + a_{nj}c_{nj} = \sum_{h=1}^{n} a_{hj}c_{hj}. \tag{4.3}$$

Example:

$$|\mathbf{A}| = \begin{vmatrix} 2 & 4 & 8 \\ 1 & 2 & 3 \\ 3 & 2 & 1 \end{vmatrix}.$$

Expansion according to the first row gives:

$$|\mathbf{A}| = 2\begin{vmatrix} 2 & 3 \\ 2 & 1 \end{vmatrix} - 4\begin{vmatrix} 1 & 3 \\ 3 & 1 \end{vmatrix} + 8\begin{vmatrix} 1 & 2 \\ 3 & 2 \end{vmatrix}$$

$$= 2(2 - 6) - 4(1 - 9) + 8(2 - 6) = -8.$$

Expansion according to the second column gives:

$$|\mathbf{A}| = -4\begin{vmatrix} 1 & 3 \\ 3 & 1 \end{vmatrix} + 2\begin{vmatrix} 2 & 8 \\ 3 & 1 \end{vmatrix} - 2\begin{vmatrix} 2 & 8 \\ 1 & 3 \end{vmatrix}$$

$$= -4(1 - 9) + 2(2 - 24) - 2(6 - 8) = -8.$$

A proof for (4.2) or (4.3) will not be given.

4.5. Definition of an Inverse Matrix

Let $\mathbf{A}$ be a square matrix. If we can find a matrix $\mathbf{B}$ of the same order as $\mathbf{A}$ such that $\mathbf{AB} = \mathbf{BA} = \mathbf{I}$, then $\mathbf{B}$ is said to be the *inverse* of $\mathbf{A}$ and is symbolized $\mathbf{A}^{-1}$. $\mathbf{A}^{-1}$, if it exists, can be found as follows.

Let $\mathbf{C}$ be the matrix of cofactors of $\mathbf{A}$ (i.e., c_{ij} is the cofactor obtained from the minor $|\mathbf{M}_{ij}|$); then

$$\mathbf{A}^{-1} = \mathbf{C}'/|\mathbf{A}|, \tag{4.4}$$

where $\mathbf{C}'$ is the transpose of $\mathbf{C}$ (or, if one prefers, $\mathbf{C}'$ is the matrix of cofactors of $\mathbf{A}'$). It is immediately seen that the inverse is undefined if $\mathbf{A}$ is not square (since then there is no determinant $|\mathbf{A}|$), and also if $|\mathbf{A}|$ is equal to zero.

4.6. Some Theorems

Inversion of a product:

$$(\mathbf{AB})^{-1} = \mathbf{B}^{-1}\mathbf{A}^{-1},$$

where it is assumed that $\mathbf{A}$ and $\mathbf{B}$ are square and that $\mathbf{A}^{-1}$ and $\mathbf{B}^{-1}$ exist. A proof of the theorem is straightforward. Let $\mathbf{AB}$ be given; then we have

$$\mathbf{AB} \cdot \mathbf{B}^{-1}\mathbf{A}^{-1} = \mathbf{A} \cdot \mathbf{I} \cdot \mathbf{A}^{-1} = \mathbf{A} \cdot \mathbf{A}^{-1} = \mathbf{I}.$$

Therefore $(\mathbf{B}^{-1}\mathbf{A}^{-1})$ is the inverse of $\mathbf{AB}$, and can be written as $(\mathbf{AB})^{-1}$. A generalization gives

$$(\mathbf{ABC})^{-1} = \mathbf{C}^{-1}\mathbf{B}^{-1}\mathbf{A}^{-1}.$$

An often useful theorem is that the inverse of a diagonal matrix is another diagonal matrix with diagonal elements that are the reciprocals of the diagonal elements of the first. As a special case, we have the unit matrix as its own inverse.

4.7. An Example

We are given a matrix

$$\mathbf{A} = \begin{pmatrix} 3 & 4 & 1 \\ 2 & 6 & 3 \\ 8 & 1 & 2 \end{pmatrix}.$$

For the value of the determinant, we use an expansion according to the first row, which gives:

$$|\mathbf{A}| = 3\begin{vmatrix} 6 & 3 \\ 1 & 2 \end{vmatrix} - 4\begin{vmatrix} 2 & 3 \\ 8 & 2 \end{vmatrix} + \begin{vmatrix} 2 & 6 \\ 8 & 1 \end{vmatrix} = 61.$$

The matrix of cofactors of **A** is:

$$\mathbf{C} = \begin{pmatrix} \begin{vmatrix} 6 & 3 \\ 1 & 2 \end{vmatrix} & -\begin{vmatrix} 2 & 3 \\ 8 & 2 \end{vmatrix} & \begin{vmatrix} 2 & 6 \\ 8 & 1 \end{vmatrix} \\ -\begin{vmatrix} 4 & 1 \\ 1 & 2 \end{vmatrix} & \begin{vmatrix} 3 & 1 \\ 8 & 2 \end{vmatrix} & -\begin{vmatrix} 3 & 4 \\ 8 & 1 \end{vmatrix} \\ \begin{vmatrix} 4 & 1 \\ 6 & 3 \end{vmatrix} & -\begin{vmatrix} 3 & 1 \\ 2 & 3 \end{vmatrix} & \begin{vmatrix} 3 & 4 \\ 2 & 6 \end{vmatrix} \end{pmatrix} = \begin{pmatrix} 9 & 20 & -46 \\ -7 & -2 & 29 \\ 6 & -7 & 10 \end{pmatrix}.$$

$\mathbf{A}^{-1}$ therefore becomes:

$$\mathbf{A}^{-1} = \mathbf{C}'/61 = \begin{pmatrix} 9/61 & -7/61 & 6/61 \\ 20/61 & -2/61 & -7/61 \\ -46/61 & 29/61 & 10/61 \end{pmatrix}.$$

It can be verified that in fact $\mathbf{A} \cdot \mathbf{A}^{-1} = \mathbf{I}$.

For large matrices these computations become extremely tedious. In fact, there are various tricks by which the computation can be facilitated, but we shall not demonstrate those here. Matrix inversion represents precisely the kind of problem for which computers have been invented.

4.8. Exercises

1. Calculate the value of the following determinants:

$$|\mathbf{A}| = \begin{vmatrix} 26 & -13 & 38 \\ 7 & 5 & 2 \\ 3 & 2 & 1 \end{vmatrix}; \quad |\mathbf{B}| = \begin{vmatrix} 55 & 100 & 16 \\ 135 & 21 & 12 \\ -80 & 79 & 3 \end{vmatrix};$$

$$|\mathbf{C}| = \begin{vmatrix} 5 & 9 & -50 \\ 55 & 50 & -125 \\ 6 & -11 & 125 \end{vmatrix}.$$

2. Write the numbers from 1 to 100 in reading order in a 10 × 10 determinant. Calculate its value by using the rules in section 4.3.

3. Prove that

$$\begin{vmatrix} 1 & a & a^2 \\ 1 & b & b^2 \\ 1 & c & c^2 \end{vmatrix} = (a - b)(b - c)(c - a).$$

4. Calculate the value of

$$|\mathbf{X}| = \begin{vmatrix} 5 & 0 & 1 & 3 \\ -3 & -2 & 0 & 4 \\ 2 & 0 & 7 & 2 \\ 7 & 3 & 11 & 13 \end{vmatrix}.$$

5. Given $\mathbf{A} = \begin{pmatrix} 1 & 1 & 1 \\ 1 & 2 & 3 \\ 1 & 4 & 9 \end{pmatrix}$, calculate $\mathbf{A}^{-1}$ and verify $\mathbf{A} \cdot \mathbf{A}^{-1} = \mathbf{I}$.

6. Given $\mathbf{B} = \begin{pmatrix} 1 & 2 & 4 \\ 1 & 3 & 9 \\ 1 & 4 & 16 \end{pmatrix}$, calculate $\mathbf{B}^{-1}$ (compare exercise 3 above).

7. Prove that if $\mathbf{A}$ is symmetric, then $\mathbf{A}^{-1}$ is also symmetric.

8. Prove that if $\mathbf{A}$ is triangular, then $\mathbf{A}^{-1}$ is also triangular.

9. Prove that $\mathbf{T}$ in exercise 13 of section 2.5 can be found as $\mathbf{T} = p(\mathbf{I} - p\mathbf{C})^{-1}\mathbf{C}$.

5

Equations

5.1. Definitions

First, we shall distinguish between homogeneous equations and non-homogeneous equations. A *homogeneous* equation is of type $\mathbf{a}'\mathbf{x} = 0$, with $\mathbf{a}$ and $\mathbf{x}$ being vectors of order $n \times 1$; $\mathbf{a}$ is the vector of coefficients, $\mathbf{x}$ the vector of unknowns. For instance, if $\mathbf{a}' = (2 \; -3)$ and $\mathbf{x}' = (x_1 \; x_2)$, then $\mathbf{a}'\mathbf{x} = 0$ gives a homogeneous equation with two unknowns: $2x_1 - 3x_2 = 0$. A *nonhomogeneous* equation has form $\mathbf{a}'\mathbf{x} = k$, with k, some constant, $\neq 0$. An example is $2x_1 - 3x_2 = 2$.

Geometrically, equations with two unknowns represent lines in the plane defined by the axes x_1 and x_2 (see figure 5.1). Similarly, an equation in three unknowns, like $3x_1 + 2x_2 + 5x_3 = 4$, represents a plane in an S_3; the space S_3 is defined by the three coordinates, and each triplet of values that satisfies the equation refers to a point located on the plane within S_3.

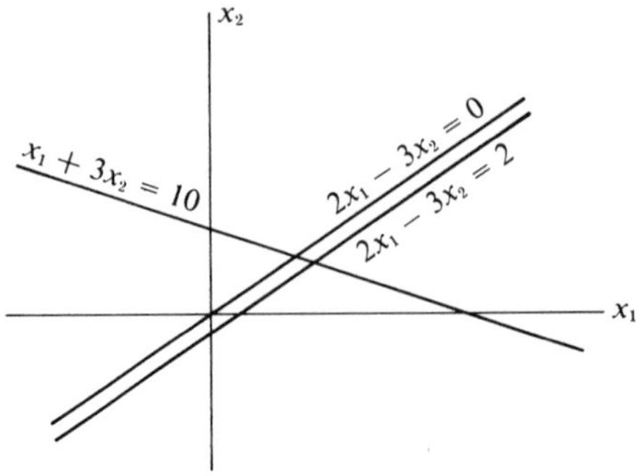

Figure 5.1 Graphs of linear equations.

Now if we generalize, we see that an equation in n unknowns will stand for an $(n - 1)$ dimensional "subspace" of an S_n. Such a subspace is called a *hyperplane*, symbolized by the letter V, with a subscript that indicates the dimensionality of the hyperplane. We then can say that a linear equation defines a V_{n-1} in an S_n. However, the S_n will also contain hyperplanes of lower order: V_{n-2}'s, V_{n-3}'s, . . . , V_2's (planes), V_1's (lines) and V_0's (points).

It is easily verified that homogeneous equations stand for hyperplanes through the origin, since the zero vector (the origin) always satisfies them. On the other hand, nonhomogeneous equations refer to hyperplanes that do not contain the origin.

5.2. Rule for the Intersection of Hyperplanes

In general, two lines in a plane will intersect in a point, if we exclude the special cases where the lines coincide or are parallel. In an S_3, two planes will intersect in a line (again, special cases excluded); a V_2 and a V_2 intersect in a V_1. Also, in an S_3, a V_2 (a plane) and a V_1 (a line) intersect in a V_0 (a point). On the other hand, two V_1's in an S_3 do not usually intersect.

The general rule is: in an S_n two hyperplanes V_{n-p} and V_{n-q} intersect in a $V_{n-(p+q)}$. Examples: a V_4 and a V_2 in an S_5 intersect in a V_1; a V_5 and a V_8 in an S_{10} intersect in a V_3. If $p + q = n$, the intersection is a V_0, a point. If $p + q > n$, there is no intersection. For instance, two lines in an S_3 do not intersect, as we know, and as the rule shows, in that it specifies an intersection of negative dimensionality (a V_{-1}). Note that the rule is only generally valid, since two lines in an S_3 may happen to intersect in a point, if they are nonparallel lines in the same plane. Also, two planes in an S_3 will in general intersect in a line, as the rule says, but there are special cases where they have no intersection (parallel planes) or where the intersection is a plane (coinciding planes). We should always bear in mind that such special cases are excluded in the formulation of the general rule.

A more general version of the rule says that a V_{n-p}, a V_{n-q}, a V_{n-r}, etc., in an S_n intersect in a $V_{n-(p+q+r+\cdots)}$. Again, special cases must be excluded.

What does the rule mean for equations? As we saw, an equation in n unknowns defines a V_{n-1} in an S_n. Two such equations define two V_{n-1}'s, with a V_{n-2} as their intersection, in general. Three equations define a V_{n-3}. If the number of equations is $n - 1$, their intersection is a V_1, a line, and if their number is n, the intersection is a point; so, in general, there will be a unique solution only for n equations in n unknowns.

5.3. Solution for n Nonhomogeneous Equations with n Unknowns

The matrix notation for n nonhomogeneous equations with n unknowns is $\mathbf{Bx} = \mathbf{k}$, with $\mathbf{B}$ an $n \times n$ matrix of coefficients, $\mathbf{x}$ the $n \times 1$ vector of unknowns, and $\mathbf{k}$ the $n \times 1$ vector of constant terms. Geometrically, the equations represent n hyperplanes in an S_n that intersect in a V_0; so there is one unique solution, giving the coordinates of the unique point of intersection (if we exclude special cases).

This solution is given by $\mathbf{x} = \mathbf{B}^{-1}\mathbf{k}$. Here it is assumed that $|\mathbf{B}| \neq 0$, since otherwise $\mathbf{B}^{-1}$ has no meaning. (One might suspect that $|\mathbf{B}| = 0$ has something to do with the special cases; this will be investigated in more detail in section 5.5.)

Example: the equations $2x_1 - 3x_2 = 2$ and $x_1 + 3x_2 = 10$ (compare figure 5.1) can be written as

$$\begin{pmatrix} 2 & -3 \\ 1 & 3 \end{pmatrix}\begin{pmatrix} x_1 \\ x_2 \end{pmatrix} = \begin{pmatrix} 2 \\ 10 \end{pmatrix}.$$

The solution is

$$\mathbf{x} = \begin{pmatrix} 2 & -3 \\ 1 & 3 \end{pmatrix}^{-1}\begin{pmatrix} 2 \\ 10 \end{pmatrix} = \begin{pmatrix} 3 & 3 \\ -1 & 2 \end{pmatrix}\begin{pmatrix} 2 \\ 10 \end{pmatrix}\Big/9 = \begin{pmatrix} 36 \\ 18 \end{pmatrix}\Big/9 = \begin{pmatrix} 4 \\ 2 \end{pmatrix}.$$

5.4. Solution for $n - 1$ Homogeneous Equations in n Unknowns

The matrix notation for $n - 1$ homogeneous equations in n unknowns is $\mathbf{Bx} = \mathbf{0}$, with $\mathbf{B}$ an $(n-1) \times n$ matrix of coefficients, $\mathbf{x}$ an $n \times 1$ vector of unknowns, and $\mathbf{0}$ an $n \times 1$ zero vector. The equations represent hyperplanes V_{n-1}, all of which pass through the origin. According to the general rule, their common intersection will be a line, which since the zero vector

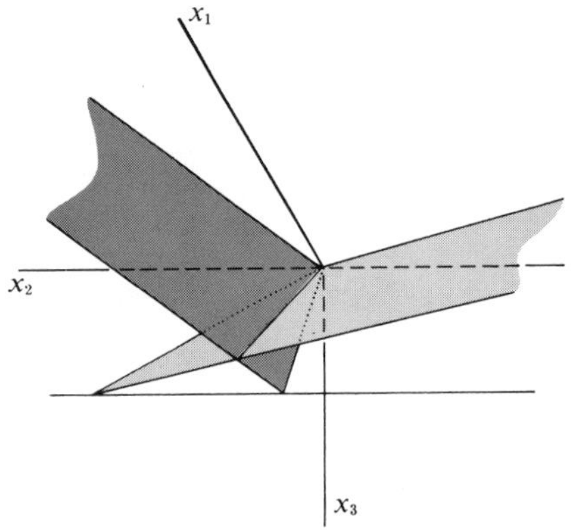

Figure 5.2 Graph of the linear equations $\begin{pmatrix} 2 & -3 & 1 \\ 4 & 1 & -2 \end{pmatrix} x = 0.$

obviously satisfies all equations, will pass through the origin. All other vectors $\mathbf{x}$ that satisfy the equations therefore will be proportional; i.e., if a non-zero vector $\mathbf{x}_i$ satisfies the equations, then $c\mathbf{x}_i$ will also satisfy them, with c an arbitrary constant. We will consider all such proportional vectors to be *one* solution.

For instance, take the following two homogeneous equations with three unknowns:

$$\begin{pmatrix} 2 & -3 & 1 \\ 4 & 1 & -2 \end{pmatrix} \mathbf{x} = 0. \tag{5.1}$$

They represent two planes in an S_3, as pictured in figure 5.2. To find a point of the intersection, add an arbitrary third equation to the set; this third equation should not be homogeneous (if it is, the unique point of intersection is the origin, which is a trivial and unacceptable solution). For instance, add the equation $2x_1 + x_2 - 3x_3 = 5$, to obtain the set

$$\begin{pmatrix} 2 & -3 & 1 \\ 4 & 1 & -2 \\ 2 & 1 & -3 \end{pmatrix} \mathbf{x} = \begin{pmatrix} 0 \\ 0 \\ 5 \end{pmatrix}, \tag{5.2}$$

with solution $\mathbf{x}' = (25/24 \quad 40/24 \quad 70/24)$. Any vector $\mathbf{x}'$ proportional to this one then will also satisfy (5.1), such as $\mathbf{x}' = (5 \quad 8 \quad 14)$, or $\mathbf{x}' = (-10 \quad -16 \quad -28)$.

Of course, we might have made a more clever choice for the third equation, like $x_3 = 1$. Instead of (5.2), we then obtain

$$\begin{pmatrix} 2 & -3 & 1 \\ 4 & 1 & -2 \\ 0 & 0 & 1 \end{pmatrix} \mathbf{x} = \begin{pmatrix} 0 \\ 0 \\ 1 \end{pmatrix}, \tag{5.3}$$

which is easier to solve. Its solution, $\mathbf{x}' = (5/14 \quad 8/14 \quad 1)$, could also have been found by substituting $x_3 = 1$ directly in (5.1). Then we have

$$\begin{pmatrix} 2 & -3 \\ 4 & 1 \end{pmatrix} \begin{pmatrix} x_1 \\ x_2 \end{pmatrix} = \begin{pmatrix} -1 \\ 2 \end{pmatrix}, \tag{5.4}$$

also with solution $x_1 = 5/14$, $x_2 = 8/14$.

5.5. Linear Dependence, and Rank of a Matrix

5.5.1. If we have m nonhomogeneous equations in n unknowns, with $m < n$, no unique solution can be identified, since the corresponding hyperplanes do not intersect in just one point but in a V_{n-m}. Of course, it is easy enough to find some vector $\mathbf{x}$ that satisfies the set: all we have to do is add $n - m$ equations to the set and solve for $\mathbf{x}$. However, this vector $\mathbf{x}$ is only one of the many possible vectors $\mathbf{x}$ that satisfy the original m equations, and so is far from unique. Similarly, for homogeneous equations, if the number of unknowns is n and the number of equations is less than $n - 1$, no unique solution can be identified.

On the other hand, if we have more than n nonhomogeneous equations in n unknowns, again no solution can be given, but now because there is no $\mathbf{x}$ at all that will satisfy the whole set. In fact, the rule says that the common hyperplane has dimensionality smaller than zero, and this means, as we saw, that the hyperplanes defined by the equations have nothing in common. There are special cases, however, which we shall now examine.

5.5.2. Parallel Hyperplanes. We are given two equations, $\mathbf{a}_1'\mathbf{x} = k_1$ and $\mathbf{a}_2'\mathbf{x} = k_2$, with $\mathbf{a}_1$ and $\mathbf{a}_2$ being $n \times 1$ vectors of coefficients. The first equation defines a hyperplane parallel to $\mathbf{a}_1'\mathbf{x} = 0$ (a hyperplane through the origin), and the second equation a hyperplane parallel to $\mathbf{a}_2'\mathbf{x} = 0$. It follows that the two hyperplanes given by the equations will be parallel themselves, if and only if $\mathbf{a}_1'\mathbf{x} = 0$ is identical to $\mathbf{a}_2'\mathbf{x} = 0$, i.e., if the vectors of coefficients are equal or proportional.

5.5.3. For the other special cases, we need treat only homogeneous equations, since a set of nonhomogeneous equations can always be transformed in a homogeneous set by introducing a dummy variable, as is easily seen in an example:

$$\begin{pmatrix} 2 & 3 \\ 5 & -1 \end{pmatrix}\mathbf{x} = \begin{pmatrix} 9 \\ 14 \end{pmatrix}. \tag{5.5}$$

Now write the equations as

$$\begin{aligned} 2x_1 + 3x_2 - 9 &= 0, \\ 5x_1 - x_2 - 14 &= 0, \end{aligned} \tag{5.6}$$

and use a dummy variable x_3, so that (5.6) can be written as

$$\begin{aligned} 2x_1 + 3x_2 + 9x_3 &= 0, \\ 5x_1 - x_2 + 14x_3 &= 0, \end{aligned} \tag{5.7}$$

where we should keep in mind that $x_3 = -1$. Therefore if $\mathbf{Bx} = \mathbf{k}$ gives nonhomogeneous equations, corresponding homogeneous equations are $(\mathbf{B} \mid \mathbf{k})\mathbf{x} = \mathbf{0}$, with the vector of constants $\mathbf{k}$ added to matrix $\mathbf{B}$ as a $(n + 1)^{\text{th}}$ column (and vector $\mathbf{x}$ augmented with an element x_{n+1}). This trick makes it possible to think of nonhomogeneous equations as just a special case of homogeneous equations.

5.5.4. Suppose now that $\mathbf{a}_1'\mathbf{x} = 0$ and $\mathbf{a}_2'\mathbf{x} = 0$ are two homogeneous equations, representing hyperplanes in an S_n with intersection V_{n-2}, following the general rule. Then it is easy to find an equation for a third hyperplane through the same intersection (i.e., so that the three hyperplanes do not have a V_{n-3} as their intersection, but form the special case with the intersection V_{n-2}). All we have to do is to construct a third equation $\mathbf{a}_3'\mathbf{x} = 0$ according to

$$\mathbf{a}_3 = c_1\mathbf{a}_1 + c_2\mathbf{a}_2 \tag{5.8}$$

with c_1 and c_2 arbitrary constants. Then any vector $\mathbf{x}$ that satisfies the first two equations (i.e., any point of the intersection V_{n-2}) also satisfies the third.

Example: we are given

$$\begin{pmatrix} 3 & -2 & 2 & -5 & 1 \\ 1 & 3 & -2 & 1 & -2 \end{pmatrix}\mathbf{x} = \mathbf{0}, \tag{5.9}$$

with a V_3 as their intersection. All points of the V_3 also satisfy

$$10x_1 + 8x_2 - 4x_3 - 6x_4 - 6x_5 = 0, \tag{5.10}$$

with (5.10) constructed as two times the first equation of (5.9) added to four times the second. Or, if we write (5.9) as

$$\mathbf{A}_2\mathbf{x} = \mathbf{0}, \tag{5.11}$$

with $\mathbf{A}_2$ as the $2 \times n$ matrix of coefficients, the condition for constructing a third equation through the common V_3 is

$$\mathbf{a}_3' = \mathbf{c}'\mathbf{A}_2. \tag{5.12}$$

The direct generalization: for a set of r equations

$$\mathbf{A}_r\mathbf{x} = \mathbf{0}, \tag{5.13}$$

with a V_{n-r} for their intersection, an $(r + 1)^{\text{th}}$ hyperplane through this V_{n-r} satisfies

$$\mathbf{a}_{r+1}' = \mathbf{c}_r'\mathbf{A}_r, \tag{5.14}$$

where $\mathbf{c}_r'$ is a $1 \times r$ vector of constants. Or, if we collect the $r + 1$ equations in one notation, $\mathbf{A}_{r+1}\mathbf{x} = 0$, then there must be a $1 \times (r + 1)$ vector $\mathbf{c}_{r+1}'$ such that

$$\mathbf{c}_{r+1}'\mathbf{A}_{r+1} = \mathbf{0}. \tag{5.15}$$

This last expression is completely equivalent to (5.14) if for the first r elements of $\mathbf{c}_{r+1}'$ we take those of $\mathbf{c}_r'$, whereas the final element of $\mathbf{c}_{r+1}'$ is set equal to -1. Note that any vector proportional to $\mathbf{c}_{r+1}'$ so defined also satisfies (5.15).

The equations defined by (5.13) are said to be *linearly independent.* Equation $\mathbf{a}_{r+1}'\mathbf{x} = 0$, so defined that (5.15) is valid, is said to be *linearly dependent* on the other r equations.

An alternative way to express the condition for linear dependence (5.15) is to form square minors $\mathbf{M}_{r+1}$ of order $r + 1$ from the matrix $\mathbf{A}_{r+1}$ by omitting $n - r - 1$ columns. Then it follows from (5.15) that for each such minor, regardless of which columns were omitted, it is true that

$$\mathbf{c}'\mathbf{M}_{r+1} = \mathbf{0}. \tag{5.16}$$

However, this implies that

$$|\mathbf{M}_{r+1}| = 0. \tag{5.17}$$

A proof of (5.17) follows from rule 7 in section 4.3. The operations described in (5.13) and (5.14) show that if $\mathbf{c}'\mathbf{A}_r$ is subtracted from the final row $\mathbf{a}_{r+1}$ in matrix $\mathbf{A}_{r+1}$, then this final row becomes zero; the same then is true for any minor $\mathbf{M}_{r+1}$. According to rule 7, the operation does not change the value of the determinant; on the other hand, a determinant with a zero row has zero value (rule 6).

Equation (5.17) gives the condition that must be satisfied if an $(r + 1)^{\text{th}}$ equation is linearly dependent on r other ones. For the original set of r linearly independent equations, it must not be true that every $|\mathbf{M}_r| = 0$, since otherwise $\mathbf{A}_r\mathbf{x} = 0$ would already contain an equation linearly dependent on a set of $r - 1$ equations.

Now the generalization. Again we start with a set of linearly independent equations, $\mathbf{A}_r\mathbf{x} = \mathbf{0}$. Add $m - r$ new equations, each of which is linearly dependent on the given set. Then for each equation we can find a vector $\mathbf{c}'$ such that $\mathbf{c}'\mathbf{A}_m = \mathbf{0}$, and all minors of order $r + 1$ or higher that can be formed out of $\mathbf{A}_m$ will have zero determinant. However, minors of order r will not all have zero determinant. Matrix $\mathbf{A}_m$ is then said to be of *rank r*. In general, the definition of the rank of a matrix is: Matrix $\mathbf{A}$ has rank r if all determinants of square minors of order $r + 1$ or higher order vanish, whereas there is at least one minor of order r with a non-zero determinant. The definition implies that the rank of an $n \times m$ matrix can never be larger than n or m, whichever is smaller.

The concept of rank makes it possible to say simply that m hyperplanes through the origin in an S_n intersect in a V_{n-r} if the rank of the matrix of coefficients of the corresponding equations is equal to r.

5.5.5. Examples. As we saw in section 5.5, in general we need $n - 1$ homogeneous equations in n unknowns to find a solution. We are now in a position to specify the condition under which they can be solved, namely, that the rank of the matrix of coefficients must be equal to $n - 1$.

If the rank is smaller, no unique solution can be identified. Also, if we have more than $n - 1$ equations, a solution can be identified only if the rank is $n - 1$.

Example: the four equations

$$\begin{pmatrix} -13 & 4 & 1 \\ 3 & 3 & 5 \\ 11 & -2 & 1 \\ 2 & 1 & 2 \end{pmatrix} \mathbf{x} = 0$$

have a solution, since the rank of the matrix is 2. The reader is invited to verify that any 3×3 minor out of the matrix of coefficients has a zero determinant, whereas 2×2 minors do not have zero determinants. The solution is $\mathbf{x}' = (1 \quad 4 \quad -3)$, or any $\mathbf{x}'$ proportional to this one.

To come back to nonhomogeneous equations $\mathbf{Ax} = \mathbf{k}$, remember that they can be reduced to the homogeneous set $(\mathbf{A} \,\vdots\, \mathbf{k})\mathbf{x} = \mathbf{0}$. If $\mathbf{A}$ is of order $n \times n$, the solution is $\mathbf{x} = \mathbf{A}^{-1}\mathbf{k}$ (section 5.3), provided that $|\mathbf{A}| \neq 0$. Since $\mathbf{A}$ is a minor out of the augmented matrix $(\mathbf{A} \,\vdots\, \mathbf{k})$, $(\mathbf{A} \,\vdots\, \mathbf{k})$ must have rank n, which is a necessary condition, but not a sufficient one, for $\mathbf{x}$ to have a solution; that is, $(\mathbf{A} \,\vdots\, \mathbf{k})$ may have rank n, but nevertheless there may be no solution for $\mathbf{x}$. Such a situation arises, for instance, with

$$3x_1 + x_2 = 5,$$

$$-3x_1 - x_2 = 2,$$

which two equations define parallel lines (section 5.5.2). We see that $(\mathbf{A} \,\vdots\, \mathbf{k})$ has rank 2, but the equations cannot be solved, since $|\mathbf{A}| = 0$. If we take the alternative homogeneous equations,

$$3x_1 + x_2 + 5x_3 = 0,$$

$$-3x_1 - x_2 + 2x_3 = 0,$$

their solution is $\mathbf{x} = (1 \quad -3 \quad 0)$, but obviously we cannot use the proportionality factor to make x_3 equal to -1, as was assumed in section 5.5.3. Apparently with nonhomogeneous equations we have to proceed with more caution. In general, it can be shown that both $\mathbf{A}$ and $(\mathbf{A} \,\vdots\, \mathbf{k})$ must have the same rank for the equations to be consistent; if this rank is r, then the corresponding hyperplanes intersect in a V_{n-r}.

5.6. Some Geometrical Implications of the Concept of Rank

Let **X** be an $m \times n$ matrix; it defines m points in an S_n. If $m > n$, and **X** has rank n, then the m points require the total S_n and are not contained in a subspace of S_n, provided that the augmented matrix $(\mathbf{X} \mid \mathbf{u})$ has rank $n + 1$, with **u** being a unit vector.

The reason for the latter addition is that if $(\mathbf{X} \mid \mathbf{u})$ also has rank n—it cannot have smaller rank, since **X** is a minor of $(\mathbf{X} \mid \mathbf{u})$—then we can select n points and solve for a vector **g** that satisfies $\mathbf{Xg} = \mathbf{u}$. Now take a vector **k** proportional to **g** in such a way that $\mathbf{k}'\mathbf{k} = 1$. Then $\mathbf{Xk} = c\mathbf{u}$, with c some constant. Geometrically this means that **k** gives direction cosines for a vector on which all points have equal projection. It follows immediately that the n points are located in a V_{n-1} perpendicular to this vector. In other words, the dimensionality of the set of points is $n - 1$, not n.

If **X** has rank r, with $r < n$, then the points are located in a subspace V_r of S_n. Here we shall be able to find $n - r$ orthogonal vectors $\mathbf{k}_i$ for which $\mathbf{Xk}_i = \mathbf{0}$, and which satisfy $\mathbf{k}_i'\mathbf{k}_i = 1$ and $\mathbf{k}_i'\mathbf{k}_j = 0$ for $i \neq j$. The remaining hyperplane that contains the points therefore must have dimensionality r. An implied condition is that the augmented matrix $(\mathbf{X} \mid \mathbf{u})$ has rank $r + 1$. If $(\mathbf{X} \mid \mathbf{u})$ had rank r, then the m points would be located in a V_{r-1} that excludes the origin.

The argument can be applied to the case where $m = n$. We then have n points in a V_{n-1} if **X** has rank n. Here it is always true that $(\mathbf{X} \mid \mathbf{u})$ also has rank n, since $(\mathbf{X} \mid \mathbf{u})$ is of the order $n \times (n + 1)$ and therefore cannot have rank $n + 1$. A familiar illustration of this case is that three points in an S_3 define a plane.

If $m < n$, and **X** has rank r, with $r < m$, then the m points are located in a V_r through the origin, provided that $(\mathbf{X} \mid \mathbf{u})$ has rank $r + 1$. If $(\mathbf{X} \mid \mathbf{u})$ has rank r, then the m points are located in a V_{r-1} that excludes the origin.

A complete example is given for the 10×3 matrix **X** in table 5.1. Matrix **X** here is of rank 2, since it has been so constructed that for each row $x_1 + 2x_2 - x_3 = 0$. In other words, there is a vector **c** such that $\mathbf{Ac} = \mathbf{0}$, with $\mathbf{c}' = (1 \quad 2 \quad -1)$. Note that if we take **k** proportional to **c** and $\mathbf{k}'\mathbf{k} = 1$, or $\mathbf{k}' = (1 \quad 2 \quad -1)/\sqrt{6}$, then **k** gives direction cosines of a vector perpendicular to the plane of points, since for the projections **y** it is true that $\mathbf{Xk} = \mathbf{0}$.

Table 5.1. A 10 × 3 matrix, **X**.

7	0	7
7	1	9
5	2	9
5	0	5
4	2	8
4	5	14
3	1	5
2	4	10
2	3	8
1	2	5

The cloud of points is pictured in figure 5.3.

An alternative picture would give the three vectors in an S_{10}. They will never require a hyperplane of dimensionality larger than 3, but here, where **X** has rank 2, they fit in a V_2, as in figure 5.4. Squared vector lengths are 198, 64, 710, respectively, and angles between them are: 55° for the pair (x_1, x_2); 29° for the pair (x_1, x_3); and 26° for the pair (x_2, x_3). The reader is invited to verify these results from the inner product matrix, $\mathbf{X}'\mathbf{X}/n$.

For completeness, we also give the ten planes in S_3 if **X** is taken as the matrix of coefficients for homogeneous equations $\mathbf{Xw} = \mathbf{0}$, with **w** as coordinates. Since **X** has rank 2, the ten planes have a $V_{m-2} = V_1$ in common; they share a common line of intersection, as is shown in figure 5.5.

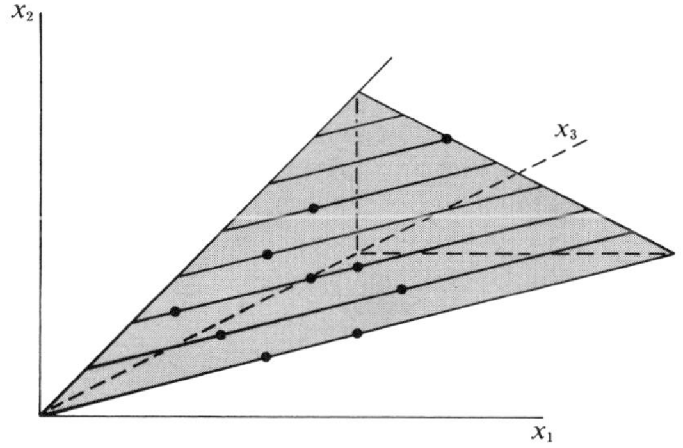

Figure 5.3 Graph of the matrix in table 5.1.

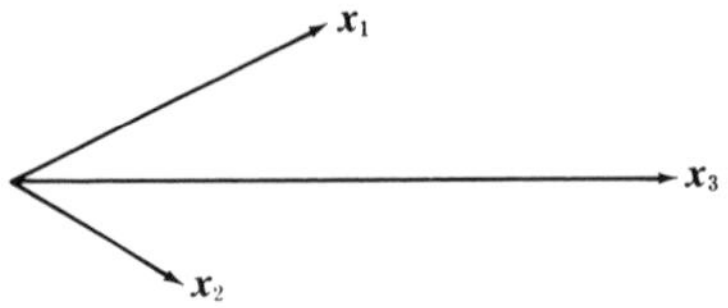

Figure 5.4
Vector representation of data in table 5.1.

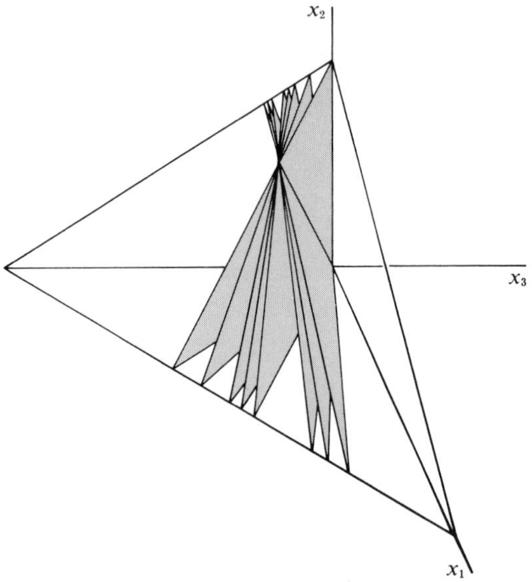

Figure 5.5 Graph of the matrix in table 5.1 as planes $\mathbf{Ax} = 0$.

5.7. Exercises

1. Given a matrix of coefficients $\mathbf{A} = \begin{pmatrix} 1 & 1 & 1 \\ 1 & 2 & 3 \\ 1 & 4 & 9 \end{pmatrix}$ for the equations $\mathbf{Ax} = \mathbf{k}$, where $\mathbf{k}' = (6 \quad 14 \quad 36)$, solve for $\mathbf{x}$ (compare exercise 5 in section 4.8).

2. Solve

$$\begin{pmatrix} 1 & -1 & -1 \\ 2 & 3 & 2 \\ 5 & 1 & 1 \end{pmatrix}\begin{pmatrix} x \\ y \\ z \end{pmatrix} = \begin{pmatrix} 0 \\ 2 \\ 6 \end{pmatrix}.$$

3. Determine the solution of

$$\begin{aligned} 5x + 7y - 2z + w &= 0; \\ 13x + 4y + 9z + 3w &= 0; \\ x + 2y - z + 7w &= 0. \end{aligned}$$

4. Investigate whether the following equations have a solution, and if so, identify it:

$$\begin{aligned} x + 3y - 4z - 5w &= 0; \\ 2x - 7y + 5z + 3w &= 0; \\ x - 3y + z + 3w &= 0; \\ 13x + y - 18z - 19w &= 0. \end{aligned}$$

5. The matrix in table 5.2 gives scores for nine persons on three tests.

Table 5.2. Scores of nine persons on three tests.

$$\begin{pmatrix} 12 & 16 & 6 \\ 8 & 3 & 8 \\ 4 & 13 & 11 \\ 7 & 4 & 2 \\ 2 & 3 & 8 \\ 8 & 6 & 2 \\ 2 & 4 & 4 \\ 9 & 10 & 12 \\ 11 & 13 & 10 \end{pmatrix}$$

First calculate a matrix **A** for deviations from the means. Then determine a matrix **X** by subtracting from each row of **A** the average for that row. The matrix **X** then gives a matrix of "profiles" for each individual. Prove that **X** has rank 2.

It follows that the rows of **X** specify points in an S_2. Therefore, it must be possible to find a rotation matrix **K** in such a way that one of the columns of **Y** = **XK** has zero elements only. Find a matrix **K** that satisfies this requirement. Make a picture of the points by using the non-zero columns of **Y** as coordinates. Also, draw in this plane the point that stands for a flat profile; draw points for rows of **A** equal to the following "extreme" profiles:

$$(6 \quad 0 \quad 0), (0 \quad 6 \quad 0), (0 \quad 0 \quad 6), (-6 \quad 0 \quad 0), (0 \quad -6 \quad 0), (0 \quad 0 \quad -6).$$

Perpendicular to the plane that pictures the profiles, one might set up an axis on which the coordinates correspond to the averages of the rows of **A**; in this way the three-dimensional space with points representing deviations from means is reconstructed.

6

Matrix Differentiation

6.1. Survey of Elementary Calculus

The reader should, I hope, already be somewhat familiar with elementary calculus. The first part of this chapter is a refresher survey of those topics from elementary calculus that are used later in this book.

6.1.1. Differentiation of Simple Rational Functions. Given a rational function $y = f(x)$, like $y = x^2 - 2x + 3$ (a parabola), a graph of the function would show a curve in the (x, y) plane. Figure 6.1 gives a sketch of how such a curve might look.

If we select a value for x, the corresponding value for y can be found from $y = f(x)$. Suppose that we now take a somewhat higher value for x, say, $x + \Delta x$. Then y will change also, in general, to $y + \Delta y$, where Δy might be negative. Then $y + \Delta y = f(x + \Delta x)$.

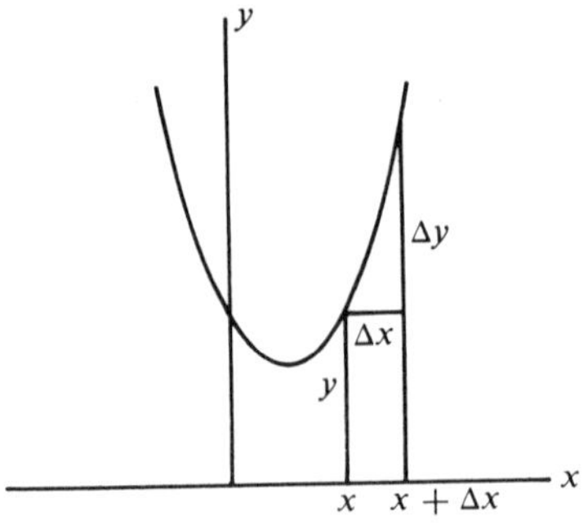

Figure 6.1
Graph of a function, $y = f(x)$.

In figure 6.1, the ratio $\Delta y/\Delta x$ represents the tangent of the angle between the x-axis and the chord from (x, y) to $(x + \Delta x, y + \Delta y)$. If Δx is small, this angle will approximate the angle between the x-axis and the tangent at point (x, y). In fact, the tangent of this latter angle can be looked on as the limit of $\Delta y/\Delta x$ if $\Delta x \to 0$.

This limit is called the *differential quotient*, and is written dy/dx. In general, it will be possible to write this quotient as a function of x, by itself. This function is symbolized y', and called the (first) *derivative* of y.

Example: $y = bx^2$. We have

$$y + \Delta y = b(x + \Delta x)^2 = bx^2 + 2bx\,\Delta x + b(\Delta x)^2.$$

But since bx^2 on the right is equal to y on the left, we can write

$$\Delta y = 2bx\,\Delta x + b(\Delta x)^2,$$

and

$$\Delta y/\Delta x = 2bx + b\,\Delta x.$$

If Δx is small, the second term at the right can be neglected, and we have: $dy/dx = 2bx$, or $y' = 2bx$.

More generally, if $y = ax^n$, then $y' = nax^{n-1}$. (6.1)

6.1.2. If u and v are functions of x, and $y = u + v$, then $y' = u' + v'$; that is, the derivative of a sum is equal to the sum of the derivatives.

Example:

$$u = 3x^2, \qquad v = 4x^3;$$
$$y = u + v = 3x^2 + 4x^3;$$
$$u' = 6x, \qquad v' = 12x^2 \quad \text{[using (6.1)]};$$
$$y' = u' + v' = 6x + 12x^2.$$

6.1.3. If u and v are functions of x, and $y = uv$, then $y' = u'v + uv'$. Example: $u = 3x^2$, $v = 4x^3$, and $y = uv$. Then

$$y' = (6x)(4x^3) + (3x^2)(12x^2) = 60x^4,$$

which could have been obtained more directly from $y = 12x^5$.

6.1.4. If u and v are two functions of x, and $y = u/v$, then $y' = (u'v - uv')/v^2$. Example: $u = 3x^2$, $v = 4x^3$;

$$y' = (24x^4 - 36x^4)/16x^6 = (-3/4)x^{-2}.$$

6.1.5. If y is a function of u, and u is a function of x, i.e., $y = f(u)$, $u = g(x)$, then $dy/dx = (dy/du) \cdot (du/dx)$.

Example: $y = u^2 - 1$, $u = 2x + 6$. Then $dy/dx = (2u) \cdot (2) = 4u = 8x + 24$. This might be compared to the other approach, where we first calculate y as a function of x:

$$y = (2x + 6)^2 - 1 = 4x^2 + 24x + 35,$$

and therefore $dy/dx = 8x + 24$.

6.1.6. Extreme Values. Let $y = f(x)$ and $y' = g(x)$. Suppose $y' = 0$ for some value x_1 of x; then the graph of $y = f(x)$ will have a horizontal tangent (parallel to the axis) at x_1. In fact, if we exclude the rather special cases of a point of inflection or a cusp (see figure 6.2), the horizontal tangent will reveal that the graph has an extreme value at that point, a minimum or a maximum. If the derivative of y' (the *second derivative* of y), symbolized y'', is positive at that point, we have a minimum, and if y'' is negative, a maximum.

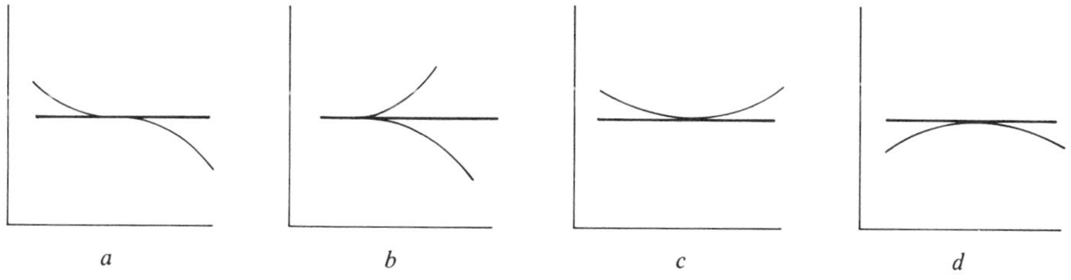

Figure 6.2 Sketch of (a) point of inflection, (b) cusp, (c) minimum, and (d) maximum.

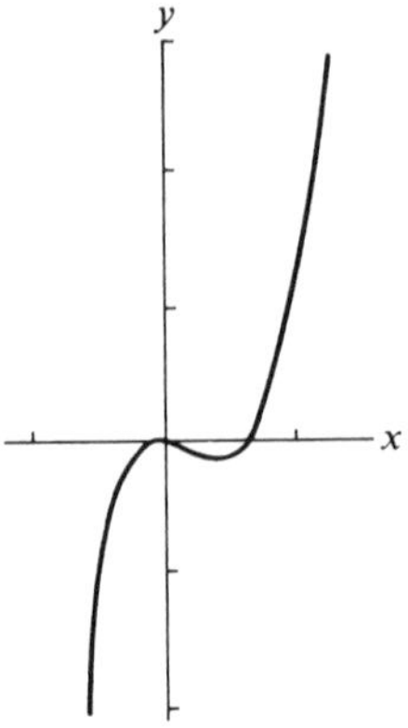

Figure 6.3
Graph of $y = 3x^3 - 2x^2$.

Example: figure 6.3 is a graph of the function $y = 3x^3 - 2x^2$. The first derivative is $y' = 9x^2 - 4x$, and it is easily solved that $y' = 0$ for $x_1 = 0$, or for $x_2 = 4/9$. For $x_1 = 0$, the corresponding y_1 is 0; and for $x_2 = 4/9$, we find $y_2 = -32/243$. Apparently, the extreme values are at these two points. To see which is which, we might calculate $y'' = 18x - 4$. Substituting x_1, we find $y'' = -4$, so that (x_1, y_1) is identified as a maximum. The other extreme value is a minimum, because $y'' = 4$ (positive) if we substitute $x_2 = 4/9$.

6.1.7. Partial Derivatives. Suppose $y = f(x_1, x_2)$, where the variables x_1 and x_2 can take independent values, for instance,

$$y = 3x_1^2 + 4x_2^2. \tag{6.2}$$

Geometrically, the function is represented as a surface in an S_3, where the S_3 is defined by the rectangular coordinates y, x_1, and x_2. For some constant value of x_2, say, $x_2 = 4$, the function is reduced to

$$y = 3x_1^2 + 64, \tag{6.3}$$

where y is expressed as a function of x_1 alone. Taken by itself, the graph of this function is a parabola, which is the intersection of the surface given by (6.2) with the plane $x_2 = 4$. If we take the derivative of (6.3), we find the slope for a tangent to the intersection. More generally, we can treat x_2 as a constant and calculate the derivative function under this assumption. Such a derivative is called a *partial derivative*, and is written $\partial y/\partial x_1$. It follows that the intersection has an extreme value at that value of x_1 for which the partial derivative is equal to zero.

An extreme value for the entire surface can be found by setting both partial derivatives equal to zero. In the example,

$$\partial y/\partial x_1 = 6x_1,$$

$$\partial y/\partial x_2 = 8x_2,$$

and it follows that y has extreme value for $6x_1 = 0$ and $8x_2 = 0$, or $x_1 = x_2 = 0$. The corresponding y can be found from (6.2); it also is equal to zero in this example. Whether the extreme value is a maximum or a minimum could be determined from further derivatives, but we shall not treat these methods here; instead, we assume it will be clear from the context what kind of extreme value we have. Moreover, in cases of doubt, one can always compute y for some values in the neighborhood of the extreme value, and thus readily discover what kind of extreme it is.

Another example:

$$y = 5x_1^2 + x_2^2 + 4x_1x_2 + 3x_2. \tag{6.4}$$

The partial derivatives are

$$\begin{aligned} \partial y/\partial x_1 &= 10x_1 + 4x_2, \\ \partial y/\partial x_2 &= 4x_1 + 2x_2 + 3. \end{aligned} \tag{6.5}$$

Setting both partials equal to zero and solving for x_1 and x_2 gives $x_1 = 3$, $x_2 = -7.5$. From (6.4) we find that the corresponding y equals -1.25. it is a minimum.

6.1.8. Differentiation with Side-Conditions. We are again given $y = f(x_1, x_2)$, but now we assume, not that x_1 and x_2 are independent, but that they are restricted by a condition expressed as $g(x_1, x_2) = 0$. Geometrically, this means we are interested in the intersection of the surface y and the surface defined by $\mathbf{g} = 0$.

Example:

$$y = 3x_1^2 + 5x_2^3, \tag{6.6}$$

$$g(x_1, x_2) = 2x_1 - 3x_2 - 1 = 0. \tag{6.7}$$

It can be shown for cases like these (without proof) that

$$dy/dx_1 = \partial y/\partial x_1 + (\partial y/\partial x_2) \cdot (dx_2/dx_1). \tag{6.8}$$

From (6.7) we can find an expression for x_2 as a function of x_1:

$$x_2 = (2x_1 - 1)/3. \tag{6.9}$$

This enables us to find a solution for (6.8):

$$dy/dx_1 = 6x_1 + (15x_2^2) \cdot (2/3) = 6x_1 + 10x_2^2. \tag{6.10}$$

In general it will not be so easy to express x_2 as a function of x_1. For instance, if the side-condition is $g(x_1, x_2) = x^4 - 2x_1^2x^2 + x_2^3 = 0$, no simple expression for x_2 is available. However, to solve (6.8) we do not really need to express x_2 as a function of x_1; all we need is the derivative dx_2/dx_1. Often this derivative can be more easily found by applying equation (6.8) to function g:

$$dg/dx_1 = \partial g/\partial x_1 + (\partial g/\partial x_2) \cdot (dx_2/dx_1). \tag{6.11}$$

Since $g = 0$, it must also be true that $dg/dx_1 = 0$, and therefore we can solve (6.11) for (dx_2/dx_1).

In our example (6.11) becomes

$$0 = 2 + (-3)(dx_2/dx_1), \tag{6.12}$$

and it follows that

$$dx_2/dx_1 = 2/3 \tag{6.13}$$

as we found earlier.

Suppose now that we want to find extreme values for y under the condition $g = 0$. Then it is necessary that $dy/dx_1 = 0$. Given further that $dg/dx_1 = 0$ (always true), and the condition $g = 0$ itself, the problem can often be solved.

Example: we use example (6.2) from section 6.1.7: $y = 3x_1^2 + 4x_2^2$. Without restriction, the corresponding surface has vertical cross-sections that are parabolas (see figure 6.4). We now set the side-condition $3x_1 - 2x_2 - 1 = 0$, which defines a vertical plane, whose intersection with the surface is a parabola. We want to determine the minimum value of this parabola. From equation (6.8), we have

$$dy/dx_1 = 6x_1 + 8x_2 \cdot (dx_2/dx_1). \tag{6.14}$$

Equation (6.11) gives

$$dg/dx_1 = 3 - 2(dx_2/dx_1) = 0. \tag{6.15}$$

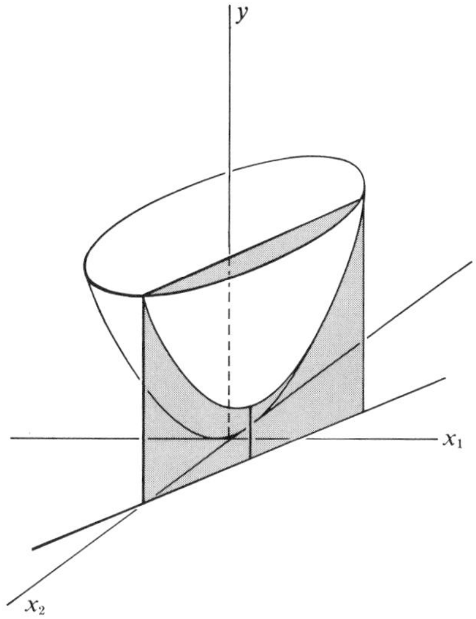

Figure 6.4 Graph of $y = 3x_1^2 + 4x_2^2$, and its intersection with the plane $3x_1 - 2x_2 - 1 = 0$.

We have finally the condition itself:

$$3x_1 - 2x_2 - 1 = 0. \tag{6.16}$$

From (6.15) we obtain

$$dx_2/dx_1 = 3/2. \tag{6.17}$$

Substituting (6.17) in (6.14), and setting (6.14) equal to zero, we find

$$0 = 6x_1 + 12x_2. \tag{6.18}$$

Then (6.16) and (6.18) are two equations with two unknowns, with solutions $x_1 = 1/4$, $x_2 = -1/8$. From the function y itself (6.2), we can solve that the corresponding y is 1/4.

6.1.9. Method of Undetermined Multipliers. Often the method described in section 6.1.8 is rather tedious, and sometimes it cannot even be applied at all. A simpler method is available, however, called the method of *undetermined multipliers* (also called Lagrange multipliers). The procedure is as follows.

For an extreme value, we set $dy/dx_1 = 0$. We know also that $dg/dx_1 = 0$. Then it must be true that

$$dy/dx_1 - \mu \, dg/dx_1 = 0 \tag{6.19}$$

for any value of μ. Using equations (6.8) and (6.11), we can rewrite (6.19) as

$$\partial y/\partial x_1 + (\partial y/\partial x_2)(dx_2/dx_1) - \mu[\partial g/\partial x_1 + (\partial g/\partial x_2)(dx_2/dx_1)] = 0, \tag{6.20}$$

or, rearranged,

$$[\partial y/\partial x_1 - \mu\ \partial g/\partial x_1] + [\partial y/\partial x_2 - \mu\ \partial g/\partial x_2](dx_2/dx_1) = 0. \tag{6.21}$$

Now it will be possible, in general, to select a value for μ so that

$$\partial y/\partial x_2 - \mu\ \partial g/\partial x_2 = 0, \tag{6.22}$$

and it follows immediately from (6.21) that

$$\partial y/\partial x_1 - \mu\ \partial g/\partial x_1 = 0. \tag{6.23}$$

Equations (6.22) and (6.23), together with the condition $g = 0$ itself, then are in general sufficient so solve for x_1, x_2, and μ.

In the second example of section 6.1.8:

$$\partial y/\partial x_1 - \mu\ \partial g/\partial x_1 = 6x_1 - \mu(3) = 0, \tag{6.24}$$

$$\partial y/\partial x_2 - \mu\ \partial g/\partial x_2 = 8x_2 - \mu(-2) = 0. \tag{6.25}$$

The condition was

$$3x_1 - 2x_2 - 1 = 0. \tag{6.26}$$

From (6.24) and (6.25) we have $\mu = 2x_1 = -4x_2$, from which it follows that $2x_1 + 4x_2 = 0$. Together with (6.26) we then can solve for x_1 and x_2: $x_1 = 1/4$, $x_2 = -1/8$. We also could solve $\mu = 1/2$, but this result is not needed.

The procedure can be simplified further by using an auxiliary function

$$F = y - \mu g. \tag{6.27}$$

Equations (6.22) and (6.23) then are simply found by taking partial derivatives

$$\begin{aligned} \partial F/\partial x_1 &= \partial y/\partial x_1 - \mu\ \partial g/\partial x_1, \\ \partial F/\partial x_2 &= \partial y/\partial x_2 - \mu\ \partial g/\partial x_2, \end{aligned} \tag{6.28}$$

which have to be set equal to zero.

Now the final point in this section is to generalize the result to functions of n variables $x_1, x_2, \ldots, x_n$. We do this without proof. Suppose

$$y = f(x_1, x_2, \ldots, x_n),$$

and there is some condition

$$g(x_1, x_2, \ldots, x_n) = 0.$$

Again we use an auxiliary function

$$F = y - \mu g,$$

and for extreme values we set its partial derivatives,

$$\partial F/\partial x_1, \partial F/\partial x_2, \ldots, \partial F/\partial x_n,$$

equal to zero. We then obtain n equations with $n + 1$ unknowns. They are the values $x_1, x_2, \ldots, x_n$, and in addition the unknown value of μ. However, $g = 0$ gives an $(n + 1)^{\text{th}}$ equation, and makes the set solvable, in general. We shall not give an example, because doing so takes too much space unless matrix notation is used. Matrix notation is the subject of the next sections.

6.2. Vector Differentiation

A function $y = \mathbf{a}'\mathbf{x}$ can be expanded as $y = a_1x_1 + a_2x_2 + \cdots + a_nx_n$; it is a function of n variables. Partial derivatives $\partial y/\partial x_1, \partial y/\partial x_2, \ldots, \partial y/\partial x_n$ are $a_1, a_2, \ldots, a_n$, which we can collect in the vector $\mathbf{a}'$. So the result can be summarized by saying $\partial y/\partial \mathbf{x} = \mathbf{a}'$, where it is understood now that $\partial y/\partial \mathbf{x}$ stands for a $1 \times n$ row vector. We might prefer to have the derivatives in a column vector; the notation would be $\partial y/\partial \mathbf{x}' = \mathbf{a}$. These operations are called *vector differentiation* (or symbolic differentiation).

6.2.1. A function $f = \mathbf{x}'\mathbf{A}\mathbf{y}$ (with $\mathbf{A}$ an $n \times m$ matrix), written out, is:

$$\begin{aligned} f = {} & x_1a_{11}y_1 + x_1a_{12}y_2 + \cdots + x_1a_{1n}y_n \\ & + x_2a_{21}y_1 + x_2a_{22}y_2 + \cdots + x_2a_{2n}y_n \\ & + \cdots \\ & + x_na_{n1}y_1 + x_na_{n2}y_2 + \cdots + x_na_{nn}y_n. \end{aligned}$$

The form $\mathbf{x}'\mathbf{A}\mathbf{y}$ is called a *bilinear form*. The partial derivative $\partial f/\partial x_1$ of the function $f = \mathbf{x}'\mathbf{A}\mathbf{y}$ is

$$\begin{aligned}\partial f/\partial x_1 &= a_{11}y_1 + a_{12}y_2 + \cdots + a_{1n}y_n \\ &= \mathbf{a}_1{}'\mathbf{y}.\end{aligned}$$

Similarly, for the other partial derivatives,

$$\begin{aligned}\partial f/\partial x_2 &= \mathbf{a}_2{}'\mathbf{y}, \\ &\cdots \\ \partial f/\partial x_n &= \mathbf{a}_n{}'\mathbf{y}.\end{aligned}$$

The result can be summarized in the notation $\partial f/\partial \mathbf{x}' = \mathbf{A}\mathbf{y}$, where the derivatives are given in a column vector. If we prefer a row vector, we should write

$$\partial f/\partial \mathbf{x} = \mathbf{y}'\mathbf{A}'.$$

Similarly, $\partial f/\partial \mathbf{y} = \mathbf{x}'\mathbf{A}$, and $\partial f/\partial \mathbf{y}' = \mathbf{A}'\mathbf{x}$.

6.2.2. The form $\mathbf{x}'\mathbf{A}\mathbf{x}$, where $\mathbf{A}$ is a symmetric matrix, is called a *quadratic form* of matrix $\mathbf{A}$. It can be written out as

$$\begin{aligned} &x_1a_{11}x_1 + x_1a_{12}x_2 + \cdots + x_1a_{1n}x_n \\ &+ x_2a_{21}x_1 + x_2a_{22}x_2 + \cdots + x_2a_{2n}x_n \\ &+ \quad \cdots \\ &+ x_na_{n1}x_1 + x_na_{n2}x_2 + \cdots + x_na_{nn}x_n. \end{aligned}$$

The function $f = \mathbf{x}'\mathbf{A}\mathbf{x}$ therefore has a partial derivative $\partial f/\partial x_1$:

$$\partial f/\partial x_1 = 2a_{11}x_1 + 2a_{12}x_2 + \cdots + 2a_{1n}x_n = 2\mathbf{a}_1{}'\mathbf{x},$$

where we have used the symmetry of $\mathbf{A}$ in that $a_{12} = a_{21}$, etc. Similarly, we find

$$\partial f/\partial x_2 = 2\mathbf{a}_2{}'\mathbf{x}, \qquad \partial f/\partial x_3 = 2\mathbf{a}_3{}'\mathbf{x}, \ldots, \qquad \partial f/\partial x_n = \mathbf{a}_n{}'\mathbf{x},$$

and the result can be summarized by writing

$$\partial f/d\mathbf{x}' = 2\mathbf{A}\mathbf{x},$$

or, in a row vector,

$$\partial f/\partial \mathbf{x} = 2\mathbf{x}'\mathbf{A}.$$

6.2.3. Extreme Values. In the same way as with ordinary equations, one can determine extreme values for matrix functions, by setting all partial derivatives equal to zero. For instance, given the function

$$y = \mathbf{x}'\mathbf{A}\mathbf{x} - \mathbf{b}'\mathbf{x},$$

where $\mathbf{y}$, $\mathbf{x}$, and $\mathbf{b}$ are $n \times 1$ vectors and $\mathbf{A}$ a symmetric $n \times n$ matrix, then

$$\partial y/\partial \mathbf{x}' = 2\mathbf{A}\mathbf{x} - \mathbf{b}. \tag{6.29}$$

We set (6.29) equal to zero, and find $2\mathbf{A}\mathbf{x} = \mathbf{b}$, which defines n non-homogeneous equations in n unknowns, which can be solved if

$$|\mathbf{A}| \neq 0 \quad \text{as} \quad \mathbf{x} = -\tfrac{1}{2}\mathbf{A}^{-1}\mathbf{b}.$$

Or take example (6.2). We there had a function $y = 3x_1^2 + 4x_2^2$. In matrix notation, this can be written as $y = \mathbf{x}'\mathbf{A}\mathbf{x}$, with $\mathbf{A}$ a symmetric 2×2 matrix:

$$\mathbf{A} = \begin{pmatrix} 3 & 0 \\ 0 & 4 \end{pmatrix}.$$

We have $\partial y/\partial \mathbf{x}' = 2\mathbf{A}\mathbf{x} = \mathbf{0}$ for the extreme values, and it follows that $\mathbf{x} = \mathbf{0}$. The example shows the advantages of matrix notation.

A more interesting example is the following. We remember (from section 5.5.4) that if $\mathbf{A}$ is of order $m \times n$ with $m > n$, then the set $\mathbf{A}\mathbf{x} = \mathbf{k}$ contains more equations than unknowns and, in general, no solution for $\mathbf{x}$ is available. However, we might be interested in finding a vector $\mathbf{x}$ that is least in error. More precisely, if we take a vector $\mathbf{x}$ and calculate $\mathbf{A}\mathbf{x} = \mathbf{b}$, say, we want the discrepancies between $\mathbf{b}$ and $\mathbf{k}$ to be small. The best over-all criterion is to minimize the sum of the squares of the discrepancies, which gives a least-squares solution for $\mathbf{A}\mathbf{x} = \mathbf{k}$.

The discrepancies are given in a vector $(\mathbf{b} - \mathbf{k})$, and the sum of the squared elements then is given by

$$f = (\mathbf{b} - \mathbf{k})'(\mathbf{b} - \mathbf{k}). \tag{6.30}$$

Using $\mathbf{b} = \mathbf{Ax}$, we can write

$$f = (\mathbf{Ax} - \mathbf{k})'(\mathbf{Ax} - \mathbf{k}) = \mathbf{x}'\mathbf{A}'\mathbf{Ax} - \mathbf{k}'\mathbf{Ax} - \mathbf{x}'\mathbf{A}'\mathbf{k} + \mathbf{k}'\mathbf{k}$$
$$= \mathbf{x}'\mathbf{A}'\mathbf{Ax} - 2\mathbf{k}'\mathbf{Ax} + \mathbf{k}'\mathbf{k}, \tag{6.31}$$

where $\mathbf{k}'\mathbf{Ax} = \mathbf{x}'\mathbf{A}'\mathbf{k}$, since both are single numbers. The problem then can be redefined by saying that we want to find a vector $\mathbf{x}$ for which f has a minimum. Setting partial derivatives equal to zero, we find

$$\partial f/\partial \mathbf{x}' = 2\mathbf{A}'\mathbf{Ax} - 2\mathbf{A}'\mathbf{k} = \mathbf{0}, \tag{6.32}$$

and therefore

$$\mathbf{A}'\mathbf{Ax} = \mathbf{A}'\mathbf{k},$$

$$\mathbf{x} = (\mathbf{A}'\mathbf{A})^{-1}\mathbf{A}'\mathbf{k}, \tag{6.33}$$

where $\mathbf{A}'\mathbf{A}$ is a symmetric matrix, the determinant of which should be different from zero.

Example: we are given the equations

$$3x_1 + 2x_2 = 11,$$
$$x_1 + 4x_2 = 7,$$
$$5x_2 = 4.$$

For $\mathbf{A}'\mathbf{A}$, we find $\mathbf{A}'\mathbf{A} = \begin{pmatrix} 10 & 10 \\ 10 & 45 \end{pmatrix}$, and $\mathbf{A}'\mathbf{k} = \begin{pmatrix} 40 \\ 70 \end{pmatrix}$. Applying (6.33), the solution is seen to be

$$\mathbf{x} = \begin{pmatrix} 10 & 10 \\ 10 & 45 \end{pmatrix}^{-1} \begin{pmatrix} 40 \\ 70 \end{pmatrix} = \left(\begin{pmatrix} 45 & -10 \\ -10 & 10 \end{pmatrix} \begin{pmatrix} 40 \\ 70 \end{pmatrix} \right) \Big/ 350$$
$$= \begin{pmatrix} 1100 \\ 300 \end{pmatrix} \Big/ 350 = \begin{pmatrix} 22 \\ 6 \end{pmatrix} \Big/ 7.$$

6.2.4. Extreme Values under a Condition. Suppose we are given some matrix function $y = f(x)$, and a condition, also in matrix form, $g(x) = 0$. Using formula (6.27) from section 6.1.9, we introduce an auxiliary function, $F = y - \mu g$. To find extreme values, partial derivatives of F have to be set equal to zero. Take the example used in sections 6.1.8 and 6.1.9. We there had $y = 3x_1^2 + 4x_2^2$, and $g(x) = 3x_1 - 2x_2 - 1 = 0$.

In matrix notation,

$$y = \mathbf{x}'\mathbf{A}\mathbf{x}, \tag{6.34}$$

$$g = \mathbf{b}'\mathbf{x} - 1 = 0, \tag{6.35}$$

where

$$\mathbf{A} = \begin{pmatrix} 3 & 0 \\ 0 & 4 \end{pmatrix}, \qquad \mathbf{b} = \begin{pmatrix} 3 \\ -2 \end{pmatrix}. \tag{6.36}$$

The auxiliary function F becomes

$$F = \mathbf{x}'\mathbf{A}\mathbf{x} - \mu(\mathbf{b}'\mathbf{x} - 1). \tag{6.37}$$

We set its partial derivatives equal to zero,

$$\partial F/\partial \mathbf{x}' = 2\mathbf{A}\mathbf{x} - \mu\mathbf{b} = \mathbf{0}, \tag{6.38}$$

and it follows that

$$\mathbf{x} = \mu\mathbf{A}^{-1}\mathbf{b}/2. \tag{6.39}$$

We multiply (6.39) with $\mathbf{b}'$, and remember that $\mathbf{b}'\mathbf{x} = 1$, from (6.35):

$$\mathbf{b}'\mathbf{x} = 1 = \mu\mathbf{b}'\mathbf{A}^{-1}\mathbf{b}/2, \tag{6.40}$$

from which it follows that

$$\mu = 2(\mathbf{b}'\mathbf{A}^{-1}\mathbf{b})^{-1}. \tag{6.41}$$

We substitute (6.41) in (6.39) to find

$$\mathbf{x} = \mathbf{A}^{-1}\mathbf{b}(\mathbf{b}'\mathbf{A}^{-1}\mathbf{b})^{-1}. \tag{6.42}$$

In our example, using (6.36), $\mathbf{b}'\mathbf{A}^{-1}\mathbf{b} = 4$, and therefore $(\mathbf{b}'\mathbf{A}^{-1}\mathbf{b})^{-1} = 1/4$. The solution for $\mathbf{x}$ then is seen to be, from (6.42),

$$\mathbf{x} = \begin{pmatrix} 1 \\ -\frac{1}{2} \end{pmatrix} \Big/ 4 = \begin{pmatrix} 1/4 \\ -1/8 \end{pmatrix},$$

which is the same result as that obtained in section 6.1.8, as it should be.

6.3. Exercises

1. Given $y = 3x^2 - 2x + 8$, find for which values of x the function has extreme values, draw a graph of the function, and determine whether the extremes are maxima or minima.

2. Given $y = 2x^3 + 3x^2 - 36x - 5$, find extreme values.
3. Given $z = 2x^2 - 3xy + y^2 + 2x - 6y - 8$, find extreme values.
4. In a 3-dimensional coordinate system (x, y, z), a cylinder $x^2 + y^2 = 1$ is given, and also a plane $x + y + z = 0$. Find the coordinates of the intersection for which z has extreme value.
5. Find a least-squares solution for the following equations:

$$\begin{aligned} 3x + y + 2z &= 2, \\ 4x \quad\; + 6z &= 4, \\ 6y - 2z &= 1, \\ 3x - y - 10z &= -6. \end{aligned}$$

6. Let x_i $(i = 1, 2, \ldots, n)$ be n observed values of a variable x. The "second moment about c" is defined as $\sum (x_i - c)^2/n$. Find the value of c for which the moment has a minimum.
7. An amateur psychologist decides to do the following experiment. He prepares five geometrical forms with areas x_i equal to 40, 45, 50, 55, and 60 square inches, respectively. He then asks a friend to give estimates of the areas. They are, in the same order, 35, 55, 51, 53, and 56 sq. in.
   (a) The five points can be represented in the x,y-plane (with y for the estimates). Fit the best straight line through the set of points. This "best" line is defined by the property that the sum of the squared distances between the observed points and the corresponding points on the line for the same value of x_i has a minimum. In other words, define

$$F(a, b) = \sum [y_i - (a + bx_i)]^2,$$

   where a and b are parameters of the straight line $y = a + bx$. Find a minimum for F; this minimum defines which values must be taken for a and b.
   (b) Note that the calculations can be speeded up considerably by taking deviations from the mean for both variables. A good suggestion here is to introduce transformed variables, $t_i = (x_i - 50)/5$ and $u_i = (y_i - 50)/5$. Repeat the first part of this exercise for t and u.
   (c) A quadratic curve is defined by the general form $y = a + bx + cx^2$. Find the best-fitting quadratic curve from the requirement that

$$F(a, b, c) = \sum (y_i - a - bx_i - cx_i)^2$$

   must reach a minimum. Hint: apply this procedure to the transformed variables t and u.
   (d) Note that the quadratic curve which is the solution in part (c) does not pass through the origin of the x,y-plane. Our amateur psychologist, however, argues that the curve should include this point since a

geometrical form with zero area will always be judged to have zero area. Find the best-fitting curve under this condition. Hint: use the transformed coordinates t and u, and realize that the curve then should include the point $(-10, -10)$. This imposes the condition $g(u, t) = a - 10b + 100c + 10 = 0$.

(e) Note, finally, that problem (d) could be solved directly by using untransformed coordinates x and y, and by fitting a curve $y = bx + cx^2$. Verify that this direct approach gives the same solution as in (d), but that the approach suggested in (d) is faster.

8. In a plane five points are given by their coordinates: (4, 5), (2, −1), (−2, 2), (1, −4), and (−5, −2). Collect the data in a 5×2 matrix $\mathbf{X}$. Let a vector through the origin be defined by its direction cosines $\mathbf{k}$ ($\mathbf{k}$ is a 2×1 vector, and $\mathbf{k}'\mathbf{k} = 1$). Then the projections of the five points on this vector will be given by $\mathbf{Xk} = \mathbf{y}$. Suppose now there is a vector $m' = (2 \quad 1 \quad 0 \quad -1 \quad -2)$. We want to identify the vector $\mathbf{k}$ so that $\mathbf{y}'\mathbf{m}/5$ has a maximum (i.e., we maximize the covariance between y and m).

NOTE: One may look on $\mathbf{m}$ as a kind of "criterion"; the problem is to find a vector in the x_1, x_2-plane such that projections on this vector "match" the criterion variable (where covariance is used as a measure of goodness of match). Note also that the solution does not change if we take as the criterion another vector proportional to $\mathbf{m}$, like $(4 \quad 2 \quad 0 \quad -2 \quad -4)$.

As an additional problem, therefore, one might identify a proportionality constant c such that the sum of the squared discrepancies between $\mathbf{y}$ and $c\mathbf{m}$ has a minimum. That is, once $\mathbf{y}$ is defined by the first part of the exercise, the next step is to determine c for which $(\mathbf{y} - c\mathbf{m})'(\mathbf{y} - c\mathbf{m})$ has a minimum.

7

Eigenvectors and Eigenvalues

7.1. Definitions

Let **A** be a square matrix of order $n \times n$. It can be shown that vectors **k** exist, such that $\mathbf{Ak} = \lambda\mathbf{k}$, with λ some scalar (the trivial solution $\mathbf{k} = 0$ is excluded). For example, if $\mathbf{A} = \begin{pmatrix} 3 & 2 \\ 1 & 4 \end{pmatrix}$, one solution for **k** is $\mathbf{k}_1 = \begin{pmatrix} 1 \\ 1 \end{pmatrix}$. Then

$$\mathbf{Ak}_1 = \begin{pmatrix} 5 \\ 5 \end{pmatrix} = 5\begin{pmatrix} 1 \\ 1 \end{pmatrix},$$

and it is seen that $\lambda_1 = 5$. Another solution is $\mathbf{k}_2 = \begin{pmatrix} 2 \\ -1 \end{pmatrix}$, for which $\mathbf{Ak}_2 = \begin{pmatrix} 4 \\ -2 \end{pmatrix}$, and $\lambda_2 = 2$.

Such vectors **k** are called *eigenvectors*, and the corresponding values λ are the *eigenvalues* of matrix **A**. They have many applications, as we shall see in later chapters.

How do we find eigenvectors? From $\mathbf{Ak} = \lambda\mathbf{k}$, it follows that $\mathbf{Ak} - \lambda\mathbf{k} = \mathbf{0}$, or

$$(\mathbf{A} - \lambda\mathbf{I})\mathbf{k} = \mathbf{0}. \tag{7.1}$$

Equation (7.1) gives n homogeneous equations with the elements of **k** as n unknowns; in addition there is the unknown value of λ. From section 5.5 we know that the equations can be solved only if the matrix of coefficients has rank smaller than n; so the determinant of the $n \times n$ matrix of coefficients $(\mathbf{A} - \lambda\mathbf{I})$ must be equal to zero:

$$|\mathbf{A} - \lambda\mathbf{I}| = 0. \tag{7.2}$$

This determinant by itself, however, gives an equation in λ only; so values for λ can be found.

In the example above, we have

$$|\mathbf{A} - \lambda\mathbf{I}| = \left|\begin{pmatrix} 3 & 2 \\ 1 & 4 \end{pmatrix} - \lambda\begin{pmatrix} 1 & 0 \\ 0 & 1 \end{pmatrix}\right| = \begin{vmatrix} 3-\lambda & 2 \\ 1 & 4-\lambda \end{vmatrix} = 0.$$

Written out, the determinant becomes

$$(3 - \lambda)(4 - \lambda) - 2 = 0, \quad \text{or} \quad \lambda^2 - 7\lambda + 10 = 0,$$

$$\text{with roots } \lambda_1 = 5, \qquad \lambda_2 = 2.$$

We then can substitute one of the roots in the equations (7.1). In our example, if we substitute λ_1, we obtain

$$\begin{pmatrix} -2 & 2 \\ 1 & -1 \end{pmatrix}\mathbf{k} = 0,$$

which gives two homogeneous equations with two unknowns; they are solvable because λ_1 has been chosen to make the rank of the matrix of coefficients equal to one. The solution is $\mathbf{k}_1 = \begin{pmatrix} 1 \\ 1 \end{pmatrix}$, where the solution is determined up to an arbitrary multiplier. If we substitute the other root for λ in (7.1), we obtain

$$\begin{pmatrix} 1 & 2 \\ 1 & 2 \end{pmatrix} \mathbf{k}_2 = 0,$$

with solution $\mathbf{k}_2 = \begin{pmatrix} 2 \\ -1 \end{pmatrix}$.

In general, the determinant $|\mathbf{A} - \lambda\mathbf{I}|$ is obtained by subtracting λ from the diagonal elements of $\mathbf{A}$. Expanded, the determinant becomes an equation of the nth degree in λ, which equation will in general have n different roots. The equation is called the *characteristic equation* of $\mathbf{A}$. It follows that a matrix $\mathbf{A}$ of order $n \times n$ will in general have n different eigenvalues and, therefore, n different eigenvectors.

These eigenvectors could be collected in a matrix of eigenvectors $\mathbf{K}$, with $\mathbf{k}_1, \mathbf{k}_2, \ldots, \mathbf{k}_n$ as columns. We then can summarize by writing, as a generalized form of $\mathbf{Ak} = \lambda\mathbf{k}$,

$$\mathbf{AK} = \mathbf{K}\Lambda \tag{7.3}$$

with Λ a diagonal matrix of eigenvalues; in our example,

$$\begin{pmatrix} 3 & 2 \\ 1 & 4 \end{pmatrix} \cdot \begin{pmatrix} 1 & 2 \\ 1 & -1 \end{pmatrix} = \begin{pmatrix} 1 & 2 \\ 1 & -1 \end{pmatrix} \cdot \begin{pmatrix} 5 & 0 \\ 0 & 2 \end{pmatrix}.$$

7.2. Canonical Form of a Matrix

Let $\mathbf{A}$ be symmetric, and suppose $\mathbf{k}_i$ and $\mathbf{k}_j$ are two eigenvectors. Then it can be shown that $\mathbf{k}_i'\mathbf{k}_j = 0$. In words, the scalar product of any two different eigenvectors is zero. Proof:

$$\mathbf{Ak}_i = \lambda_i\mathbf{k}_i; \tag{7.4}$$

therefore, using the fact that $\mathbf{A}$ is symmetric,

$$\mathbf{k}_i'\mathbf{A} = \lambda_i\mathbf{k}_i'. \tag{7.5}$$

Now we can write

$$\mathbf{k}_i'\mathbf{Ak}_j = \mathbf{k}_i'(\mathbf{Ak}_j) = \mathbf{k}_i'(\lambda_j\mathbf{k}_j) = \lambda_j\mathbf{k}_i'\mathbf{k}_j, \tag{7.6a}$$

but also

$$\mathbf{k}_i'\mathbf{A}\mathbf{k}_j = (\mathbf{k}_i'\mathbf{A})\mathbf{k}_j = \lambda_i\mathbf{k}_i'\mathbf{k}_j; \tag{7.6b}$$

it follows that

$$\lambda_j\mathbf{k}_i'\mathbf{k}_j = \lambda_i\mathbf{k}_i'\mathbf{k}_j. \tag{7.7}$$

Equation (7.7) implies that either $\lambda_i = \lambda_j$, or $\mathbf{k}_i'\mathbf{k}_j = 0$. But since the roots of the characteristic equation will in general be different, the first alternative can be excluded, and it must be true that $\mathbf{k}_i'\mathbf{k}_j = 0$. It follows, for a symmetric matrix, that the matrix of eigenvectors $\mathbf{K}$ will satisfy $\mathbf{K}'\mathbf{K} = \mathbf{D}$, with $\mathbf{D}$ being a diagonal matrix whose diagonal elements are equal to $\mathbf{k}_i'\mathbf{k}_i$.

We have seen, however, that eigenvectors are determined only up to an arbitrary multiplier. We can always select a multiplier such that $\mathbf{k}_i'\mathbf{k}_i = 1$. The eigenvector is then said to be *normalized* to one, and we have $\mathbf{K}'\mathbf{K} = \mathbf{I}$.

From equation (7.3) we know that we can write $\mathbf{AK} = \mathbf{K}\Lambda$, and it follows that

$$\mathbf{K}'\mathbf{A}\mathbf{K} = \mathbf{K}'\mathbf{K}\Lambda = \Lambda. \tag{7.8}$$

Λ is sometimes called the diagonal *canonical form* of matrix $\mathbf{A}$. It also follows, if we postmultiply (7.3) by $\mathbf{K}'$, that

$$\mathbf{A} = \mathbf{K}\Lambda\mathbf{K}', \tag{7.9}$$

where we use the property that $\mathbf{KK}' = \mathbf{I}$.

Example: We are given a matrix $\mathbf{A} = \begin{pmatrix} 134 & 12 \\ 12 & 141 \end{pmatrix}$; to solve for its eigenvectors, we expand the determinant,

$$\begin{vmatrix} 134 - \lambda & 12 \\ 12 & 141 - \lambda \end{vmatrix} = \lambda^2 - 275\lambda + 18{,}750.$$

The two roots are found to be $\lambda_1 = 150$, $\lambda_2 = 125$. Substituting the first root in an equation of type (7.1), we have

$$\begin{pmatrix} -16 & 12 \\ 12 & -9 \end{pmatrix}\mathbf{k}_1 = \mathbf{0}, \quad \text{and} \quad \mathbf{k}_1 = \begin{pmatrix} 3 \\ 4 \end{pmatrix},$$

up to an arbitrary multiplier. If we take $\mathbf{k}_1 = \begin{pmatrix} .6 \\ .8 \end{pmatrix}$ we see that $\mathbf{k}_1'\mathbf{k}_1 = 1$. To find $\mathbf{k}_2$, we substitute λ_2 in equation (7.2), and find $\begin{pmatrix} 9 & 12 \\ 12 & 16 \end{pmatrix}\mathbf{k}_2 = \mathbf{0}$, with solution $\mathbf{k}_2 = \begin{pmatrix} 4 \\ -3 \end{pmatrix}$. Normalized to one, this eigenvector becomes $\mathbf{k}_2 = \begin{pmatrix} .8 \\ -.6 \end{pmatrix}$.

Equation (7.3) now can be written as

$$\begin{pmatrix} 134 & 12 \\ 12 & 141 \end{pmatrix}\begin{pmatrix} .6 & .8 \\ .8 & -.6 \end{pmatrix} = \begin{pmatrix} .6 & .8 \\ .8 & -.6 \end{pmatrix}\begin{pmatrix} 150 & 0 \\ 0 & 125 \end{pmatrix},$$

and for equation (7.8) we find

$$\begin{pmatrix} .6 & .8 \\ .8 & -.6 \end{pmatrix}\begin{pmatrix} 134 & 12 \\ 12 & 141 \end{pmatrix}\begin{pmatrix} .6 & .8 \\ .8 & -.6 \end{pmatrix} = \begin{pmatrix} 150 & 0 \\ 0 & 125 \end{pmatrix}.$$

Equation (7.9) is identified as

$$\begin{pmatrix} 134 & 12 \\ 12 & 141 \end{pmatrix} = \begin{pmatrix} .6 & .8 \\ .8 & -.6 \end{pmatrix}\begin{pmatrix} 150 & 0 \\ 0 & 125 \end{pmatrix}\begin{pmatrix} .6 & .8 \\ .8 & -.6 \end{pmatrix}.$$

Note: In practice it will be difficult to find eigenvalues by solving the characteristic equation, which is an equation of nth degree and therefore rather troublesome if n is large, or even if n is larger than 2. Appendix 1 describes a different method.

7.3. Asymmetric Matrices

If $\mathbf{A}$ is of order $n \times n$ but not symmetric, equation (7.8) will not be valid. In fact, we must distinguish between left and right eigenvectors, since solutions for $\mathbf{Ap} = \lambda\mathbf{p}$ will not satisfy $\mathbf{p}'\mathbf{A} = \lambda\mathbf{p}'$; $\mathbf{p}$ is a *right eigenvector*, and the *left eigenvector* would be $\mathbf{q}$, for which $\mathbf{q}'\mathbf{A} = \lambda\mathbf{q}'$.

Equation (7.3), for right eigenvectors, can be written as

$$\mathbf{AP} = \mathbf{P}\boldsymbol{\Lambda}; \tag{7.10}$$

the similar equation for left eigenvectors is

$$\mathbf{Q}'\mathbf{A} = \boldsymbol{\Lambda}\mathbf{Q}'. \tag{7.11}$$

It can be shown that the matrix of eigenvalues Λ is the same in both (7.10) and (7.11). Let $\mathbf{p}_i$ be one of the right eigenvectors of $\mathbf{A}$. From (7.10) we have $\mathbf{A}\mathbf{p}_i = \lambda_i \mathbf{p}_i$, and therefore

$$(\mathbf{A} - \lambda_i \mathbf{I})\mathbf{p}_i = 0. \tag{7.12}$$

These n homogeneous equations can only be solved if the determinant $|\mathbf{A} - \lambda\mathbf{I}| = 0$. The latter expression defines the characteristic equation, with n solutions for the eigenvalues λ. With left eigenvalues, we have $\mathbf{q}_j'\mathbf{A} = \lambda_j \mathbf{q}_j$, or

$$\mathbf{q}_j'(\mathbf{A} - \lambda_j \mathbf{I}) = 0. \tag{7.13}$$

The condition for solvability is again that $|\mathbf{A} - \lambda\mathbf{I}| = 0$, which gives the identical characteristic equation.

Another theorem can be derived from (7.10) and (7.11). Postmultiplying (7.10) by $\mathbf{P}^{-1}$ and premultiplying (7.11) by $\mathbf{Q}'^{-1}$, we have

$$\mathbf{A} = \mathbf{P}\Lambda\mathbf{P}^{-1} = \mathbf{Q}'^{-1}\Lambda\mathbf{Q}'. \tag{7.14}$$

This is the equivalent of (7.9) for a nonsymmetric matrix. Equation (7.8) also has an equivalent:

$$\mathbf{P}^{-1}\mathbf{A}\mathbf{P} = \mathbf{Q}'\mathbf{A}\mathbf{Q}'^{-1} = \Lambda. \tag{7.15}$$

The equations (7.14) and (7.15) suggest some intimate connection between $\mathbf{P}$ and $\mathbf{Q}'^{-1}$ and between $\mathbf{Q}'$ and $\mathbf{P}^{-1}$. In fact, it is easily shown that rows of $\mathbf{P}^{-1}$ contain left eigenvectors of $\mathbf{A}$. The proof follows immediately if (7.10) is both pre- and postmultiplied by $\mathbf{P}^{-1}$. We then have

$$\mathbf{P}^{-1}\mathbf{A}\mathbf{P}\mathbf{P}^{-1} = \mathbf{P}^{-1}\mathbf{P}\Lambda\mathbf{P}^{-1},$$

and therefore

$$\mathbf{P}^{-1}\mathbf{A} = \Lambda\mathbf{P}^{-1}. \tag{7.16}$$

Remember, however, that eigenvectors are determined only up to an arbitrary scalar constant. Therefore, given a solution for $\mathbf{P}$ and a solution for $\mathbf{Q}'$, it is not necessarily true that $\mathbf{Q}' = \mathbf{P}^{-1}$. But the scalars can be fixed to produce this identity, and here, with $\mathbf{P}\mathbf{Q}' = \mathbf{I}$, equation (7.14) reduces to

$$\mathbf{A} = \mathbf{P}\Lambda\mathbf{Q}'. \tag{7.17}$$

To illustrate, and as another perhaps useful exercise, take a matrix

$$\mathbf{A} = \begin{pmatrix} 8 & -2 & 0 \\ 0 & 8 & -2 \\ -12 & 22 & -4 \end{pmatrix}.$$

To find eigenvalues, we develop the determinant

$$\begin{vmatrix} 8-\lambda & -2 & 0 \\ 0 & 8-\lambda & -2 \\ -12 & 22 & -4-\lambda \end{vmatrix} = 0,$$

which results in the characteristic equation

$$\lambda^3 - 12\lambda^2 + 44\lambda - 48 = 0.$$

Trial and error shows that $\lambda = 2$ is one solution. The other solutions then are easily found by dividing the equation by $(\lambda - 2)$ and solving the resulting quadratic equation. The three roots are $\lambda_1 = 6$, $\lambda_2 = 4$, and $\lambda_3 = 2$.

To find a first right eigenvector, we substitute $\lambda_1 = 6$ in the set of equations (7.12):

$$\begin{pmatrix} 2 & -2 & 0 \\ 0 & 2 & -2 \\ -12 & 22 & -10 \end{pmatrix} \mathbf{p}_1 = \mathbf{0},$$

with solution $\mathbf{p}_1' = (1 \quad 1 \quad 1)$.

By a similar process the other eigenvectors can be found. The right eigenvector matrix $\mathbf{P}$ appears to be

$$\mathbf{P} = \begin{pmatrix} 1 & 1 & 1 \\ 1 & 2 & 3 \\ 1 & 4 & 9 \end{pmatrix}.$$

For a first left eigenvector, we substitute $\lambda_1 = 6$ in the set of equations (7.13) to obtain

$$\mathbf{q}_1' \begin{pmatrix} 2 & -2 & 0 \\ 0 & 2 & -2 \\ -12 & 22 & -10 \end{pmatrix} = \mathbf{0},$$

from which it is found that $\mathbf{q}_1' = (6 \quad -5 \quad 1)$. A similar procedure for the other two eigenvectors results, finally, in the left eigenvector matrix

$$\mathbf{Q}' = \begin{pmatrix} 6 & -5 & 1 \\ -3 & 4 & -1 \\ 2 & -3 & 1 \end{pmatrix}.$$

It must be understood that columns of $\mathbf{P}$ and rows of $\mathbf{Q}'$ are determined up to arbitrary constants. Note that we might fix these constants in the solution for $\mathbf{Q}'$ to make $\mathbf{Q}'$ equal to $\mathbf{P}^{-1}$. Then $\mathbf{Q}'$ has to be written as

$$\mathbf{Q}' = \begin{pmatrix} 3 & -2.5 & .5 \\ -3 & 4 & -1 \\ 1 & -1.5 & .5 \end{pmatrix},$$

and equation (7.17) can be specified as

$$\mathbf{A} = \begin{pmatrix} 1 & 1 & 1 \\ 1 & 2 & 3 \\ 1 & 4 & 9 \end{pmatrix} \begin{pmatrix} 6 & & \\ & 4 & \\ & & 2 \end{pmatrix} \begin{pmatrix} 3 & -2.5 & .5 \\ -3 & 4 & -1 \\ 1 & -1.5 & .5 \end{pmatrix}.$$

7.4. Some Theorems

7.4.1. If $\mathbf{k}$ is an eigenvector of $\mathbf{A}$, then $\mathbf{k}$ is also an eigenvector of $\mathbf{A}^2$, $\mathbf{A}^3$, etc. Proof: $\mathbf{A}^2\mathbf{k} = \mathbf{A}(\mathbf{A}\mathbf{k}) = \mathbf{A}(\lambda\mathbf{k}) = \lambda(\mathbf{A}\mathbf{k}) = \lambda^2\mathbf{k}$. It follows that λ^2 is an eigenvalue of $\mathbf{A}^2$ (and λ^3 an eigenvalue of $\mathbf{A}^3$, etc.).

7.4.2. If $\mathbf{k}$ is an eigenvector of $\mathbf{A}$, then $\mathbf{k}$ is also an eigenvector of $\mathbf{A}^{-1}$ (assuming that $|\mathbf{A}| \neq 0$). Proof: given $\mathbf{A}\mathbf{k} = \lambda\mathbf{k}$; multiply both sides by $\mathbf{A}^{-1}$ to obtain $\mathbf{k} = \lambda\mathbf{A}^{-1}\mathbf{k}$, and divide by λ to get $\mathbf{A}^{-1}\mathbf{k} = \lambda^{-1}\mathbf{k}$. We see that the eigenvalues of $\mathbf{A}^{-1}$ are the reciprocals of the eigenvalues of $\mathbf{A}$. In general, $\mathbf{A}^{-1}\mathbf{K} = \mathbf{K}\mathbf{\Lambda}^{-1}$.

7.4.3. If $\mathbf{A}$ is symmetric, then the eigenvectors of $\mathbf{A}$ are also eigenvectors of $\mathbf{A}'\mathbf{A}$. The proof follows from 7.4.1, since $\mathbf{A}'\mathbf{A} = \mathbf{A}^2$.

7.4.4. The sum of the eigenvalues of a matrix is equal to the sum of the diagonal elements of the matrix (the trace of the matrix). We shall give no proof, but the reader may verify the theorem in the examples given above.

7.4.5. If **A**, a matrix of order $n \times n$, has rank r, then **A** has $n - r$ eigenvalues equal to zero. This follows from the definition of linear dependence. In fact, if **A** is of rank r ($r < n$), we will be able to find non-zero vectors **c** such that $\mathbf{Ac} = \mathbf{0}$. But these vectors satisfy the definition of an eigenvector, with eigenvalue equal to zero.

7.4.6. The Eckhart-Young Theorem. Let **A** be a symmetric matrix, of order $n \times n$. Suppose we want to find a matrix **B** of rank 1 in such a way that the sum of the squared discrepancies between the elements of **A** and the corresponding elements of **B** has a minimum. It can be shown that the solution is $\mathbf{B} = \mathbf{k}_1\lambda_1\mathbf{k}_1'$, where λ_1 is the largest eigenvalue of **A**, and $\mathbf{k}_1$ the corresponding eigenvector, normalized to one (Schönemann *et al.* 1965).

The theorem can be generalized. Suppose we take the first r largest eigenvalues and the corresponding eigenvectors. The eigenvectors are collected in an $n \times r$ matrix $\mathbf{K}^*$ (an incomplete version of **K**), and the eigenvalues in a diagonal matrix Λ^*. It is assumed that the eigenvectors are normalized to one. Then $\mathbf{K}^*\Lambda^*\mathbf{K}^{*\prime}$ is an $n \times n$ matrix of rank r (the rank of a product cannot be larger than the smallest rank in the factors), and is a least-squares solution for the approximation of **A** by a matrix of rank r. It is assumed, here, that the eigenvalues are all positive.

7.4.7. If **A** is of rank r by itself, and we take the r eigenvectors for which the eigenvalues are different from zero (compare 7.4.5) collected in a matrix $\mathbf{K}^*$ (of order $n \times r$), then $\mathbf{A} = \mathbf{K}^*\Lambda^*\mathbf{K}^{*\prime}$. This follows from the Eckhart-Young theorem, but is also immediately clear from equation (7.9) itself, since eigenvectors for which the eigenvalue is zero do not play a role in this equation.

As an example, take the 3×3 matrix

$$\mathbf{A} = \begin{pmatrix} 2.284 & -.570 & -.856 \\ -.570 & 1.143 & .714 \\ -.856 & .714 & .571 \end{pmatrix}.$$

The matrix of (nonnormalized) eigenvectors is

$$\begin{pmatrix} 2 & 2 & 1 \\ -1 & 3 & -2 \\ -1 & 1 & 4 \end{pmatrix},$$

with corresponding eigenvalues 3, 1, and 0, respectively.

The normalized matrix of eigenvectors, then, is

$$\begin{pmatrix} .816 & .535 & .218 \\ -.408 & .802 & -.436 \\ -.408 & .267 & .873 \end{pmatrix}.$$

According to our last theorem, **A** can be reconstructed from $\mathbf{K}^*\Lambda^*\mathbf{K}^{*\prime}$, where $\mathbf{K}^*$ contains the first two normalized eigenvectors only. This gives

$$\begin{pmatrix} .816 & .535 \\ -.408 & .802 \\ -.408 & .267 \end{pmatrix} \begin{pmatrix} 3 & \\ & 1 \end{pmatrix} \begin{pmatrix} .816 & -.408 & -.408 \\ .535 & .802 & .267 \end{pmatrix}, \tag{7.18}$$

and it can be verified that (7.18) is equal to matrix **A**.

To illustrate the Eckhart-Young theorem, we may approximate **A** (which is a matrix of rank 2, since one eigenvalue is zero) by a matrix of rank 1, and take

$$\begin{pmatrix} .816 \\ -.408 \\ -.408 \end{pmatrix} \cdot (3) \cdot (.816 \quad -.408 \quad -.408) = \begin{pmatrix} 2 & -1 & -1 \\ -1 & .5 & .5 \\ -1 & .5 & .5 \end{pmatrix},$$

where it is easily seen that this matrix in fact has rank 1. The discrepancies between the last matrix and **A** itself are

$$\begin{pmatrix} .284 & .430 & .144 \\ .430 & .643 & .214 \\ .144 & .214 & .071 \end{pmatrix}.$$

But since this "residual" depends on the second eigenvector only, it must be equal to

$$\begin{pmatrix} .535 \\ .802 \\ .267 \end{pmatrix} \cdot (1) \cdot (.535 \quad .802 \quad .267),$$

as in fact it is (rounding errors cause small discrepancies).

7.4.8. Let **A** be a symmetric matrix for which we can write $\mathbf{A} = \mathbf{K}\Lambda\mathbf{K}'$. Then we can find a matrix $\mathbf{A}^{\frac{1}{2}}$ for which $\mathbf{A}^{\frac{1}{2}} \cdot \mathbf{A}^{\frac{1}{2}} = \mathbf{A}$.

The matrix $\mathbf{A}^{\frac{1}{2}} = \mathbf{K}\Lambda^{\frac{1}{2}}\mathbf{K}'$. A constructive proof is

$$\mathbf{A}^{\frac{1}{2}} \cdot \mathbf{A}^{\frac{1}{2}} = \mathbf{K}\Lambda^{\frac{1}{2}}\mathbf{K}' \cdot \mathbf{K}\Lambda^{\frac{1}{2}}\mathbf{K} = \mathbf{K}\Lambda^{\frac{1}{2}}\Lambda^{\frac{1}{2}}\mathbf{K}' = \mathbf{K}\Lambda\mathbf{K}' = \mathbf{A}.$$

Similarly we can define a matrix $\mathbf{A}^{-\frac{1}{2}}$ that satisfies $\mathbf{A}^{-\frac{1}{2}} \cdot \mathbf{A}^{-\frac{1}{2}} = \mathbf{A}^{-1}$ (or $\mathbf{A}^{-\frac{1}{2}} \cdot \mathbf{A}^{\frac{1}{2}} = \mathbf{I}$); it is $\mathbf{A}^{-\frac{1}{2}} = \mathbf{K}\Lambda^{-\frac{1}{2}}\mathbf{K}'$.

7.5. Illustration of Application

As an illustration of how eigenvectors can be useful, we take the following example. In S_2 we have five points (cf. exercise 6.3.8), coordinates of which are given in matrix $\mathbf{X}'$:

$$\mathbf{X}' = \begin{pmatrix} 4 & 2 & -2 & 1 & -5 \\ 5 & -1 & 2 & -4 & -2 \end{pmatrix},$$

where each column of $\mathbf{X}'$ specifies one point. Coordinates are given as deviations from the mean. The variance-covariance matrix is

$$\mathbf{X}'\mathbf{X}/n = \begin{pmatrix} 50 & 20 \\ 20 & 50 \end{pmatrix} \Big/ 5 = \begin{pmatrix} 10 & 4 \\ 4 & 10 \end{pmatrix}.$$

Our problem is to find a vector in the plane such that the projections of the points on that vector have maximum variance.

Let the vector be given by its direction cosines **k**, for which $\mathbf{k}'\mathbf{k} = 1$. The projections are $\mathbf{Xk} = \mathbf{y}$. The variance of the projections is $\mathbf{y}'\mathbf{y}/n = \mathbf{k}'\mathbf{X}'\mathbf{Xk}/n$. We want a maximum for this expression under the condition that $\mathbf{k}'\mathbf{k} = 1$.

Using the results in section 6.2.4, we take an auxiliary function

$$F = \mathbf{k}'(\mathbf{X}'\mathbf{X}/n)\mathbf{k} - \mu(\mathbf{k}'\mathbf{k} - 1). \tag{7.19}$$

Its partial derivatives are set equal to zero:

$$\partial F/\partial \mathbf{k}' = 2(\mathbf{X}'\mathbf{X}/n)\mathbf{k} - 2\mu\mathbf{k} = 0. \tag{7.20}$$

It follows that

$$(\mathbf{X}'\mathbf{X}/n)\mathbf{k} = \mu\mathbf{k}. \tag{7.21}$$

In words, (7.21) shows that $\mathbf{k}$ is an eigenvector of the variance-covariance matrix. It is readily seen, from mere inspection, that one eigenvector is $\mathbf{k}_1 = \begin{pmatrix} 1 \\ 1 \end{pmatrix}$ with eigenvalue 14, and the other eigenvector is $\mathbf{k}_2 = \begin{pmatrix} 1 \\ -1 \end{pmatrix}$ with eigenvalue 6. The eigenvectors here are not yet normalized to one. If we do that, we find $\mathbf{K} = \begin{pmatrix} 1 & 1 \\ 1 & -1 \end{pmatrix} \Big/ \sqrt{2}$.

Now, if we look back at equation (7.21), and multiply both sides by $\mathbf{k}'$, we find

$$\mathbf{k}'(\mathbf{X}'\mathbf{X}/n)\mathbf{k} = \mu\mathbf{k}'\mathbf{k} = \mu, \tag{7.22}$$

since $\mathbf{k}'\mathbf{k} = 1$.

But the expression at the left of (7.22) is the variance of the projections; it follows that the eigenvalue is equal to this variance. Apparently, for maximum variance we have to take the largest eigenvalue. The corresponding eigenvector $\mathbf{k}_1$ has both elements equal to $1/\sqrt{2}$, and therefore makes equal angles with the original coordinates. To avoid misunderstanding, we should note that the latter result is not generally true; it happens here because the variances of the original coordinates are equal (to 10).

7.6. Normalization of Eigenvectors

It should be realized that normalization of eigenvectors to one, as we have preferred so far, is just a convention. It is an easy convention, but not more than that. Instead, it might be often more elegant to normalize eigenvectors to their corresponding eigenvalue, so that $\mathbf{k}'\mathbf{k} = \lambda$. To avoid confusion, let us write $\mathbf{f}$ for such an eigenvector normalized to λ. The relation between $\mathbf{F}$ and $\mathbf{K}$ then can be written as $\mathbf{F} = \mathbf{K}\Lambda^{\frac{1}{2}}$, or $\mathbf{K} = \mathbf{F}\Lambda^{-\frac{1}{2}}$.

Using $\mathbf{F}$, some expressions can be simplified, but others become somewhat more involved. For instance, equation (7.9) is simplified to

$$\mathbf{A} = \mathbf{F}\mathbf{F}', \tag{7.23}$$

since $\mathbf{A} = \mathbf{K}\Lambda\mathbf{K}' = (\mathbf{K}\Lambda^{\frac{1}{2}})(\Lambda^{\frac{1}{2}}\mathbf{K}') = \mathbf{F}\mathbf{F}'$.

Another simple equation is

$$\mathbf{F}'\mathbf{F} = \Lambda, \tag{7.24}$$

since $\mathbf{F}'\mathbf{F} = \Lambda^{\frac{1}{2}}\mathbf{K}'\mathbf{K}\Lambda^{\frac{1}{2}} = \Lambda^{\frac{1}{2}}\mathbf{I}\Lambda^{\frac{1}{2}} = \Lambda$.

On the other hand, equation (7.8) would now read

$$\Lambda = \mathbf{K}'\mathbf{A}\mathbf{K} = \Lambda^{\frac{1}{2}}\mathbf{F}'\mathbf{A}\mathbf{F}\Lambda^{\frac{1}{2}}. \tag{7.25}$$

On the whole, it is often very convenient, in derivations, to use the property $\mathbf{K}'\mathbf{K} = \mathbf{K}\mathbf{K}' = \mathbf{I}$, which is a good reason for normalizing eigenvectors to one. The point here is that doing so is *not essential.*

7.7. Exercises

1. Determine the characteristic equation of the matrix

$$\mathbf{A} = \begin{pmatrix} 1.00 & .80 & .50 \\ .80 & .60 & .40 \\ .50 & .40 & .25 \end{pmatrix},$$

and determine the roots. Find the eigenvector corresponding to the largest root. (NOTE: One of the roots is negative, which implies that **A** cannot be a variance-covariance matrix. A suggestion why this is so is found in section 7.5, where we saw that eigenvalues represent variances of projections. But variances cannot be negative. A matrix that has no negative eigenvalues is called a positive semidefinite matrix, or Gramian matrix.)

2. Find the vector on which the projections of the following set of ten points in an S_2 have maximim variance: (5, 1), (8, 3), (1, 4), (4, 3), (3, 2), (5, 0), (7, 5), (8, 4), (0, 2), and (9, 6).

3. Calculate $\mathbf{A}^{\frac{1}{2}}$ and $\mathbf{A}^{-\frac{1}{2}}$ for the matrix $\mathbf{A} = \begin{pmatrix} 8 & 2 \\ 2 & 5 \end{pmatrix}$.

4. In an experiment on learning, the performance is measured over three consecutive time-intervals. Results, for 5 subjects, are

$$\begin{pmatrix} 15 & 24 & 21 \\ 23 & 32 & 35 \\ 9 & 12 & 9 \\ 13 & 19 & 25 \\ 21 & 12 & 27 \end{pmatrix}.$$

We are interested in the *form* of the learning curves only, not their height, and therefore subtract the mean of each row from the elements in that row. The resulting matrix is called **X** (with **Xu** = **0**). We want to express each curve as a weighted sum of basic curves:

$$x_i' = a_{i1}y_1' + a_{i2}y_2' + \cdots.$$

Here y_j' is a basic learning curve and a_{ij} are numbers representing weights. In general form: $\mathbf{X} = \mathbf{A}\mathbf{Y}'$. Find a solution for **A** and $\mathbf{Y}'$. [Hint: There is no unique solution, unless further conditions are imposed. Suggestion: Assume that $\mathbf{Y}'\mathbf{Y} = \mathbf{I}$ (the basic curves are independent and normalized), and also that $\mathbf{A}'\mathbf{A}$ is diagonal (the vectors of weights are independent). These assumptions define a unique solution.] Prove that two basic curves are sufficient.

8

The Multinormal Distribution

Later in this book we will sometimes need to refer to the multinormal distribution. This short chapter summarizes what we will need to know about it for later chapters. We begin from the normal distribution.

8.1. Definition

A variable x, with mean $\bar{x}$ and variance σ^2 is said to follow a normal distribution if its density function is

$$y = \frac{1}{\sigma\sqrt{2\pi}} \exp\left(-\tfrac{1}{2}(x - \bar{x})^2/\sigma^2\right). \tag{8.1}$$

Our main problem will be to generalize to a function of more variables. The first step is the binormal distribution, a function of two variables,

x_1 and x_2. Both will be normal by themselves, so that we can write

$$f(x_1) = \frac{1}{\sigma_1\sqrt{2\pi}} \exp\left(-\tfrac{1}{2}(x_1 - \bar{x}_1)^2/\sigma_1^2\right), \tag{8.2}$$

$$f(x_2) = \frac{1}{\sigma_2\sqrt{2\pi}} \exp\left(-\tfrac{1}{2}(x_2 - \bar{x}_2)^2/\sigma_2^2\right). \tag{8.3}$$

The binormal distribution is, without proof,

$$g(x_1, x_2) = \frac{1}{2\pi\sigma_1\sigma_2\sqrt{1 - r^2}} \exp\left\{-\frac{1}{2(1 - r^2)}\left[\frac{(x_1 - \bar{x}_1)^2}{\sigma_1^2} - \frac{2r(x_1 - \bar{x}_1)(x_2 - \bar{x}_2)}{\sigma_1\sigma_2} + \frac{(x_2 - \bar{x}_2)^2}{\sigma_2^2}\right]\right\} \tag{8.4}$$

where r stands for the product-moment correlation coefficient. Functions (8.2) and (8.3) are the marginal distributions of (8.4). Function (8.4) can be pictured as a surface with ordinate g above the plane formed by x_1 and x_2, as sketched in figure 8.1.

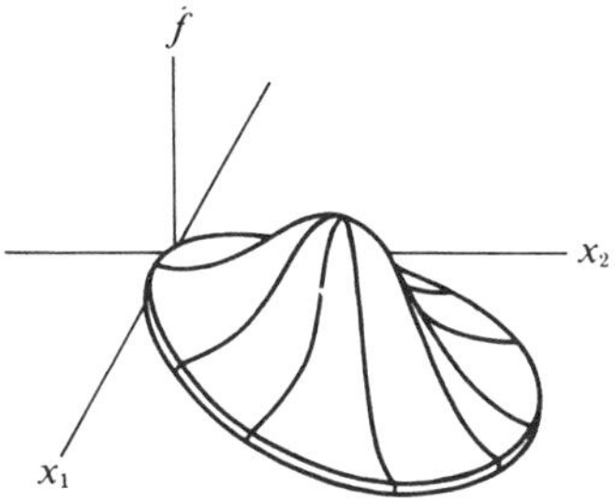

Figure 8.1
Bivariate normal distribution.

Our first concern is to find a more convenient notation to replace (8.4). First let us look at the expression within the square brackets:

$$\frac{(x_1 - \bar{x}_1)^2}{\sigma_1^2} - \frac{2r(x_1 - \bar{x}_1)(x_2 - \bar{x}_2)}{\sigma_1\sigma_2} + \frac{(x_2 - \bar{x}_2)}{\sigma_2^2}\,. \tag{8.5}$$

It can be shown that this expression can be replaced by the more compact notation

$$(\mathbf{x} - \bar{\mathbf{x}})'\mathbf{A}^{-1}(\mathbf{x} - \bar{\mathbf{x}}), \tag{8.6}$$

where **A** is the variance-covariance matrix

$$\mathbf{A} = \begin{pmatrix} \sigma_1^2 & r\sigma_1\sigma_2 \\ r\sigma_1\sigma_2 & \sigma_2^2 \end{pmatrix}, \tag{8.7}$$

and where $(\mathbf{x} - \bar{\mathbf{x}})'$ is the row vector $((x_1 - \bar{x}_1)\ (x_2 - \bar{x}_2))$.

For the determinant $|\mathbf{A}|$ we find

$$|\mathbf{A}| = \sigma_1^2\sigma_2^2(1 - r^2), \tag{8.8}$$

and therefore, for the inverse matrix $\mathbf{A}^{-1}$,

$$\mathbf{A}^{-1} = \begin{pmatrix} \sigma_2^2 & -r\sigma_1\sigma_2 \\ -r\sigma_1\sigma_2 & \sigma_1^2 \end{pmatrix} \Big/ \sigma_1^2\sigma_2^2(1 - r^2). \tag{8.9}$$

Using (8.9), the identity between (8.5) and (8.6) is easily shown to be true.

Further, we meet in (8.4) the expression $1/(2\pi\sigma_1\sigma_2\sqrt{1 - r^2})$. From (8.8) we derive that this expression is equal to

$$(2\pi)^{-1}\,|\mathbf{A}|^{-\frac{1}{2}}. \tag{8.10}$$

Using (8.6) and (8.10), we find that we can rewrite (8.4) as

$$g(x_1, x_2) = (2\pi)^{-1}\,|\mathbf{A}|^{-\frac{1}{2}} \exp\,[-\tfrac{1}{2}(\mathbf{x} - \bar{\mathbf{x}})'\mathbf{A}^{-1}(\mathbf{x} - \bar{\mathbf{x}})]. \tag{8.11}$$

An alternative to (8.11) is to take the variables in standard form, so that both $\bar{x}_1$ and $\bar{x}_2$ are zero; then for **A** we can write **R**, the correlation matrix. Expression (8.11) then becomes

$$g(x_1, x_2) = (2\pi)^{-1}\,|\mathbf{R}|^{-\frac{1}{2}} \exp\,(-\tfrac{1}{2}\mathbf{x}'\mathbf{R}^{-1}\mathbf{x}). \tag{8.12}$$

8.2. The Equiprobability Contour

We now shall define an *equiprobability contour*, as the set of points (x_1, x_2) for which g is some constant. Obviously, for g = constant, we must have

$$(\mathbf{x} - \bar{\mathbf{x}})'\mathbf{A}^{-1}(\mathbf{x} - \bar{\mathbf{x}}) = \text{constant}, \tag{8.13}$$

since all other elements in (8.11) are constants already. On the other hand, expression (8.13) is an equation in the second degree in x_1 and x_2, and therefore represents a conic section—specifically, an ellipse—since x_1^2 and x_2^2 will have positive coefficients.

8.3. Generalization

The final problem is to generalize (8.11) and (8.12) to the multinormal case, where g is a function of n variables $x_1, x_2, \ldots, x_n$, for which we shall use the notation $g(x^{(n)})$. It can be shown that the distribution then has the same general form as in (8.11) and (8.12), with the understanding that $\mathbf{A}$ (or $\mathbf{R}$) is now an $n \times n$ matrix, and $(\mathbf{x} - \bar{\mathbf{x}})$ an $n \times 1$ vector. Also, the exponent of the factor 2π becomes equal to $-\frac{1}{2}n$. We write the function as

$$g(x^{(n)}) = (2\pi)^{-\frac{1}{2}n}\,|\mathbf{A}|^{-\frac{1}{2}} \exp\,[-\tfrac{1}{2}(\mathbf{x} - \bar{\mathbf{x}})'\mathbf{A}^{-1}(\mathbf{x} - \bar{\mathbf{x}})]. \qquad (8.14)$$

An equiprobability contour now becomes a hyperellipsoid, defined by

$$(\mathbf{x} - \bar{\mathbf{x}})'\mathbf{A}^{-1}(\mathbf{x} - \bar{\mathbf{x}}) = \text{constant}. \qquad (8.15)$$

The standard version of (8.14) is

$$g(x^{(n)}) = (2\pi)^{-\frac{1}{2}n}\,|\mathbf{R}|^{-\frac{1}{2}} \exp\,(-\tfrac{1}{2}\mathbf{x}'\mathbf{R}^{-1}\mathbf{x}). \qquad (8.16)$$

PART II

Techniques of Multivariate Analysis

9

An Overview of Multivariate Techniques

9.1. Introduction

Multivariate techniques start from a multivariate data matrix. As was explained in Chapter 1, this matrix is a table that gives the results of a number of observations on a number of variables simultaneously. Without any loss of generality, we shall adopt the convention that columns of the table refer to variables, rows to a set of observations.

Multivariate analysis comes down to operations on columns (sometimes on rows, too) of the data matrix. What operations have to be performed depends on the specific model that inspires the analysis. Therefore, there is no cookbook in which one can look up the appropriate technique for a given type of matrix. The analysis does not depend so much on the nature of the matrix as on the nature of the specific questions asked about the variables and their interrelations.

In a general sense, these questions will have to do with the explanation of variables. We shall want to identify variables that are determiners of other variables and to specify how effective they are as determiners. Explanations of this type tend to grow into explanatory networks, once the number of variables starts increasing.

Explanation, as we use the term here, focuses on the variance of variables. A variable that assumes a constant value throughout needs no explanation, in this conception. What we want to explain is *why* the variable varies, and a variable is explained to the extent that its variance can be attributed to an identifiable source.

This source is a variable by itself. Sometimes it can be identified with an observed variable. A trivial example is that the observed variable "income" is a source of the variation in the observed variable "tax to be paid." But often we shall also want to refer to unobserved variables. A simple example might help here to explain the concept of "unobserved variable." First of all, to say that the entries in a column of the data matrix are "values" assumed by a "variable" is by itself a theoretical decision, an interpretation. It is a rather trivial theoretical decision, but nevertheless it is one. What we observe, therefore, are the values taken by the variable or a sample of values, and strictly speaking not the variable itself. But we shall continue to use the term "observed variable," for short.

The simplest use of the term "unobserved variable" is when we assume that an observed variable is subject to error and is thus not perfectly reliable. This assumption means that there are two sources of variation for the observed variable: one source is what we measure (whatever this is), and the other source is a random component that is added to what we measure. The first source will be called the systematic component, and the second the error component. Both components can be conceptualized as variables. Notice that *both* these variables are unobserved: we do not know what values they take. All we know is that the observed value is the sum of two components, but we are not able to reconstruct these two additive portions. Notice also that the unobserved variables are theoretical constructions, in that they follow from a theory or a certain interpretation of the observed variable.

In a formal sense, it is true that the two components "explain" the observed variable, since we identify two sources of variation. On the other hand, this explanation is still extremely formal and has no substantive meaning at all. It becomes a bit more meaningful if we develop reliability theory further, toward procedures for estimating reliability. First, the reliability of a variable is defined as the ratio of the variance that can be attributed to the systematic component to the variance of the observed variable itself. An estimate of the reliability can be obtained if we have another observed variable for which it can be safely assumed that it depends on the same systematic component and differs only in the error component. Then the joint dependence on one systematic component is quantitatively revealed in the correlation between the two variables.

We can then say that the correlation between the two variables is explained because of their joint dependence on an identical, unobserved variable. This idea can be extended to a larger number of observed variables; then we have begun to develop a model for factor analysis (many observed variables dependent on a common unobserved source). The substantive explanatory value of the unobserved variable has now increased considerably.

Once we have come this far, it is not a drastic step to assume that an observed variable may depend on several different systematic sources. This assumption would lead to a further development of factor analysis (many observed variables dependent on several unobserved sources of variation), but it is obviously also related to multiple regression theory. In multiple regression we try to identify how much an observed variable depends on other observed variables. Very often we shall very reasonably assume that the latter measure different things; it follows that the observed variable we want to explain can be interpreted as depending on several different sources. These sources are indicated by observed variables, which in their turn have systematic and error components and have a pattern of intercorrelations between them. It follows that we must be prepared to unravel relations further if we really want to understand the underlying network.

This may suffice to give an impression of how we shall use the concept of "unobserved variable." We shall now change over to a more systematic overview of multivariate models.

9.2. Overview of Multivariate Models

9.2.1. Pictorial Overview. For a first overview we shall use pictures to illustrate what multivariate methods can do. In these pictures a variable is indicated by a circle with the variable name in it. If variable x is a determiner of variable y, this fact will be represented by an arrow that starts at x and points to y. Figures 9.1 to 9.8 give the illustrations.

First we have a look at *multiple regression analysis*. Here it is assumed that one variable (variable y in figure 9.1) is to be explained in terms of other observed variables x (only two of them are given in figure 9.1.) Usually the explanation will not be complete, in that the variables in set x do not account for the total variance of y. Then we complete the picture by adding an unobserved variable (e_y, in the figure) that is interpreted to account for all the variance in y not covered by the set x. This unobserved variable might be an error source, or a systematic component independent of x, or a mixture of these two.

A next type of analysis is obtained if the variables in the set x can be ordered. Most commonly, this is an ordering in time. If x_1 is prior in time to x_2, it follows that x_1 can be a determiner of x_2, but not the other way round. Therefore we can have arrows only from variables with lower subscript to variables with higher subscript, given that the subscripts reflect order in time. This type of analysis is called *path analysis*, and is illustrated in figure 9.2. Note that we assume incomplete determination everywhere, so that we have to add an unobserved variable for each variable in the series (except the first). Note also that path analysis is nothing but a repeated application of multiple regression analysis (applied to each variable in succession, with the prior variables as its determiners).

Figure 9.3 illustrates another extension of multiple regression analysis. We now have several variables y that we want to explain in terms of a set of variables x. We then may apply multiple regression analysis to each y in turn, as is suggested by figure 9.3 (with unobserved components for the y set omitted). However, the situation may be simplified if we assume unobserved variables intermediating between the set x and the set y as shown in figure 9.4. This technique is called *canonical correlation analysis*. It assumes that there are unobserved variables dependent on the set x and determining the set y. A more elaborate version is given in figure 9.5, where we have a sequence of intermediate unobserved variables. The idea now is that there are unobserved variables χ completely dependent on the set x. Each variable χ is then a determiner of an unobserved variable η, which, however, is not completely dependent on the χ, so we add an unobserved variable e_η. Finally, the set η is interpreted as determiners of the observed variables in set y (again, since the determination is not complete, we should add variables e_{y_1}, e_{y_2}, etc., but they have been omitted in the figure).

Figure 9.6 illustrates the *factor analysis* model. It assumes that a set of observed variables x can be interpreted as dependent on a set of unobserved variables f (with, in addition, specific unobserved components,

e_{x_1}, e_{x_2}, etc., not drawn in the figure). Note that factor analysis is hidden in canonical analysis. In fact, in figure 9.5 the set η operates as a set of factors of the observed variables x. One may even change the canonical model into a kind of double factor analysis by simply reversing the direction of the arrows connecting the set x with the set χ. This reversal, of course, is a clear change in conceptualization, but computationally it is a very minor operation. What results is figure 9.7, where we can see better that the χ's are factors of the observed variables x, while the η's are factors of the set y. Each factor χ is related to one factor η; the nature of the relation between these pairs is left implicit, so we have drawn connections without arrowheads.

In the models described so far, there is no place for mutual determination. That is, in following a path along arrows and in the direction of the arrows, we cannot come back to a point where we were before. Models with this property are called "recursive" models. They are contrasted with *nonrecursive models* (this negative expression is somewhat unfortunate), in which we can follow a path and sometimes return to the point where we started. The simplest example is mutual influence between two variables: x determines y and y determines x. Another simple example is the intransitive triangle: x determines y, y determines z, z determines x. Nonrecursive features can be built into earlier models, as is shown in figure 9.8. Here the canonical model of figure 9.4 is made nonrecursive by allowing for a back and forth relation between the two unobserved variables.

So much for this first overview. It not only illustrates standard multivariate techniques, but also shows some of the relations between them. Apart from the relations explicitly mentioned above, note that 9.3 and 9.6 are comparable: factor analysis is comparable to repeated multiple regression analysis with unobserved variables as determiners. Figure 9.2 also can be interpreted as illustrating factor analysis, with x_1 and e_2, e_3, and e_4 as factors. Family relationships will become even more salient in the next overview.

9.2.2. An Overview in Terms of Operations on Matrices. As was remarked in section 9.1, multivariate analysis comes down basically to operations on columns of the data matrix. We shall illustrate this with the help of figure 9.9. First, in figure 9.9a we have the *multiple regression* situation: one variable y, corresponding to the last column of the data matrix, is explained in terms of several observed variables x at the left.

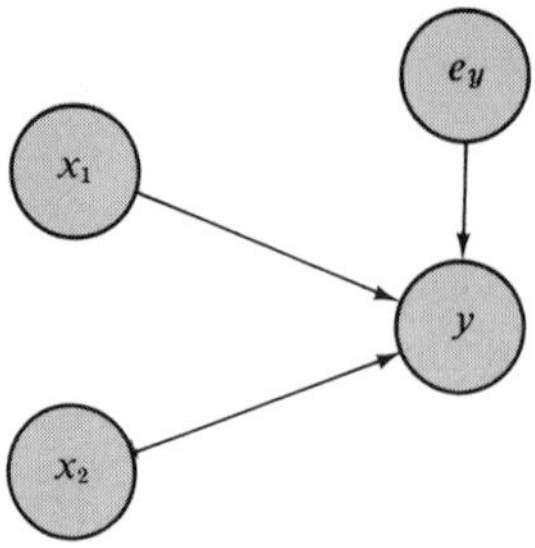

Figure 9.1
Multiple correlation.

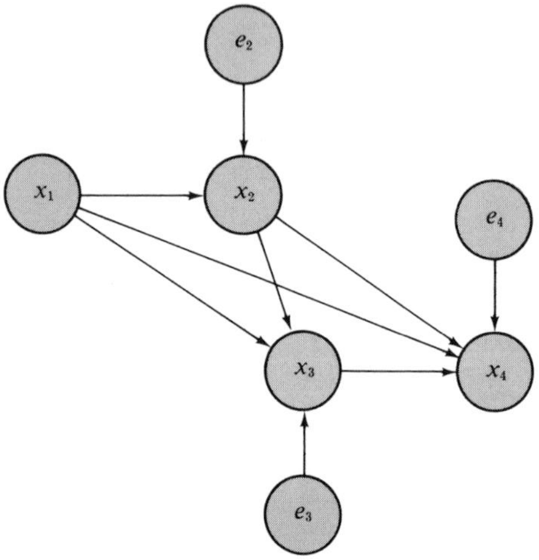

Figure 9.2 Path analysis.

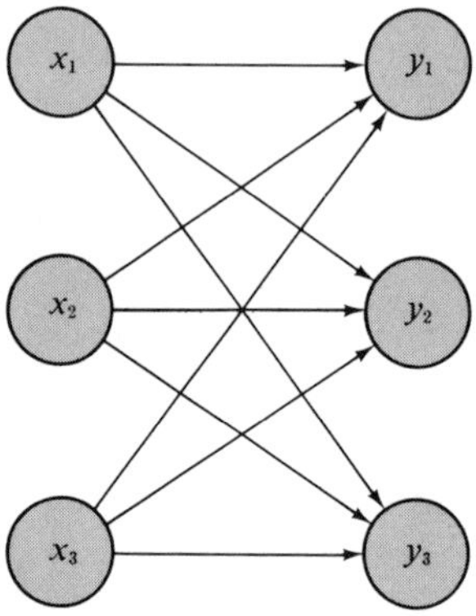

Figure 9.3 Repeated multiple correlation.

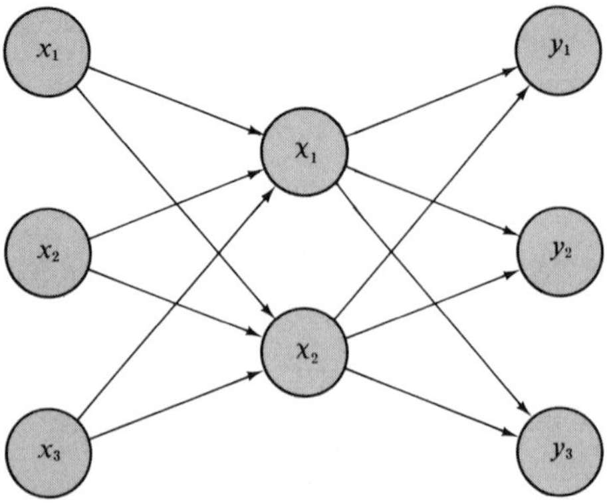

Figure 9.4 Canonical correlation.

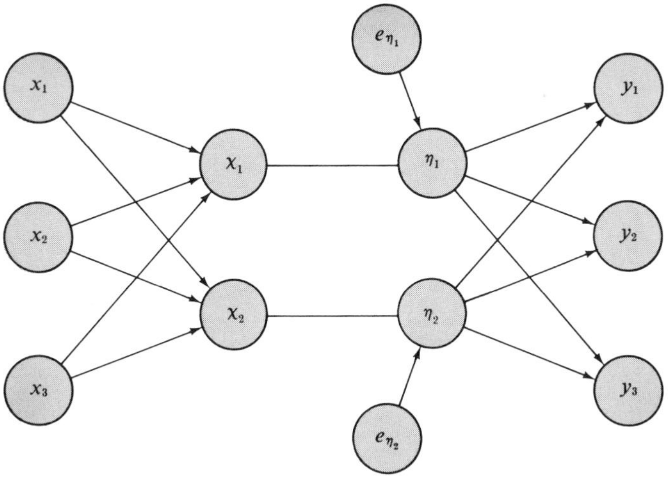

Figure 9.5 Canonical correlation.

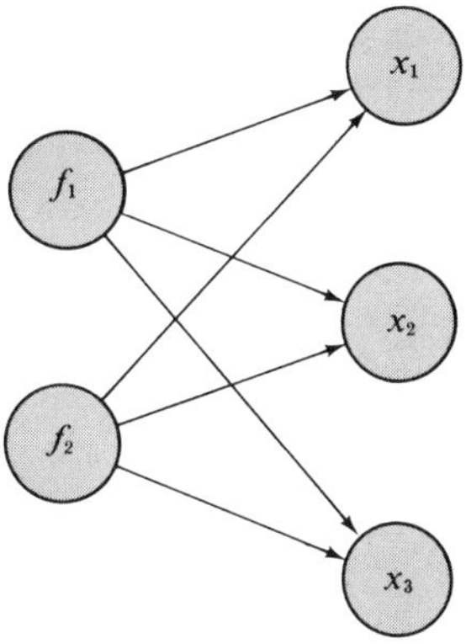

Figure 9.6 Factor analysis.

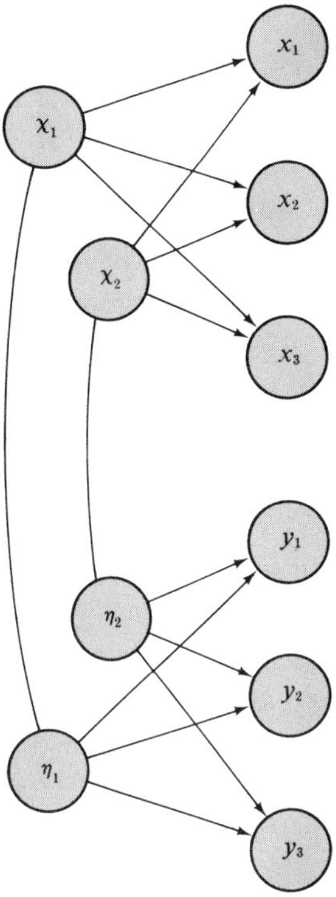

Figure 9.7 Canonical correlation as a double factor analysis.

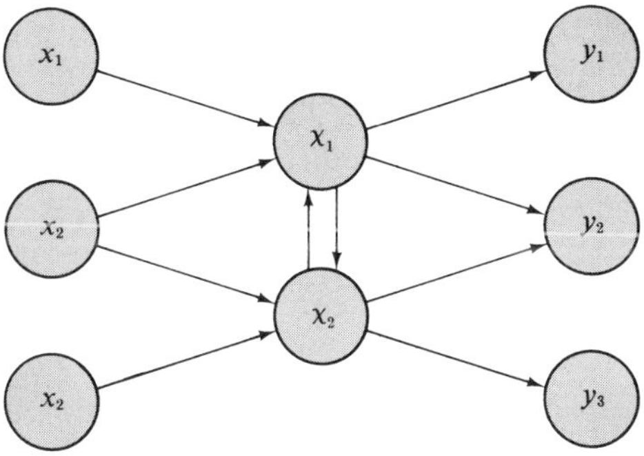

Figure 9.8 Canonical correlation with nonrecursive property added.

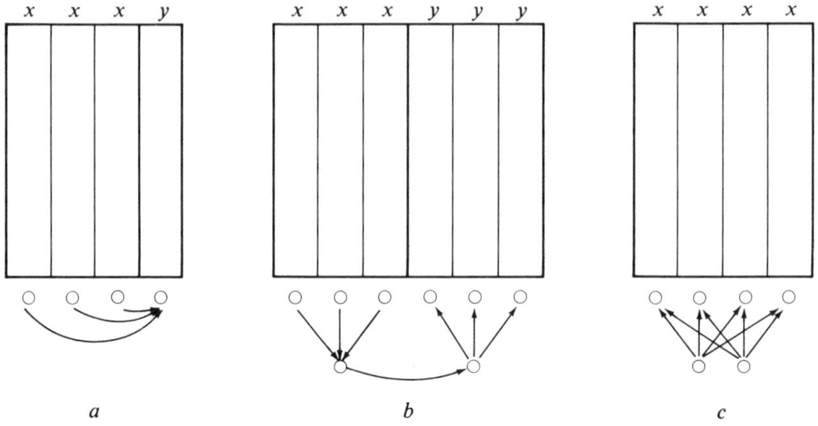

Figure 9.9 Schematic representation of basic operations in multivariate analysis.

For *path analysis*, we would order the variables from left to right and repeat multiple correlation analysis for each variable in turn.

Figure 9.9b illustrates *canonical analysis:* a distinction is made between two sets of observed variables x and y. Variables in set x are determiners of variables in set y, and we use intermediating unobserved variables to canalize the influence of set x on set y (only one chain is drawn in figure 9.9b). Finally, in figure 9.9c we have the *factor analysis* model, where one set of observed variables is related to unobserved determinants.

Nothing new has been introduced so far. However, with figure 9.9 we can illustrate a number of new techniques as well. The first one is *discriminant analysis*. This technique can be applied if the observations are divided into subgroups, where the objective is to find a linear compound of the observed variables such that the subgroups are optimally differentiated on this compound. First, take the situation with two subgroups. The original data matrix will be a matrix with columns x for the observed variables, but this matrix can be divided according to the rows: some observations (rows) refer to subgroup I and the other observations to subgroup II. This can be illustrated with figure 9.9a if we use variable y as a dummy variable that codes for the subgroups; i.e., we agree that $y = 1$ for observations from subgroup I, and $y = 0$ for observations from subgroup II (the choice of numbers 1 and 0 for the coding is arbitrary). Column y then gives the division into the subgroups. Discriminant analysis, then, is nothing but multiple regression analysis with the dummy variable taken as the variable we want to explain.

When there are more than two subgroups, we will need more than one column for the coding of the groups. For instance, with three subgroups we could use two dummies, y_1 and y_2, and code as follows: $y_1 = 1$ for observations from group I and $y_1 = 0$ otherwise; $y_2 = 1$ for observations from subgroup II and $y_2 = 0$ otherwise. Taking pairs (y_1, y_2), we see that the first subgroup is coded as (1, 0), the second as (0, 1), and the third as (0, 0). The situation can be pictured in figure 9.9b with two variables y for the dummies. This shows a formal identity between canonical analysis and discriminant analysis.

It is interesting to note that figures 9.9a and 9.9b also cover *analysis of variance*. First, take simple analysis of variance, where one dependent variable is measured under different experimental conditions (called treatments). For instance, irregularity of heartbeat is measured for male and female subjects, and under two levels of environmental stress. This design can be coded as follows: use x_1 as a dummy to code stress conditions ($x_1 = 1$ for high stress, $x_1 = 0$ for low stress), and use x_2 as a dummy to code sex ($x_2 = 1$ for male subjects, $x_2 = 0$ for females). Values of the dependent variable are registered in the y column. We then have the analysis of variance model, and it is seen to be formally identical to the multiple regression model.

Suppose there is more than one independent variable (we measure irregularity of heartbeat y_1, but also the galvanic skin response y_2). This type of design is called *multivariate analysis of variance* (MANOVA). The appropriate figure is 9.9b, with variables in set x to code the experimental conditions, and variables in set y for the dependent variables. MANOVA apparently is formally equivalent to canonical analysis.

Of course, we can also introduce dummy variables in figure 9.9c. They could code division in subgroups, experimental conditions, etc. If we maintain the factor analysis model strictly, it does not matter whether variables are dummies or not. But the dummies may have special status (for instance, we may require that the observed variables must be explained by factors that also differentiate between various subgroups), and then we obtain a special kind of factor analysis (in this example, *canonical discriminant factor analysis*).

It may seem perverse to ask what happens if all the variables are dummies, but there are good answers to this question. For instance, for figure 9.9a we would have the standard chi-square situation. The textbook example is: some cows are inoculated and some are not; all of them are exposed to a virus; and it is registered which cows become ill. Code

$x_1 = 1$ for inoculated cows and $x_1 = 0$ otherwise, code $y = 1$ for ill cows and $y = 0$ otherwise, and we have figure 9.9a. More complicated examples of chi-square tests could easily be accommodated in one of the figures.

Both overviews clearly show that multivariate techniques do not fall into separate categories as special techniques for special situations, but instead form a close family. Very often they are interchangeable, even in computation. In fact, multivariate analysis can be reduced, in computation, to a limited number of standard subroutines (Beaton 1968). This underlines the basic view that the emphasis should not be on distinctions between techniques but on the specification of the model.

In our approach this computational identity will not be exploited. In Chapter 1 I suggested that there are a limited number of basic tricks; this should be sufficient for the reader to obtain at least some idea about the kind of computational uniformity we are talking about. Our approach will be to specify the model for each class of techniques, and then develop these techniques directly from the formalization of the model.

10

Simple Regression and Correlation

10.1. Introduction

This short chapter will show how regression and correlation can be expressed and treated in matrix notation. We shall first consider regression in which one variable, x_1, is given fixed values. The other variable, x_2, is the dependent variable; i.e., its values are observed and are therefore subject to error. In the general sense, the problem is to find an expression for x_2 as a function of x_1. Regression theory can be further generalized to include problems where one variable depends on several fixed variables, but this falls outside our scope.

The most popular class of functions in regression analysis are polynomials: $x_2 = a + bx_1 + cx_1^2 + dx_1^3 + \cdots$. However, we shall concentrate on the simplest form only, where the function is linear: $x_2 = a + bx_1$. If this function fits the data perfectly, we could draw a straight line through the points that represent the data in the (x_1, x_2)-plane. However, to the extent x_2 is subject to error (and to the extent the linear function is in error), points will deviate from a straight line. We measure these deviations along the x_2 coordinate axis.

In correlation, both x_1 and x_2 are observed variables, in that data are collected as (x_1, x_2) pairs. For instance, it is a problem in regression if we take children of fixed ages and observe their heights, whereas it is a problem in correlation if we sample children and for each child observe its height and its weight. The latter observations can be represented as a cloud of points in the (x_1, x_2)-plane. In this cloud we might draw a line that runs, roughly, through the averages of x_2 for given observed values of x_1. If this line is a straight line, it is also called a regression line. However, whereas for regression the line would represent a linear function, it does not for correlation. This becomes immediately clear by realizing that a different line runs through average values of x_1 for given values of x_2: there are two regression lines, and two different lines cannot stand for one linear relationship between variables. It is better, therefore, to consider regression lines in a correlation problem in a more modest sense, as showing how values of x_2 can be empirically predicted from given values of x_1, and vice versa.

10.2. Regression

Let x_1 be the independent variable that takes fixed values, and x_2 the dependent variable. Without loss of generality, we may assume that x_1 is measured in deviations from the mean. This implies that the sum of the x_1 values is zero, or, in matrix notation, that $\mathbf{u}'\mathbf{x}_1 = 0$, where $\mathbf{x}_1$ is the column vector of n fixed values for the independent variable, and where $\mathbf{u}'$ is a row vector with unit elements.

The regression line will have the general form $x_2 = a + bx_1$. Our first problem is to translate this equation into matrix notation; i.e., given the $n \times 1$ vector $\mathbf{x}_1$, we want to specify a vector $\hat{\mathbf{x}}_2$ in such a way that for each pair of corresponding elements the linear equation holds. The solution is

$$\hat{\mathbf{x}}_2 = a\mathbf{u} + b\mathbf{x}_1, \tag{10.1}$$

where a and b are scalar constants, and $\mathbf{u}$ is the unity vector.* Realize that the vectors $\mathbf{x}_1$ and $\hat{\mathbf{x}}_2$ define n points in the (x_1, x_2)-plane that are located

* The treatment of regression can be easily extended to cover polynomials of degree higher than one. For instance, if the curve is of second degree, equation (10.1) has to be replaced by $\hat{\mathbf{x}}_2 = a\mathbf{u} + b\mathbf{x}_1 + c\mathbf{x}_1^{(2)}$, where the notation $\mathbf{x}_1^{(2)}$ stands for a column vector with elements equal to the squares of the elements of $\mathbf{x}_1$. Exercise 7 in Chapter 6 gives an example.

on the regression line. In contrast, the vectors $\mathbf{x}_1$ and $\mathbf{x}_2$ (with $\mathbf{x}_2$ the $n \times 1$ vector of observed values of variable x_2) specify n points that represent observed data and that may deviate from the regression line. In fact, the difference vector $(\mathbf{x}_2 - \hat{\mathbf{x}}_2)$ gives these deviations (distances of the observed points from the line, measured along the x_2-axis).

The basic problem is, of course, to find the values of a and b. They determine where the line must be drawn in the cloud of points. Then we first have to define a criterion by which the best-fitting line can be singled out. The classical criterion is the least-squares criterion, which says that the sum of the squared deviations must be minimized. Let this sum be called f; then, given that the deviations are collected in the vector $(\mathbf{x}_2 - \hat{\mathbf{x}}_2)$, we have

$$\begin{aligned} f &= (\mathbf{x}_2 - \hat{\mathbf{x}}_2)'(\mathbf{x}_2 - \hat{\mathbf{x}}_2) \\ &= (\mathbf{x}_2 - b\mathbf{x}_1 - a\mathbf{u})'(\mathbf{x}_2 - b\mathbf{x}_1 - a\mathbf{u}), \end{aligned} \tag{10.2}$$

where we have used equation (10.1) to eliminate $\hat{\mathbf{x}}_2$.

We can expand f as

$$f = \mathbf{x}_2'\mathbf{x}_2 + b^2\mathbf{x}_1'\mathbf{x}_1 + a^2n - 2b\mathbf{x}_1'\mathbf{x}_2 - 2a\mathbf{u}'\mathbf{x}_2 + 2ab\mathbf{u}'\mathbf{x}_1, \tag{10.3}$$

where we have used $\mathbf{u}'\mathbf{u} = n$. Further, we know that the last term on the right is equal to zero and can be omitted. To find an extreme value, we set partial derivatives of f equal to zero:

$$\partial f/\partial a = 2an - 2\mathbf{u}'\mathbf{x}_2 = 0, \tag{10.4}$$

$$\partial f/\partial b = 2b\mathbf{x}_1'\mathbf{x}_1 - 2\mathbf{x}_1'\mathbf{x}_2 = 0. \tag{10.5}$$

From (10.4) it follows that $a = \mathbf{u}'\mathbf{x}_2/n = \bar{x}_2$, the mean of variable x_2. From (10.5) we have

$$b = \mathbf{x}_1'\mathbf{x}_2/\mathbf{x}_1'\mathbf{x}_1. \tag{10.6}$$

The latter form can be expressed as the ratio of the covariance and the variance of x_1, since $\mathbf{x}_1'\mathbf{x}_1/n = \sigma_1^2$ (because x_1 is taken in deviations from the mean) and $\mathbf{x}_1'\mathbf{x}_2/n$ is the covariance. Note that the latter term gives the covariance, even if x_2 is not taken as deviations from the mean, because

$$\mathbf{x}_1'(\mathbf{x}_2 - \bar{x}_2\mathbf{u}) = \mathbf{x}_1'\mathbf{x}_2 - \bar{x}_2 \cdot \mathbf{u}'\mathbf{x}_1 = \mathbf{x}_1'\mathbf{x}_2.$$

For the sum of the squared distances f, we obtain, by substituting in (10.3) the solution for a and b:

$$\begin{aligned} f &= \mathbf{x}_2'\mathbf{x}_2 + (\mathbf{x}_1'\mathbf{x}_2)^2/\mathbf{x}_1'\mathbf{x}_1 + n\bar{x}_2^2 - 2(\mathbf{x}_1'\mathbf{x}_2)^2/\mathbf{x}_1'\mathbf{x}_1 - 2n\bar{x}_2^2 \\ &= (\mathbf{x}_2'\mathbf{x}_2 - n\bar{x}_2^2) - (\mathbf{x}_1'\mathbf{x}_2)^2/\mathbf{x}_1'\mathbf{x}_1 \end{aligned} \tag{10.7}$$

If we divide both sides by n, we find an expression for the *variance about the regression line*, symbolized $\sigma_{2.1}^2$; it is a kind of residual variance (a measure for the scatter of the points about the line):

$$\sigma_{2.1}^2 = (\mathbf{x}_2'\mathbf{x}_2/n - \bar{x}_2^2) - (\mathbf{x}_1'\mathbf{x}_2)^2/n\mathbf{x}_1'\mathbf{x}_1. \tag{10.8}$$

Note, however, that the first term between parentheses on the right is just the variance of $x_2 : \sigma_2^2$. The other term can be written as

$$(\mathbf{x}_1'\mathbf{x}_2)^2/n\mathbf{x}_1'\mathbf{x}_1 = (\text{cov})^2/\sigma_1^2. \tag{10.9}$$

It follows that (10.8) can be re-arranged to read:

$$\sigma_2^2 = \sigma_{2.1}^2 + (\text{cov})^2/\sigma_1^2. \tag{10.10}$$

This equation (10.10) shows that the variance of x_2 can be divided into two portions. The first one is the residual variance about regression, and the second one is the portion of the variance of x_2 that can be attributed to (or "explained by") regression on x_1.

For an illustration, let us take data about an index of total real-product output (sum of labor and capital services in constant prices for U.S.

Table 10.1. Output of private domestic economy for 1951–1965 (from Jorgerson and Griliches 1967).

Year	Index number	Year	Index number
1951	.852	1959	1.069
1952	.873	1960	1.096
1953	.917	1961	1.115
1954	.904	1962	1.189
1955	.981	1963	1.240
1956	.999	1964	1.307
1957	1.013	1965	1.387
1958	1.000		

private domestic product) during 15 years. Variable x_1 is time, and the index is x_2. Data are given in table 10.1.

From equation (10.4), we find $a = \bar{x}_2 = 1.063$. The slope is found from (10.6) as $9.756/280 = .035$. For the variance of x_2, we find $\mathbf{x}_2'\mathbf{x}_2/n - \bar{x}_2^2 = .0240$. The "explained" variance is $(\mathbf{x}_1'\mathbf{x}_2)^2/n\mathbf{x}_1'\mathbf{x}_1 = .0227$, and it follows that the residual variance $\sigma^2_{2.1} = .0013$. In other words, one might say that 95% of the variance of x_2 can be explained by the regression on time (see figure 10.1).

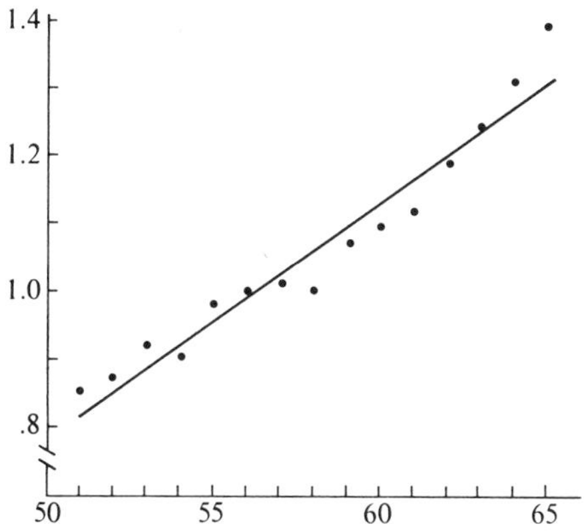

Figure 10.1
Regression line for data in table 10.1.

10.3. Correlation

A *correlation* problem is, loosely speaking, a doubled regression problem, since now we want to fit two lines: a regression line for the regression of x_2 on x_1, but also a regression line for the regression of x_1 on x_2.

Let us assume that both variables are given in standard scores, which will simplify notations considerably. First, for the regression of x_2 on x_1, formula (10.6) now can be simplified to

$$b_1 = \mathbf{x}_1'\mathbf{x}_2/n, \tag{10.11}$$

since $\mathbf{x}_1'\mathbf{x}_1 = n\sigma_1^2 = n$. Here $\mathbf{x}_1'\mathbf{x}_2/n$ is the covariance between standardized variables, called the correlation coefficient. For the residual variance, equation (10.8), we now write

$$\sigma^2_{2.1} = 1 - r^2, \tag{10.12}$$

since the first term on the right of (10.8) was the variance of x_2 (equal to one, for a standardized variable), and the second term reduces to r^2. The alternative for (10.10) becomes

$$1 = \sigma_{2.1}^2 + r^2, \tag{10.13}$$

which shows that r^2 is the portion of the variance of x_2 explained by regression (sometimes called "*coefficient of determination*"), and $1 - r^2$ the portion of the variance that is left "unexplained."

Similar formulas now obtain for the second regression line. From considerations of symmetry alone, it follows that

$$b_2 = \mathbf{x}_2'\mathbf{x}_1/n = r, \tag{10.14}$$

where b_2 is the slope of the second regression line with the coordinate axes reversed. It follows that the two regression lines are symmetric about the bisector of the axes. Also, the residual variances for both lines are equal.

10.4. Binormal Distribution

For a binormal distribution, another interpretation becomes possible. The distribution is given by

$$g(x_1, x_2) = (2\pi)^{-1}|\mathbf{R}|^{-\frac{1}{2}} \exp\left(-\tfrac{1}{2}\mathbf{x}'\mathbf{R}^{-1}\mathbf{x}\right). \tag{10.15}$$

Suppose we take a constant value for x_1, say, $x_1 = c$. The exponent can then be written as:

$$\begin{aligned} -\tfrac{1}{2}\mathbf{x}'\mathbf{R}^{-1}\mathbf{x} &= -\frac{1}{2}\,\frac{(c\ x_2)\begin{pmatrix} 1 & -r \\ -r & 1 \end{pmatrix}\begin{pmatrix} c \\ x_2 \end{pmatrix}}{1 - r^2} \\ &= -\frac{1}{2}\,\frac{c^2 - 2rcx_2 + x_2^2}{1 - r^2} \\ &= -\frac{1}{2}\,\frac{(x_2 - rc)^2 + c^2(1 - r^2)}{1 - r^2} \\ &= -\frac{1}{2}\,\frac{(x_2 - rc)^2}{1 - r^2} - \tfrac{1}{2}c^2. \end{aligned} \tag{10.16}$$

Substituting this in (10.15), we find:

$$g(x_1, x_2) = \frac{1}{2\pi\sqrt{1-r^2}} \exp\left(-\frac{1}{2}\frac{(x_2 - rc)^2}{1-r^2} - \tfrac{1}{2}c^2\right)$$

$$= \frac{1}{\sqrt{2\pi}} \exp\left(-\tfrac{1}{2}c^2\right)$$

$$\times \frac{1}{\sqrt{2\pi}\sqrt{1-r^2}} \exp\left(-\frac{1}{2}\frac{(x_2 - rc)^2}{1-r^2}\right). \quad (10.17)$$

It is seen from (10.17) that $g(x_1, x_2)$ is decomposed into two factors. The first one is $f_1(c)$, the density element for $x_1 = c$. The second one gives a normal distribution for x_2, with mean rc and variance $1 - r^2$, which is just the conditional distribution $f_2(x_2 \mid c)$. Thus, for a given value of $x_1 = c$, values of x_2 are normally distributed with mean located on the regression line and scatter equal to the variance about the regression line.

In this sense we can use the regression line for prediction: i.e., for a given value of x_1, the best prediction of x_2 is the mean of the conditional distribution, which is the point on the regression line. Variance about regression, then, is a measure for the prediction error.

10.5. Vector Model

As we have seen before (section 3.1), in the vector model we have two vectors in an S_n, with length $\rho_1 = \sigma_1$, $\rho_2 = \sigma_2$, respectively, and the angle between them equal to $\cos^{-1}(r)$. It follows that the projection of vector $\mathbf{x}_2$ on $\mathbf{x}_1$ will be equal to $r\sigma_2$. This shows that vector $\mathbf{x}_2$ can be decomposed in two components, one in the direction of $\mathbf{x}_1$ (the component of "explained variance") and one in a direction perpendicular to $\mathbf{x}_1$ (the residual variance, independent from x_1), and similarly for $\mathbf{x}_1$. The picture is very much the same as in components of forces in classical mechanics (see figure 10.2).

Computational techniques are given for an example. Table 10.2 gives the production (in thousands) of trucks and buses combined, and of jeeps, in Brazil for the period 1957 to 1964. Let $\mathbf{x}_1$ and $\mathbf{x}_2$ stand for the raw data. The vector of means becomes $\mathbf{u}'\mathbf{X}/n = (522 \quad 128)/8 = \bar{\mathbf{x}}$.

Sums of squares and the cross-products are

$$\mathbf{X}'\mathbf{X} = \begin{pmatrix} 27568 & 8894 \\ 8894 & 2174 \end{pmatrix},$$

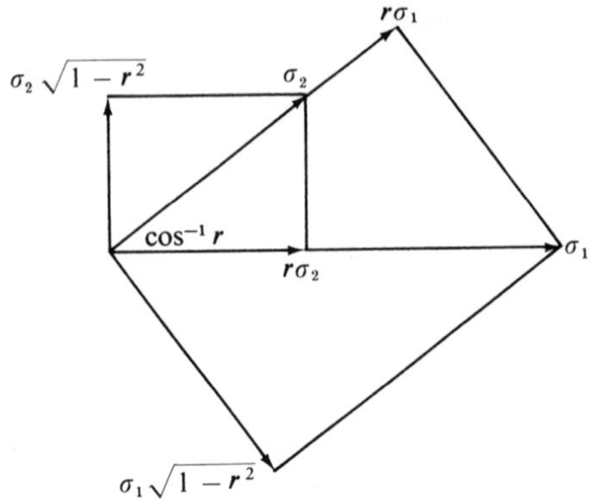

Figure 10.2 Decomposition of correlated vectors.

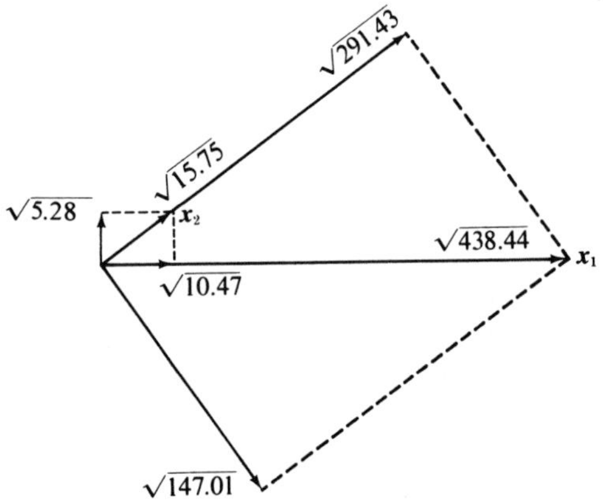

Figure 10.3 Vector representation of data in table 10.2.

Table 10.2. Vehicle output in Brazil, in thousands (Munk 1969).

Year	Trucks and buses	Jeeps
1957	2	9
1958	45	14
1959	66	18
1960	76	20
1961	73	18
1962	94	22
1963	74	14
1964	73	13

and the variance-covariance matrix becomes

$$\mathbf{X}'\mathbf{X}/n - \bar{\mathbf{x}} \cdot \bar{\mathbf{x}}' = \begin{pmatrix} 438.44 & 67.75 \\ 67.75 & 15.75 \end{pmatrix}.$$

The correlation coefficient then is found as $67.75/\sqrt{(15.75)(438.44)} =$ .814. Its square, the coefficient of determination, is .665. The "explained" variance for x_1 therefore is $(.665)(438.44) = 291.43$, and the residual variance 147.01. Similarly, the explained variance for $x_2 = (.665)(15.75) =$ 10.47, and the residual portion 5.28. The vector model is pictured, on scale, in figure 10.3.

11

Multiple and Partial Correlation

11.1. The Simplest Case: Three Variables

11.1.1. Multiple Correlation. We are given an $n \times 3$ matrix $\mathbf{X}$, with $\mathbf{X}$ standardized, so that $\mathbf{X}'\mathbf{X}/n = \mathbf{R}$. The matrix could be pictured as n points in an S_3. The multiple-correlation problem is to find in this space a plane

$$\hat{\mathbf{x}}_3 = b_1\mathbf{x}_1 + b_2\mathbf{x}_2 \tag{11.1}$$

such that the deviations $(\mathbf{x}_3 - \hat{\mathbf{x}}_3)$ are small. The standard criterion is to minimize the sum of the squares of these deviations. The plane is a *regression plane*, and each deviation measures the distance from a point to this plane along the x_3-coordinate. The plane defined in (11.1) is only one of three possible planes, but we shall not treat the other two, because the reasoning is completely similar (just interchange columns of $\mathbf{X}$), and because in applications we generally have a natural way to select one variable (x_3) as dependent on the other two. For instance, x_3 might be a respondent's occupational level, x_1 is his father's occupational level, and x_2 his own level of education; we shall be interested to know, then, how x_3 depends on x_1 and x_2. Or x_1 and x_2 might be measures in two different intelligence tests, and x_3 a measure of school proficiency three years later; the problem would be to find how x_3 could be predicted from x_1 and x_2.

For the formal approach, we agree that $\mathbf{X}_{12}$ is the matrix obtained from $\mathbf{X}$ by omitting the third column, and that $\mathbf{R}_{12}$ is the partition of $\mathbf{R}$ defined by $\mathbf{X}_{12}'\mathbf{X}_{12}/n$. An alternative for (11.1) is

$$\hat{\mathbf{x}}_3 = \mathbf{X}_{12}\mathbf{b}, \tag{11.2}$$

where $\hat{\mathbf{x}}_3$ now stands for an $n \times 1$ vector, and $\mathbf{b}$ is a 2×1 vector. The sum of squared deviations is

$$\begin{aligned} f &= (\mathbf{x}_3 - \hat{\mathbf{x}}_3)'(\mathbf{x}_3 - \hat{\mathbf{x}}_3) = (\mathbf{x}_3 - \mathbf{X}_{12}\mathbf{b})'(\mathbf{x}_3 - \mathbf{X}_{12}\mathbf{b}) \\ &= \mathbf{x}_3'\mathbf{x}_3 + \mathbf{b}'\mathbf{X}_{12}'\mathbf{X}_{12}\mathbf{b} - 2\mathbf{b}'\mathbf{X}_{12}'\mathbf{x}_3 \end{aligned} \tag{11.3}$$

An extreme value is found by taking the derivative with respect to $\mathbf{b}'$, and setting it equal to zero, which gives

$$\partial f/\partial \mathbf{b}' = 2\mathbf{X}_{12}'\mathbf{X}_{12}\mathbf{b} - 2\mathbf{X}_{12}'\mathbf{x}_3 = \mathbf{0}$$

or

$$\mathbf{X}_{12}'\mathbf{X}_{12}\mathbf{b} = \mathbf{X}_{12}'\mathbf{x}_3. \tag{11.4}$$

If we divide by n, we have

$$\mathbf{R}_{12}\mathbf{b} = \mathbf{r}_3, \tag{11.5}$$

where $\mathbf{r}_3 = \mathbf{X}_{12}'\mathbf{x}_3/n$ (the vector of correlations between the two predictors and the third variable). It follows that the *regression weights* $\mathbf{b}$ are given by

$$\mathbf{b} = \mathbf{R}_{12}^{-1}\mathbf{r}_3, \tag{11.6}$$

assuming $|\mathbf{R}_{12}| \neq 0$.

If we now divide (11.3) by n, we obtain an expression for the residual variance about the regression plane (sum of squared deviations divided by n), symbolized $\sigma_{3.12}^2$. Substituting it in (11.6) in that form, the result is

$$\begin{aligned} \sigma_{3.12}^2 &= (\mathbf{x}_3'\mathbf{x}_3/n) + \mathbf{r}_3'\mathbf{R}_{12}^{-1}\mathbf{R}_{12}\mathbf{R}_{12}^{-1}\mathbf{r}_3 - 2\mathbf{r}_3'\mathbf{R}_{12}^{-1}\mathbf{r}_3 \\ &= (\mathbf{x}_3'\mathbf{x}_3/n) - \mathbf{r}_3'\mathbf{R}_{12}^{-1}\mathbf{r}_3 \\ &= 1 - \mathbf{r}_3'\mathbf{R}_{12}^{-1}\mathbf{r}_3. \end{aligned} \tag{11.7}$$

As we found for simple correlation, the unit variance of x_3 can be divided into two portions:

$$1 = \sigma_{3.12}^2 + \mathbf{r}_3'\mathbf{R}_{12}^{-1}\mathbf{r}_3. \tag{11.8}$$

The first term at the right side of (11.8) gives the residual variance about the regression plane, and the second term is the part of the variance of x_3 that is explained by regression. The *multiple-correlation coefficient* $R_{3.12}$ is defined as the square root of the latter term; i.e.,

$$R_{3.12} = (\mathbf{r}_3'\mathbf{R}_{12}^{-1}\mathbf{r}_3)^{\frac{1}{2}}, \tag{11.9}$$

which can also be expressed as

$$R_{3.12} = (\mathbf{b}'\mathbf{r}_3)^{\frac{1}{2}}.$$

For example, we are given a correlation matrix,

$$\mathbf{R} = \begin{pmatrix} 1.00 & .70 & .53 \\ .70 & 1.00 & .17 \\ .53 & .17 & 1.00 \end{pmatrix}.$$

We take x_3 as dependent on x_1 and x_2. For $\mathbf{b}$, we find

$$\mathbf{b} = \begin{pmatrix} 1.00 & .70 \\ .70 & 1.00 \end{pmatrix}^{-1} \begin{pmatrix} .53 \\ .17 \end{pmatrix} = \begin{pmatrix} 1.00 & -.70 \\ -.70 & 1.00 \end{pmatrix} \begin{pmatrix} .53 \\ .17 \end{pmatrix} \Big/ .51$$

$$= \begin{pmatrix} .411 \\ -.201 \end{pmatrix} \Big/ .51 = \begin{pmatrix} .81 \\ -.39 \end{pmatrix}.$$

The squared multiple-correlation coefficient becomes $R_{3.12}^2 = \mathbf{b}'\mathbf{r}_3 = .36$, and $R_{3.12}$ itself is .60.

11.1.2. Vector Model. In the vector model $\mathbf{X}/\sqrt{n}$ is pictured by three vectors in an S_n, where $\mathbf{R}$ gives cosines of the angles between them. Of course, the three vectors can always be fitted into a three-dimensional subspace of an S_n. Since $\hat{\mathbf{x}}_3 = \mathbf{X}_{12}\mathbf{b}$, we see that $\hat{\mathbf{x}}_3$ is linearly dependent on $\mathbf{x}_1$ and $\mathbf{x}_2$, and therefore the vector $\hat{\mathbf{x}}_3$ must be located in the (x_1, x_2)-plane. It can be shown that this vector actually is the projection of vector $\mathbf{x}_3$ on the (x_1, x_2)-plane. To prove this, it is sufficient to show that the cosine of the angle between $\mathbf{x}_3$ and $\hat{\mathbf{x}}_3$ is equal to the length of $\hat{\mathbf{x}}_3$.

First, we know that the length of $\hat{\mathbf{x}}_3$ is the square root of the variance of $\hat{\mathbf{x}}_3$, which is equal to

$$\hat{\mathbf{x}}_3'\hat{\mathbf{x}}_3/n = \mathbf{b}'\mathbf{R}_{12}\mathbf{b} = \mathbf{r}_3'\mathbf{R}_{12}^{-1}\mathbf{r}_3 = R^2_{3.12}.$$

In words, the length of vector $\hat{\mathbf{x}}_3$ equals the multiple-correlation coefficient.

The second step is to find the cosine of the angle between $\mathbf{x}_3$ and $\hat{\mathbf{x}}_3$. We know that the inner product $\mathbf{x}_3'\hat{\mathbf{x}}_3/n$ is equal to the length of $\mathbf{x}_3$ times the length of $\hat{\mathbf{x}}_3$ times the cosine of the angle between them. Now $\mathbf{x}_3$ has unit length, $\hat{\mathbf{x}}_3$ has length $R_{3.12}$, and the inner product is

$$\mathbf{x}_3'\hat{\mathbf{x}}_3/n = \mathbf{x}_3'\mathbf{X}_{12}\mathbf{b}/n = \mathbf{r}_3'\mathbf{b} = R^2_{3.12}.$$

It follows that the cosine of the angle is also equal to the multiple-correlation coefficient. Therefore, $\hat{\mathbf{x}}_3$ is the projection of $\mathbf{x}_3$. Furthermore, $\hat{\mathbf{x}}_3$ gives the vector in the (x_1, x_2)-plane that makes the smallest angle with $\mathbf{x}_3$, which, of course, agrees with the notion that $\hat{\mathbf{x}}_3$ is the linear compound of $\mathbf{x}_1$ and $\mathbf{x}_2$ that has maximum correlation with $\mathbf{x}_3$.

The result is pictured in figure 11.1. One may wonder whether vector **b** is also represented in this picture somehow. In fact, it is. To see how, think of $\hat{\mathbf{x}}_3$ as a kind of resulting "force," as in classical mechanics, with components in the directions of $\mathbf{x}_1$ and $\mathbf{x}_2$ according to a parallelogram of forces. Then $\mathbf{b}_1$ and $\mathbf{b}_2$ are these components. In figure 11.1 $\mathbf{b}_2$ is opposite

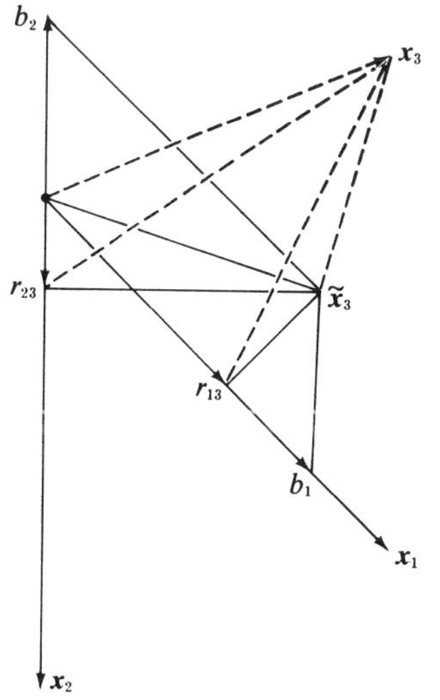

Figure 11.1 Vector representation of multiple correlation. Solid lines are in the (x_1, x_2)-plane; vector $\mathbf{x}_3$ is not in this plane. Vector $\tilde{\mathbf{x}}_3$ is the projection of $\mathbf{x}_3$ on the (x_1, x_2)-plane.

in direction to $\mathbf{x}_2$ ($\mathbf{b}_2$ is negative). Vector $\mathbf{b}$ will have two positive values only if $\mathbf{x}_3$ has a projection inside the acute angle formed by $\mathbf{x}_1$ and $\mathbf{x}_2$.

Sometimes one prefers an equation for $\hat{\mathbf{x}}_3$ as a weighted sum of $\mathbf{x}_1$ and $\mathbf{x}_2$ such that $\hat{\mathbf{x}}_3$ also has unit variance. The weights then are often indicated by the symbols $\boldsymbol{\beta}$:

$$\hat{\mathbf{x}}_3 = \beta_1 \mathbf{x}_1 + \beta_2 \mathbf{x}_2.$$

In the vector model, this just means that $\hat{\mathbf{x}}_3$ is extended to unit length, i.e., multiplied by $R_{3\cdot 12}$. It follows that

$$\boldsymbol{\beta} = \mathbf{b}/R_{3.12}.$$

11.1.3. Partial Correlation. A *partial correlation* (written $r_{12\cdot 3}$) refers to the correlation between x_1 and x_2 after the portion in both that can be predicted from a third variable x_3 has been eliminated. Or, approached differently, vector $(\mathbf{x}_1 - r_{13}\mathbf{x}_3)$ gives the deviations of x_1 from the regression line for regression of x_1 on x_3, and similarly $(\mathbf{x}_2 - r_{23}\mathbf{x}_3)$ gives the deviations of x_2 from the regression line for regression of x_2 on x_3. These deviations are both uncorrelated with x_3. The partial correlation is the correlation between these deviations; it represents correlation between x_1 and x_2 insofar as they are independent of x_3.

The concept of partial correlation gives rise to several interesting methodological problems. One of them is the so-called *spurious correlation*, the classical example of which is that incomes of reverends in Boston is correlated with the price of Jamaica rum, if the correlation is calculated from pairs of observations taken every ten years. The point, of course, is that this correlation is purely artificial, since both variables depend on the trend in monetary value. Another example is: the more firemen involved in extinguishing a fire, the larger the damage; again, both variables are dependent upon a third, the extent of the fire. The number of divorces increases with the number of cars, the number of patients in psychiatric hospitals increases with the number of families having color TV, etc. In these cases, a third variable can be found that will cause the partial correlation between the first two to vanish. However, it will not *always* vanish, since two variables may both be partly independent of a third variable, and in addition have some intrinsic relationship to each other. It is even possible that the partial correlation is larger than the raw correlation. A hypothetical example is that we do not find a correlation between the amount of rainfall and the amount of wheat produced, measured over consecutive years, whereas the partial correlation after elimination of the effect of daily temperature is positive. That is, for years with equal temperature, there is a correlation between rainfall and amount of wheat produced, but this relation is contaminated by variation in temperature, since higher rainfall accompanies lower temperature, which is disadvantageous for wheat production.

Now we turn to the formal approach. As we saw, the partial correlation is defined as the correlation between $\mathbf{x}_1 - r_{13}\mathbf{x}_3$ and $\mathbf{x}_2 - r_{23}\mathbf{x}_3$. First, the covariance between these deviation variables is

$$(\mathbf{x}_1 - r_{13}\mathbf{x}_3)'(\mathbf{x}_2 - r_{23}\mathbf{x}_3)/n$$
$$= (\mathbf{x}_1{}'\mathbf{x}_2/n) - (r_{13}\mathbf{x}_3{}'\mathbf{x}_2/n) - (r_{23}\mathbf{x}_1{}'\mathbf{x}_3/n) + (r_{13}r_{23}\mathbf{x}_3{}'\mathbf{x}_3/n)$$
$$= r_{12} - r_{13}r_{23}. \qquad (11.10)$$

To find the correlation, we still should divide by the square root of the product of the variances. The first variance is

$$(\mathbf{x}_1 - r_{13}\mathbf{x}_3)'(\mathbf{x}_1 - r_{13}\mathbf{x}_3)/n = (\mathbf{x}_1{}'\mathbf{x}_1/n) - (2r_{13}\mathbf{x}_1{}'\mathbf{x}_3/n) + (r_{13}{}^2\mathbf{x}_3{}'\mathbf{x}_3/n)$$
$$= 1 - r_{13}{}^2. \qquad (11.11)$$

Similarly, we find for the other variance:

$$(\mathbf{x}_2 - r_{23}\mathbf{x}_3)'(\mathbf{x}_2 - r_{23}\mathbf{x}_3)/n = 1 - r_{23}{}^2. \qquad (11.12)$$

It follows that the partial correlation is

$$r_{12.3} = (r_{12} - r_{13}r_{23})/\sqrt{(1 - r_{13}{}^2)(1 - r_{23}{}^2)}. \qquad (11.13)$$

How is the partial correlation represented in the vector model? First, take a plane through the origin perpendicular to $\mathbf{x}_3$, and project $\mathbf{x}_1$ and $\mathbf{x}_2$ on this plane (figure 11.2). These projections represent components of

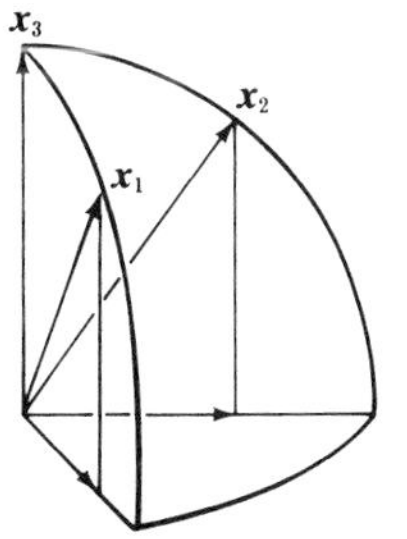

Figure 11.2 Vector representation of partial correlation. The multiple-correlation coefficient would be equal to the projection of $\mathbf{x}_3$ on the (x_1, x_2)-plane (not pictured). The partial correlation corresponds with the cosine of the angle between the projections of x_1 and x_2 on the plane perpendicular to $\mathbf{x}_3$.

variance independent of $\mathbf{x}_3$. Since $\mathbf{x}_1$ has length 1, and since its projection on $\mathbf{x}_3$ is equal to r_{13}, it follows from Pythagoras's theorem that its projection on the plane must be equal to $\sqrt{1 - r_{13}^2}$. Similarly, $\mathbf{x}_2$ has a projection on the plane with length $\sqrt{1 - r_{23}^2}$. The partial correlation is the correlation between the two variables represented by these vectors, and therefore is given as the cosine of the angle between the projections. Note, just for the fun of it, that if $\mathbf{x}_1$ and $\mathbf{x}_2$ are already orthogonal to $\mathbf{x}_3$, then the multiple correlation is zero—$\mathbf{x}_3$ has zero projection on the (x_1, x_2)-plane—and the partial correlation $r_{12\cdot 3}$ is equal to r_{12} itself. The reverse of the latter is not true, however: $\mathbf{x}_1$ and $\mathbf{x}_2$ might both be correlated with $\mathbf{x}_3$ and still have a partial correlation equal to r_{12}.

11.2. The General Case

The general case is a straightforward generalization of the three-variable situation. However, to simplify, we shall change the notation slightly, and agree that the matrix $\mathbf{X}$ of order $n \times m$ gives observations on "predictors" only, and that there is an $(m + 1)^{\text{th}}$ variable y for which we want to find the regression equation

$$\hat{\mathbf{y}} = \mathbf{Xb}. \tag{11.14}$$

All variables are assumed to be standardized, $\mathbf{R} = \mathbf{X}'\mathbf{X}/n$, and for $\mathbf{X}'\mathbf{y}/n$ we shall write $\mathbf{r}$.

Equation (11.14) defines a V_m in the S_{m+1} if we take the picture with n points in the S_{m+1}. Again we can set up a function

$$f = (\mathbf{y} - \mathbf{Xb})'(\mathbf{y} - \mathbf{Xb}),$$

and find an extreme value (minimum) for f. The result becomes

$$\mathbf{b} = \mathbf{R}^{-1}\mathbf{r}, \tag{11.15}$$

whereas we can find that the variance of y can be divided in a portion $\sigma_{y.12\cdots m}^2$ for variance about the regression hyperplane, and a portion representing explained variance, equal to $R_{y.12\cdots m}^2 = \mathbf{r}'\mathbf{R}^{-1}\mathbf{r}$, defined as the squared *multiple-correlation coefficient*. Since all derivations are directly analogous to the three-dimensional case, we leave them as an exercise. Also, in the vector model, we would have that $\hat{\mathbf{y}}$ is the projection of $\mathbf{y}$ on the V_m defined by $\mathbf{X}$, and that this projection has length equal to the multiple-correlation coefficient, which is also equal to the cosine of the angle between **y** and $\hat{\mathbf{y}}$. Again, regression weights **b** are pictured by decomposing $\hat{\mathbf{y}}$ according to a "hyperparallelogram of forces." The derivations suppose $|\mathbf{R}| \neq 0$. If $|\mathbf{R}| = 0$, how do we proceed? The answer is that the m **x**-vectors apparently fit into a hyperplane of lower dimensionality than m, since **R** has rank lower than m; therefore we can omit the x-variables that are linearly dependent on other x-variables.

For partial correlations also, the situation is completely analogous to the three-variable case. In the vector model, imagine a hyperplane in the space perpendicular to vector **y**, with projections of the **x**-vectors on this plane. The plane will be a V_m, and any vector $\mathbf{x}_i$ has a projection on it with length $\sqrt{1 - r^2_{yx_i}}$, whereas angles between projections have cosines equal to partial correlations. Formula (11.13) remains valid, with obvious changes, for any correlation $r_{x_ix_j.y}$.

However, we also can define higher-order partial correlations. They are symbolized $r_{x_ix_j.x_kx_l}$, or, for short, $r_{ij.kl}$, and represent the correlations between the deviations of x_i from the regression plane for the regression of x_i on x_k and x_l, on the one hand, and the deviations from x_j from the regression plane for the regression of x_j on x_k and x_l, on the other hand. In practice, if such a correlation must be calculated, one may calculate $r_{ij.k}$, $r_{il.k}$, and $r_{jl.k}$ first, and calculate $r_{ij.kl}$ from

$$r_{ij.kl} = \frac{r_{ij.k} - r_{il.k}r_{il.k}\,,}{\sqrt{1 - r^2_{il.k}}\,\sqrt{1 - r^2_{jli.k}}} \tag{11.16}$$

which is just formula (11.13) applied to correlations after x_k has already been partialled out. The partial correlation $r_{ij.kl}$ is called a *second-order* partial correlation. Higher-order partial correlations like $r_{ij.klm}$, etc., can be calculated by applying formula (11.13) repeatedly to partial correlations of the preceding level of order. In section 15.4.2, we shall see that there are other, simpler techniques.

11.3. Example

For a numerical ilustration, take the following three variables (Blau and Duncan 1967):

x = father's education level;
y = son's education level;
z = son's final occupational status.

(This example will be discussed in much more detail in the next chapter.) Correlations between the variables are given in Table 11.1. Obviously, it is reasonable to impose a definite time order on the variables, in that the father's education precedes the son's education, and the latter precedes the son's final occupational status. We are interested in seeing what is the regression of z on the two "earlier" variables.

Table 11.1. Correlations for an example.

	x	y	z
x	1	.453	.322
y	.453	1	.596
z	.322	.596	1

Regression weights $\mathbf{b}$ are calculated from (11.6) as

$$\mathbf{b} = \begin{pmatrix} 1 & .453 \\ .453 & 1 \end{pmatrix}^{-1} \begin{pmatrix} .322 \\ .596 \end{pmatrix} = \begin{pmatrix} 1 & -.453 \\ -.453 & 1 \end{pmatrix} \begin{pmatrix} .322 \\ .596 \end{pmatrix} \Big/ .795$$

$$= \begin{pmatrix} .052 \\ .450 \end{pmatrix} \Big/ .795 = \begin{pmatrix} .065 \\ .566 \end{pmatrix}.$$

The multiple correlation is found from $R_{z.xy}{}^2 = \mathbf{b}'\mathbf{r} = .359$, and $R_{z.xy} = .599$. It appears that $R_{z.xy}$ is hardly larger than r_{zy}; apparently the prediction of z from y is not made much better if we include x as a second

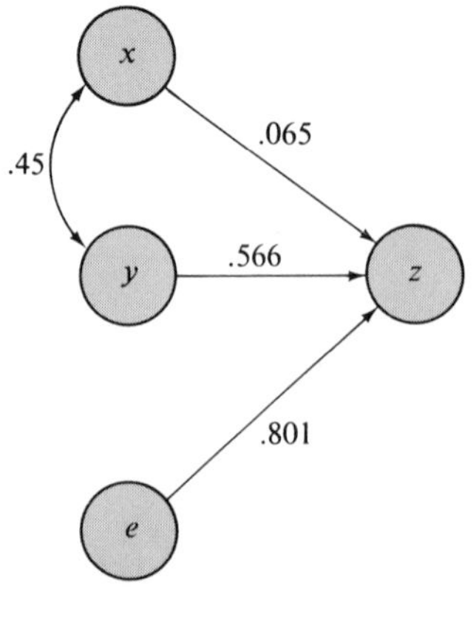

Figure 11.3
Path diagram
for multiple correlation.

predictor. This also appears true of the partial correlation $r_{yz.x}$. It is calculated from (11.13) as

$$r_{yz.x} = (.596 - (.453)(.322))/(\sqrt{1 - .453^2}\ \sqrt{1 - .322^2}) = .533.$$

So $r_{yz.x}$ is a bit smaller than r_{yz} itself, which shows that very little of the correlation between y and z can be explained by their common dependence on x.

The regression equation is $\hat{z} = .065x + .566y$, where $\hat{z}$ is the component of z that is dependent on x and y. Its variance is $R_{z.xy}^2 = .359$. We could however, also write an equation

$$z = .065x + .566y + \hat{e} \qquad (11.17)$$

in which the component $\hat{e}$ is added to stand for the component of z that is independent of x and y. Its variance is $1 - R_{z.xy}^2 = .641$. Since x, y, and z in equation (11.17) are standardized variables, it is perhaps more elegant to make the same assumption for the fourth variable, and write

$$z = .065x + .566y + .801e, \qquad (11.18)$$

where e, in contrast to $\hat{e}$, is a standardized variable. Its contribution to the variance of z now appears as a coefficient in equation (11.18), equal to the square root of the variance contributed by it.

The result can be pictured as in figure 11.3. Here the variables are given as nodes of a network. Single-headed arrows show how z depends on x, y, and e. Numbers written next to the arrows refer to equation (11.18). The double-headed arrow shows that x and y are correlated (the direction of the dependence is left implicit). This type of model is discussed in more detail in the next chapter.

12

Partial-Correlation Analysis and Path Analysis

12.1. Partial-Correlation Analysis

Partial-correlation analysis has found applications in sociology, where it is used to detect causal relations between variables.

First of all, let us state explicitly that correlations never *prove* causal relations. All they can do is to be consistent with some causal theory. Also, if they are consistent with some theory, one can always find a different theory and show that the observed correlations are consistent with this other theory, too. There is nothing in this that should disturb the reader. In fact, formal analysis of data is only one way to arrive at an understanding of the real world, and formal analysis is not sufficient by itself. It should be guided by what is reasonably known about the content to which the data refer. Most researchers would agree that one cannot make a blind diagnosis from formal data alone.

The point can be illustrated by looking back to the variables used in the example of section 11.3. Suppose it were true that the father's education is a determiner of the son's education, and that the latter in its turn determines the son's occupational status. If this were all, there would be only an indirect link between the father's education and the son's occupation level, and we should find that the partial correlation between these variables, if the son's education is taken out, vanishes. Suppose this were in fact found. The conclusion then is that the data are consistent with the theory. However, from a formal point of view, the data are also consistent with the theory that the son's occupational status determines his education and that the latter causes his father's education, or with the theory that the son's education is a common cause of the father's education and his own occupational level. Both theories are silly, but this is revealed, not by the pattern of correlations, but by looking at what the data refer to. The crucial point here is what we know about the time-order of the variables. The father's education precedes the son's education in time, and the son's education comes before the son's job. This ordering excludes a number of theories.

Use of partial-correlation techniques in order to tease out the nature of causal relations between observed variables was suggested by Lazarsfeld (1955). The method was presented later in more detail by Simon (1957) and Blalock (1961, 1963). What it comes down to is that the researcher postulates a model of causal relations, draws a diagram, so to speak, where he indicates, by means of arrows going from one variable to another, how causal influences might run. The model is then "tested" by trying to derive from it which partial correlations should vanish and which should not. The derivations to be tested might be extended to higher-order partials, and could include hypotheses about multiple-regression weights as well. For instance, if x_a is a determiner of x_b, and x_b in its turn influences x_c, whereas x_a has no direct effect on x_c, then $r_{ac \cdot b}$ should vanish. Also, the multiple correlation $R_{c \cdot ab}$ should not be larger than the zero-order correlation r_{bc}.

The latter suggests how the procedure can be systematized. In fact, a zero-order correlation indicates that we can find a linear equation $\hat{x}_c = p_{bc}x_b$ where $\hat{x}_c$ stands for the component of x_c that is dependent on x_b. We know, from formula (10.11), that for standardized variables $p_{bc} = r_{bc}$. One might complete the equation by adding a component e_c that is independent of x_b to it:

$$x_c = p_{bc}x_b + p_{ec}e_c,$$

where e_c is also a standardized variable. If we include x_a as another predictor of x_c, the equation can be extended to

$$x_c = p_{ac}x_a + p_{bc}x_b + p_{ec}e_c,$$

where e_c now stands for a residual component independent of both x_b and x_a. Compare equation (11.18). If the theory were true that x_a affects x_c only indirectly, via x_b, we should find that the additional contribution of x_a in the latter equation is negligible. So the theory can be tested by inspecting the numerical value of p_{ac}.

The following sections show how such notions can be handled more systematically.

12.2. Path Analysis

12.2.1. Path analysis was developed by Sewall Wright (1934, 1954, 1960a and b; Tukey 1954) as a technique for dealing with observed interrelated variables for which it can be assumed that there are several "ultimate" variables that completely determine them. The ultimate variables may be observed variables themselves (like "income" as a determiner of "amount of tax paid"), or they may be hypothetical underlying variables (like "intelligence" as a determiner of observed scores in various tests).

Figure 12.1 pictures a situation in which x_1 and x_2 are observed ultimate variables, and x_u and x_v latent (unobserved, hypothetical, synthetic)

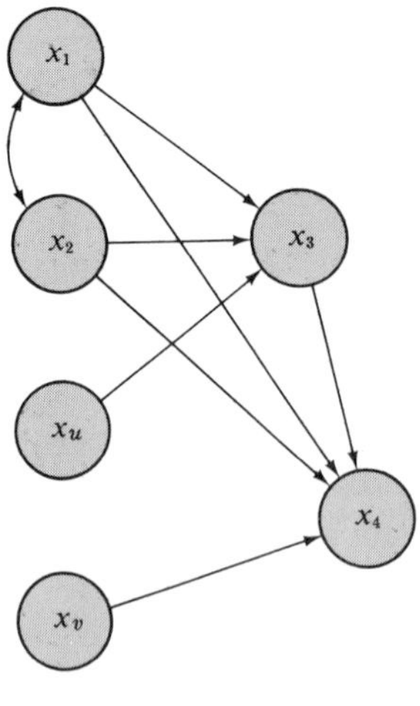

Figure 12.1
Illustration of path diagram.

ultimate variables. The ultimate variables completely determine the observed variables x_3 and x_4, as is indicated by the one-headed arrows. Note that x_4 is not only directly dependent on the ultimate variables x_1 and x_2, but is also indirectly dependent on them through the intermediary x_3. In Figure 12.1 a double-headed arrow is drawn between the ultimate variables x_1 and x_2 to indicate that they are interdependent, but that the direction of the influence is not made explicit. On the other hand, x_u and x_v are independent of each other (no arrow), and are also independent of x_1 and x_2.

Path analysis aims to specify linear equations that are equivalent to such a diagram. Any variable with single-headed arrows pointing to it can be expressed as a function of the variables from which the arrows leave. For instance, it appears from figure 12.1 that we can write an equation

$$x_3 = b_{31}x_1 + b_{32}x_2 + b_{3u}x_u, \tag{12.1}$$

showing that x_3 is completely dependent on x_1, x_2, and x_u.

Similarly, we have

$$x_4 = b_{41}x_1 + b_{42}x_2 + b_{43}x_3 + b_{4v}x_v. \tag{12.2}$$

In the latter equation, the term $b_{43}x_3$ could be eliminated, however, because x_3 can be replaced by the right side of equation (12.1). We then obtain an equation for x_4 in terms of only ultimate factors.

The coefficients b in equations like (12.1) and (12.2) are called *path coefficients*. The equations themselves will be called *structural equations.*

12.2.2. We shall assume henceforth that variables x_i are expressed in standard form. The expectation of terms like x_i^2 will then be equal to unity, and the expectation of x_ix_j equal to the correlation coefficient r_{ij}. The assumption is by no means essential, but it simplifies the formal treatment greatly.

The problem, of course, is to identify the coefficients b in the equations if the correlations between variables are known. To show how this can be done, take equation (12.1), and imagine that it stands for a vector equation. I.e., $\mathbf{x}_1$, $\mathbf{x}_2$, and $\mathbf{x}_3$ are $n \times 1$ vectors of observed values (standardized), whereas $\mathbf{x}_u$ stands for an $n \times 1$ vector of unobserved "error" values (also standardized); the coefficients b are scalar multipliers. Now multiply both sides of (12.1) by $\mathbf{x}_1'$:

$$\mathbf{x}_1'\mathbf{x}_3 = b_{31}\mathbf{x}_1'\mathbf{x}_1 + b_{32}\mathbf{x}_1'\mathbf{x}_2 + b_{3u}\mathbf{x}_1'\mathbf{x}_u. \tag{12.3}$$

If we divide by n (number of observations), we see immediately that the equation reduces to

$$r_{13} = b_{31} + b_{32}r_{12}, \tag{12.4}$$

where we take advantage of the fact that $r_{1u} = 0$, since x_u was assumed to be independent of x_1. Similarly, we can multiply both sides of equation (12.1) by $\mathbf{x}_2'$ to obtain, again after division by n,

$$r_{23} = b_{31}r_{12} + b_{32}. \tag{12.5}$$

Equations (12.4) and (12.5) are now seen to form two nonhomogeneous equations with two unknowns, b_{31} and b_{32}. Since they can be solved, equation (12.1) is identified except for the coefficient b_{3u}.

However, if we multiply (12.1) by $\mathbf{x}_u'$, we see that it is true that

$$r_{3u} = b_{3u}, \tag{12.6}$$

and multiplication of (12.1) by $\mathbf{x}_3'$ gives

$$1 = b_{31}r_{13} + b_{32}r_{23} + b_{3u}r_{3u}, \tag{12.7}$$

so that we can solve for b_{3u} by taking the square root of

$$1 - b_{31}r_{13} - b_{32}r_{23}.$$

This identifies equation (12.1) completely. In a similar manner we could proceed with equation (12.2), but instead we will turn to a somewhat more general approach.

12.2.3. We shall describe the general approach with an example where we have four variables. We assume that variable subscripts indicate time order. Thus, if $i < j$, variable i can be a determiner of variable j, but variable j cannot be a determiner of variable i. This leads to the recursive model described in Chapter 9 and illustrated in figure 9.2.

If we list the relations, we obtain:

x_1 is a determiner of x_2, x_3, and x_4;

x_2 is determined by x_1, and is a determiner of x_3 and x_4;

x_3 is determined by x_1 and x_2, and is a determiner of x_4;

x_4 is determined by x_1, x_2, and x_3.

It follows that the linear relations can be built up in a triangular layout:

$$\begin{aligned} x_2 &= b_{21}x_1; \\ x_3 &= b_{31}x_1 + b_{32}x_2; \\ x_4 &= b_{41}x_1 + b_{42}x_2 + b_{43}x_3. \end{aligned} \tag{12.8}$$

This model, however, is hardly realistic, since it implies complete determination of all variables by x_1. Therefore we add unobserved variables e_i in the equations, where e_i stands for a component of x_i that is independent of the preceding x-variables. (Here e_i could be a random error variable, or a specific systematic component of x_i, or both.) The equations then become:

$$\begin{aligned} x_2 &= b_{21}x_1 + b_{2_e}e_2; \\ x_3 &= b_{31}x_1 + b_{32}x_2 + b_{3_e}e_3; \\ x_4 &= b_{41}x_1 + b_{42}x_2 + b_{43}x_3 + b_{4_e}e_4. \end{aligned} \tag{12.9}$$

Here e_2 is a variable postulated to explain variance in x_2 not determined by x_1, so that x_2 depends partly on x_1 and partly on this specific component e_2. In general, we assume that a latent variable e_i is independent of all preceding variables x_h $(h < i)$, and we assume also that all e-variables are uncorrelated. A variable e_i might be called an *exogenous* component; it stands for a source of variation not dependent on other variables in the system. In the sense of path analysis as described in section 12.2.1, the e's are ultimate factors, together with x_1; x_1 itself is also exogenous. Note that the set of equations (12.9) implies that x_2, x_3, and x_4 are not ultimate factors, but *dependent* or *endogenous* variables—this can be seen if, in the equation for x_3, for instance, we eliminate the term with x_2 in it by substituting for x_2 the right side of the first equation; then x_3 becomes dependent on x_1, e_2, and e_3. Similarly in the third equation we could first eliminate the term with x_3 and then the term with x_2, so that x_4 comes to be a linear function of x_1, e_2, e_3, and e_4. Now think of all the equations (12.9) as vector equations; i.e., the x's stand for column vectors of n observed values, and the e's stand for column vectors of n unobserved values. The rule for finding a solution for the b's is: take one of the equations, that for $\mathbf{x}_i$ (to make the rule general), and multiply both sides by $\mathbf{x}_g'$ for all values of $g < i$. Divide by n, and the result is equations of the form:

$$r_{ig} = b_{i1}r_{1g} + b_{i2}r_{2g} + \cdots + b_{i,i-1}r_{i-1,g}, \tag{12.10}$$

where we take advantage of the agreement that all correlations r_{g,e_i} are zero.

To illustrate the rule, take the first equation of set (12.9). Multiplication of both sides by $\mathbf{x}_1'$ gives, after division by n,

$$r_{12} = b_{21}.$$

This solves for b_{21}. Now take the second equation, that for $\mathbf{x}_3$. First multiply both sides by $\mathbf{x}_1'$ to obtain

$$r_{13} = b_{31} + b_{32}r_{12}. \tag{12.11a}$$

Then multiply both sides by $\mathbf{x}_2'$:

$$r_{23} = b_{31}r_{12} + b_{32}. \tag{12.11b}$$

The two resulting equations (12.11) are two nonhomogeneous equations in two unknowns, which can be solved for the b's. Go on with the equation for $\mathbf{x}_4$ to obtain:

$$\begin{aligned} r_{14} &= b_{41} \quad\quad\;\; + b_{42}r_{12} + b_{43}r_{13}; \\ r_{24} &= b_{41}r_{12} + b_{42} \quad\quad\;\; + b_{43}r_{23}; \\ r_{34} &= b_{41}r_{13} + b_{42}r_{23} + b_{43}. \end{aligned} \tag{12.12}$$

We obtain three nonhomogeneous equations in three unknowns, and can solve for the b's. In general, for the ith equation, the rule will result in i nonhomogeneous equations in i unknowns, from which we can solve for the i values of b_i. The latter requires that the determinant of coefficients be different from zero. This determinant is the determinant $|\mathbf{R}_{i-1}|$, where $\mathbf{R}_{i-1}$ is the matrix of correlations between the first $i - 1$ observed variables. In general, this requirement will be met. If it is not met, this would mean that we can find an observed variable in the set (or more than one) that is completely dependent on the others. This variable then is redundant and might be omitted without loss of information. The event is rather unlikely in applied work.

The remaining task is to identify the coefficients b_{ie}. They are readily found from the equation for $\mathbf{x}_i$. Multiply both sides by $\mathbf{e}_i'$:

$$r_{i,e_i} = b_{i,e}, \tag{12.13}$$

from which we see that b_{ie} is always equal to the correlation between a variable and its own e-component. Secondly, multiply the equation by $\mathbf{x}_i'$, to find

$$1 = b_{i1}r_{1i} + b_{i2}r_{2i} + \cdots + b_{i,i-1}r_{i,i-1} + b_{ie}r_{i,e_i}. \qquad (12.14)$$

But since we know from (12.13) that the last term at the right of (12.14) is equal to b_{ie}^2, we can solve b_{ie} as the square root of $1 - \sum b_{ig}r_{gi}$.

12.3. Numerical Illustration

We now turn to an illustration of the technique. The data are taken from Blau and Duncan (1967), in whose study five variables are described that refer to the process of occupational stratification:

x_1, father's level of education;
x_2, father's occupational status;
x_3, son's educational level;
x_4, occupational level of son's first job;
x_5, occupational level of son's later job.

The matrix of intercorrelations between these five variables is given in table 12.1. Application of the procedure described in section 12.2 results in the following solution for the structural equations:

$$\begin{aligned} x_2 &= .516x_1 + .857e_2; \\ x_3 &= .310x_1 + .278x_2 + .859e_3; \\ x_4 &= .025x_1 + .215x_2 + .433x_3 + .818e_4; \\ x_5 &= -.014x_1 + .121x_2 + .399x_3 + .280x_4 + .752e_5. \end{aligned} \qquad (12.15)$$

This solution is pictured as a diagram in figure 12.2.

12.4. Relation to Multiple-Correlation Technique

It should be noted that path analysis in the form described so far is just an application of multiple-correlation techniques. The formal equivalence is directly seen from sets of equations like (12.11) and (12.12). Take the set (12.11), for instance. It can be written as

$$\mathbf{R}_2\mathbf{b}_3 = \mathbf{r}_3, \qquad (12.16)$$

Table. 12.1. Correlations between five variables related to social stratification.

	x_1	x_2	x_3	x_4	x_5
x_1	1.000	.516	.453	.332	.322
x_2		1.000	.438	.417	.405
x_3			1.000	.538	.596
x_4				1.000	.541
x_5					1.000

where $\mathbf{R}_2$ stands for the matrix of correlations between the first two variables x_1 and x_2, $\mathbf{b}_3$ stands for the vector with elements b_{31} and b_{32}, and $\mathbf{r}_3$ represents the vector of correlations r_{13} and r_{23}. The solution for $\mathbf{b}_3$ is found to be $\mathbf{b}_3 = \mathbf{R}_2^{-1}\mathbf{r}_3$. But this is exactly the same equation as (11.15) for multiple regression weights. In fact, if we omit the e-components at the right side of equations (12.9), we would retain only the component that is dependent on earlier x-variables, and the equations would appear to be ordinary regression equations. The variance of the retained component then must be equal to the squared multiple-correlation coefficient, and the

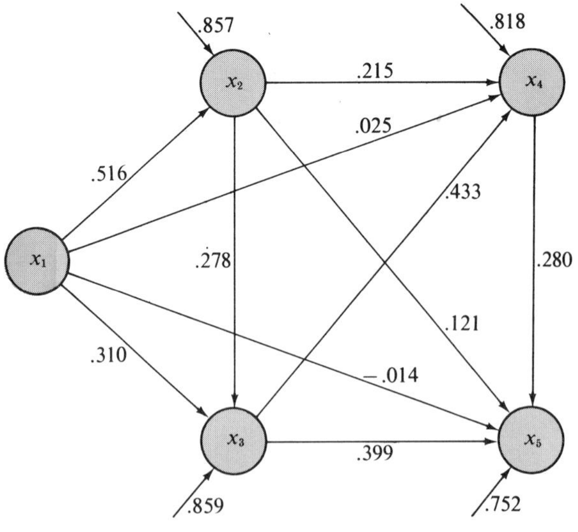

Figure 12.2 Path diagram for five sociological variables.

variance of the e-component equal to the residual variance about regression. So path analysis as described so far is nothing but multiple-regression analysis, successively applied to the regression of each later variable on the earlier ones.

This is also true for another version of path analysis. To see how, look back at the example in section 12.3. We find for the last two equations of (12.15) that the coefficients of x_1 in them are very small, so small that a statistical test would reveal that they are not significantly different from zero. The first thing that might come to mind, then, is just to omit these two terms from the equations. However, doing so would be wrong, not only because the variance on the right side then would no longer equal one, but also because the values for the coefficients in the retained terms are not the most efficient ones we might obtain. Instead of simply omitting a term, we should repeat the analysis, but with variable x_1 omitted from it. We thus obtain a new solution for the equations for x_4 and x_5 without an x_1-term in them:

$$\begin{aligned} x_4 &= .225x_2 + .439x_3 + .819e_4; \\ x_5 &= .115x_2 + .395x_3 + .281x_4 + .752e_5. \end{aligned} \qquad (12.17)$$

Actually, this still is multiple-regression analysis. Another variation is dependence analysis, which we shall describe in the next section.

12.5. Dependence Analysis

Dependence analysis is a technique suggested by Boudon (1965, 1967, 1968). Its only difference from "classical" path analysis is that, in setting up structural equations like (12.9), it is assumed as part of the theoretical model that some coefficients are equal to zero. For instance, in dependence analysis the equation for x_4 might from the very beginning have been assumed to be

$$x_4 = b_{42}x_2 + b_{43}x_3 + b_{4e}e_4. \qquad (12.18)$$

If we then apply the rule for solving for b's, we would obtain the following set of equations to take the place of (12.12):

$$\begin{aligned} r_{14} &= b_{42}r_{12} + b_{43}r_{13}; \\ r_{24} &= b_{42} \quad\;\; + b_{43}r_{23}; \\ r_{34} &= b_{42}r_{23} + b_{43}. \end{aligned} \qquad (12.19)$$

The set (12.19) gives three nonhomogeneous equations in two unknowns, which means that the set is overdetermined, and no solution can be identified unless the small miracle happens that the equations satisfy the special conditions for solvability given in section 5.5.4. One approach, then, might be to take a least-squares solution as the least bad compromise (section 6.2.3). Equations (12.19) become, in the example of 12.3:

$$\begin{aligned} .332 &= .516b_{42} + .453b_{43}; \\ .417 &= b_{42} + .438b_{43}; \\ .538 &= .438b_{42} + b_{43}; \end{aligned} \qquad (12.20)$$

with solution

$$\mathbf{b} = \begin{pmatrix} 1.458 & 1.110 \\ 1.110 & 1.397 \end{pmatrix}^{-1} \begin{pmatrix} .824 \\ .871 \end{pmatrix} = \begin{pmatrix} .291 \\ .376 \end{pmatrix},$$

which is different from the result in (12.17), where the two values for b_4 were found to be .225 and .439. It is the latter solution that must be preferred, because the structural equation is better estimated statistically (but in this book statistical issues are avoided), and because the procedure outlined in section 12.4 safeguards against the danger of self-fulfilling prophecy. In fact, if we first identify the complete recursive model, and if we then recalculate equations only if it appears that in the complete system some coefficients are not statistically different from zero, then we can be sure the recalculated equations are fair and acceptable. On the other hand, if we set some coefficients *a priori* equal to zero, we always can find the best equation on that assumption, but the assumption might be false to start with and untenable in the light of the available data. We then would arrive at equations that satisfy the *a priori* model and be happy with them, without even being aware of the fact that the model itself is inconsistent with the data. This is what we mean by "self-fulfilling prophecy."

12.6. Partially Implicit Recursive Model

Another variety of path analysis is to leave parts of the complete model implicit. This was suggested by our first introduction of the technique in section 12.2.1, where x_1 and x_2 were taken as ultimate factors and their interdependence was not made explicit. If this is the model we want, the solution is very simple: we omit from system (12.9) the equation for x_2 and solve for the others. If we also are not interested in the equation for x_3 (i.e., x_1, x_2, and x_3 are taken as ultimate factors with unexplicitated interdependence), then we skip calculations for b_3, too.

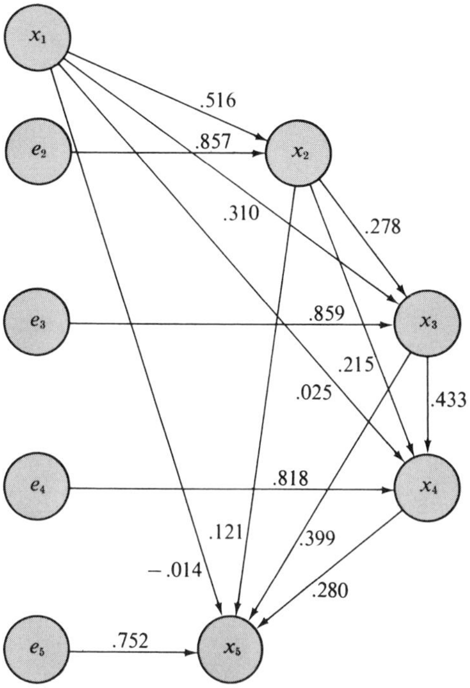

Figure 12.3 Alternative drawing of figure 12.2.

Note that the latent variables x_u and x_v in figure 12.1 are nothing but the exogenous variables e_3 and e_4. To show this better, we have redrawn figure 12.2 in such a way that the *e*'s look more like ultimate factors: see figure 12.3.

In section 12.2.1 we also started from the idea that ultimate factors explain the other variables completely. In the equation for x_4 in (12.9), however, referring this equation to figure 12.1, we see that x_4 is expressed as dependent not only on ultimate factors x_1, x_2, and e_4, but also on x_3, which is not an ultimate factor. To find an equivalent equation in terms of only ultimate factors is simple: we just eliminate the term $b_{43}x_3$ in the equation for x_4, by substituting for it what we have found in the equation for x_3. The result is

$$x_4 = b_{41}x_1 + b_{42}x_2 + b_{43}(b_{31}x_1 + b_{32}x_2 + b_{3_e}e_3) + b_{4_e}e_4$$
$$= (b_{41} + b_{43}b_{31})x_1 + (b_{42} + b_{43}b_{32})x_2 + b_{43}b_{3_e}e_3 + b_{4_e}e_4, \quad (12.21)$$

which is an equation in ultimate factors, only.

Equation (12.21) illustrates a general "principle" of path analysis. Look at the coefficient for x_1 in (12.21): it is $b_{41} + b_{43}b_{31}$. Now imagine a walk through the diagram, following directions of arrows: there is one direct path from x_1 to x_4 on which we meet coefficient b_{41}, but there is also an indirect path via x_3 on which we first meet b_{31} and then b_{43}; multiplying them as we walk along, we have $b_{43}b_{31}$ for the indirect path. The general principle is: to find a coefficient in the equation in terms of ultimate factors, walk along all possible paths from the ultimate factor to the dependent variable, multiple path coefficients that are met along each path, and add over all paths.

12.7. Concrete Path Coefficients

For some applications, one might prefer to have path coefficients that apply to variables which are not first standardized. For instance, imagine a study of the determinants of growth of plants. One wants to know to what extent variation in stem length y_3 (measured in inches) can be contributed to variation in the usual range of temperature y_1 (measured in degrees Fahrenheit) and variation of humidity y_2 (percentage of water in the air). These y-variables are measured in real units as deviations from the mean. The relation with standardized variables is: $y_i = x_i s_i$, where s_i is the standard deviation. Now take the standard equation:

$$x_3 = b_{31}x_1 + b_{32}x_2 + b_{3_e}e_3. \quad (12.22)$$

The corresponding equation in terms of real units is

$$y_3 = (s_3/s_1)b_{31}y_1 + (s_3/s_2)b_{32}y_2 + s_3 b_{3_e}e_3, \quad (12.23)$$

from which we see that $(s_3/s_1)b_{31}$ and $(s_3/s_2)b_{32}$ are the new path coefficients for variables measured in real units. They are sometimes called *concrete path coefficients* (Turner and Stevens 1959). Their use will be defended on the grounds that standard path coefficients can be misleading. For instance, suppose that b_{31} and b_{32} are about equal. Equation (12.22) then suggests that temperature and humidity are about equally effective determiners of stem length. If, however, s_2 is much larger than s_1, the "real" effect of temperature is much larger. In other words, the small variation in temperature is as effective as the large variation in humidity. However, concrete path coefficients are equally "misleading" if one reverses the argument. Suppose that concrete path coefficients are about equal. This suggests that temperature has the same effect as humidity. But since variation in temperature is small, its real contribution to stem length is also small. Of course, the truth is that it does not matter whether we standardize variables or not, since this is a matter of convention. If one procedure seems misleading compared to the other, it is just because of human frailty.

12.8. Remaining Issues

Although path analysis seems a rather elegant and effective technique for describing causal networks, there are a few problems that make it advisable to be cautious in interpreting results of path analysis. In fact, path coefficients, since they are multiple-regression weights used in a special way, have all the disadvantages of the latter. For example, although such weights specify a best equation from the available data, they tend to be rather vulnerable because they capitalize on chance elements in the data. The classical solution to this problem for prediction purposes is to compute weights from one set of data, and test their merits by applying the equations to a fresh set of data to see how well they work in that new environment. Very often the result is found to be somewhat disappointing, in that the ultimate predictive power is much less than was first suggested. A similar *cross-validation* might be recommended for path analysis, if the numerical results are to be taken seriously.

Although multiple-regression equations produce the best solution for empirical prediction, they do not necessarily give a clear understanding of the underlying process. Take the following hypothetical example, where, for convenience, we take the same variables as were used in the earlier numerical illustration of the technique:

x_1 = father's education level;
x_2 = son's education level;
x_3 = occupational status of son's first job;
x_4 = occupational status of son's final job.

Suppose the correlations were as given in table 12.2. Classical path analysis gives the following results:

$$x_2 = .8x_1 + .6e_2;$$
$$x_3 = .6x_1 + .8e_3; \qquad (12.24)$$
$$x_4 = -x_1 + .8x_2 + .6x_3 + .734e_4.$$

Our interest is especially in the last equation of (12.24), which seems to suggest that the son's final job status depends on his educational level and

Table 12.2. A hypothetical correlation matrix.

	x_1	x_2	x_3	x_4
x_1	1.000	.800	.600	.000
x_2		1.000	.480	.288
x_3			1.000	.384
x_4				1.000

his first job status, but is inversely related to his father's education. A naive sociologist might be tempted to conclude that there is a sort of compensatory mechanism, a rebellion of a later generation against the first; sons want to be different from their fathers. However, the model from which the correlation matrix in table 12.2 was derived is quite different; its equation for x_4 was

$$x_4 = .48e_2 + .48e_3 + .734e_4. \qquad (12.25)$$

In other words, we started from the assumption that the son's final job status is completely independent of x_1, and that it depends on his own education and his own first job status precisely to the extent the latter two variables are also independent of x_1. The resulting equation for x_4 in (12.24) may then appear surprising; however, its form results from the fact that the term x_1 in it is used as a *suppressor*. This means the following: if we predict x_4 from x_2 and x_3, as is assumed in the path-analysis model, then x_2 and x_3 do in fact contain the relevant components e_2 and e_3 and to this extent they are useful predictors. However, x_2 and x_3 also contain the component x_1, which is irrelevant in our assumptions. In (12.24), then, the term with x_1 is used to "suppress" the irrelevant components in x_2 and x_3; i.e., x_2 and x_3 do contain a portion x_1 in them, and this portion is subtracted again in the equation by adding the negative term with x_1.

The suppressor phenomenon is well-known in psychological research. It illustrates the fact that multiple-regression equations can be ambiguous, and that they do not always give proper insight if weights are taken at face value. The difficulty arises from the fact that the predictors in the equation are interdependent among themselves, and therefore cannot be interpreted afterward as if they were independent determiners.

We shall come back to path analysis later, to show that it can be interpreted as a special variety of factor analysis (section 15.4), to show how it can be combined with factor analysis and canonical analysis (sections 14.3 and 15.5), and finally to show a solution for nonrecursive models (Chapter 16).

13

Factor Analysis

13.1. General Introduction

The objective of factor analysis might best be explained in terms of vector spaces. The point of departure is a matrix **X**, an $n \times m$ matrix of observed values on m variables. It is assumed that the m variables are standardized, so that $\mathbf{X}'\mathbf{X}/n = \mathbf{R}$ (the $m \times m$ correlation matrix).

We remember that the picture of $\mathbf{X}/\sqrt{n}$ results in a bundle of m vectors in an S_n, all of them with unit length, and with angles between them that have cosines equal to correlations. It will be clear that the m vectors can always be fitted into a subspace V_m of an S_n, since it never requires more then m dimensions to picture relations between m vectors.

However, it may happen that the m vectors can be fitted into a subspace of even lower dimensionality. If so, we would be able to find vectors in the V_m on which the x-vectors have zero projections. Suppose there is a set of $m - p$ such vectors; then we would be able to fit the n x-vectors into a V_p. The advantage is obvious: without any loss of information we then could describe the positions of the x-vectors in terms of p coordinates instead of m coordinates.

In practice, we cannot expect to find such a V_p with p considerably smaller than m. However, perhaps p can be made small if we pay a certain price for it, by allowing a little distortion of the vector bundle as we force it to fit into the V_p. In other words, we might find vectors in V_m on which the projections are not actually zero, but are small enough to be neglected. For instance, imagine three vectors in a V_m located like three ribs of an opened umbrella. The projections of the ribs on the stick of the umbrella are small; therefore the angles between the ribs are not changed very much if we force the ribs into a flat plane (damage to the umbrella is left out of consideration).

Factor analysis generalizes the idea to higher-dimensional spaces. The x-vectors in a V_m are ribs of a hyperumbrella that has $m - p$ orthogonal sticks on which the ribs have sufficiently small projections. It then is possible to force the ribs into a V_p without too much distortion of the angles between the ribs.

13.2. Basic Procedure

Since the x-vectors have unit length, the sum of the squared projections of each one on an orthogonal coordinate system in a V_m is equal to unity, according to the generalized Pythagorean theorem. Therefore, the total sum of the squared projections for all m vectors together equals m. It follows immediately that if there are vectors in a V_m on which the sum of the squared projections is relatively small, then there must be other vectors on which the sum of the squared projections is relatively large. This defines the strategy we shall follow: we shall first try to find a vector in the V_m such that the sum of the squared projections of the x-vectors on it is a maximum. Then we shall try to find a second direction in the V_m orthogonal to the first, again with a maximum sum of squared projections on it. And so on, until after the pth direction the total sum of squared projections equals m (which implies zero projections on the remaining V_{m-p}), or until the total sum of squared projections is sufficiently near to m (which implies near-zero projections on the remaining hyperplane).

For the formal approach, let us define an $n \times 1$ vector **y** in the V_m. We want **y** to be located in the V_m (in order to be able to determine projections of x-vectors on it). This condition implies that **y** is linearly dependent on the x-vectors, and therefore that we can find an $m \times 1$ vector **g** such that $\mathbf{Xg} = \mathbf{y}$. Further, we remember that in the picture of $\mathbf{X}/\sqrt{n}$ the x-vectors have unit length. We want the same to be true for **y**. This implies that $\mathbf{y}'\mathbf{y}/n = 1$, or

$$\mathbf{g}'\mathbf{X}'\mathbf{X}\mathbf{g}/n = \mathbf{g}'\mathbf{R}\mathbf{g} = 1. \tag{13.1}$$

Under these conditions, the inner product $\mathbf{X}'\mathbf{y}/n$ gives projections of x-vectors on the y-vector, for, since all vectors have unit length, the cosine of any angle between an x-vector and the y-vector equals the projection of the x-vector on the y-vector. Let these projections be called **f**. Then we have

$$\mathbf{f} = \mathbf{X}'\mathbf{y}/n = \mathbf{X}'\mathbf{X}\mathbf{g}/n = \mathbf{R}\mathbf{g}. \tag{13.2}$$

Our objective is to obtain a maximum for the sum of the squared projections by a suitable choice of **g**. In other words, we want to maximize $\mathbf{f}'\mathbf{f} = \mathbf{g}'\mathbf{R}^2\mathbf{g}$ under the condition $\mathbf{g}'\mathbf{R}\mathbf{g} = 1$. Using the method of undetermined multipliers, we can see directly that we have to take the partial derivatives of the auxiliary function

$$\mathbf{g}'\mathbf{R}^2\mathbf{g} - \lambda(\mathbf{g}'\mathbf{R}\mathbf{g} - 1)$$

and set these equal to zero. The result is

$$2\mathbf{R}^2\mathbf{g} - 2\lambda\mathbf{R}\mathbf{g} = \mathbf{0},$$

or, rearranging and dividing by 2,

$$\mathbf{R}^2\mathbf{g} = \lambda\mathbf{R}\mathbf{g}. \tag{13.3}$$

But since $\mathbf{Rg} = \mathbf{f}$, we can also write

$$\mathbf{R}\mathbf{f} = \lambda\mathbf{f}, \tag{13.4}$$

which shows that **f** is an eigenvector of **R**. This eigenvector should be normalized to the eigenvalue λ, for, by premultiplying (13.3) by $\mathbf{g}'$, it is seen that

$$\mathbf{g}'\mathbf{R}^2\mathbf{g} = \lambda\mathbf{g}'\mathbf{R}\mathbf{g} = \lambda, \tag{13.5}$$

where we use the condition $\mathbf{g}'\mathbf{R}\mathbf{g} = 1$. However, $\mathbf{g}'\mathbf{R}^2\mathbf{g} = \mathbf{f}'\mathbf{f}$, and therefore

$$\mathbf{f}'\mathbf{f} = \lambda. \tag{13.6}$$

It follows also that the eigenvalue λ directly gives the sum of the squared projections. Obviously, in order to have a maximum for this sum, we should take the solution for the eigenvector with largest eigenvalue.

The vector $\mathbf{g}$ is also the eigenvector of $\mathbf{R}$, but normalized to a different value. This follows from equation (13.3) if we write

$$\mathbf{R}^2\mathbf{g} - \lambda\mathbf{R}\mathbf{g} = 0,$$

and therefore

$$\mathbf{R}(\mathbf{R}\mathbf{g} - \lambda\mathbf{g}) = \mathbf{0},$$

so that $\mathbf{R}\mathbf{g} - \lambda\mathbf{g} = \mathbf{0}$. Since $\mathbf{g}'\mathbf{R}\mathbf{g} = 1$, we have $\mathbf{g}'\mathbf{R}\mathbf{g} = \lambda\mathbf{g}'\mathbf{g} = 1$, and

$$\mathbf{g}'\mathbf{g} = \lambda^{-1}, \tag{13.7}$$

so that $\mathbf{g}$ is the eigenvector of $\mathbf{R}$ normalized to the inverse of the eigenvalue. We therefore can also write

$$\mathbf{g} = \mathbf{f}\lambda^{-1}. \tag{13.8}$$

However, we should at this point look at other solutions of equation (13.4). In general, $\mathbf{R}$ will have m eigenvectors with m eigenvalues different from zero (assuming that $\mathbf{R}$ has rank m; if $\mathbf{R}$ has lower rank, say rank p with $p < m$, then there are $n - p$ zero eigenvalues—cf. section 7.4.5). Let $\mathbf{F}$ be the $m \times m$ matrix of eigenvectors (each column of $\mathbf{F}$ gives an eigenvector). Equation (13.4) then can be generalized to

$$\mathbf{R}\mathbf{F} = \mathbf{F}\Lambda, \tag{13.9}$$

where, for convenience, we shall agree that the eigenvalues given in the diagonal matrix Λ are ordered in decreasing order. Since $\mathbf{R}$ is symmetric, the eigenvectors are orthogonal, and since we agreed that the eigenvectors in $\mathbf{F}$ should be normalized to their eigenvalues, we have

$$\mathbf{F}'\mathbf{F} = \Lambda. \tag{13.10}$$

On the other hand, since $\mathbf{R}\mathbf{F} = \mathbf{F}\Lambda$, and $\Lambda = \mathbf{F}'\mathbf{F}$, we can write $\mathbf{R}\mathbf{F} = \mathbf{F}\mathbf{F}'\mathbf{F}$, or $(\mathbf{R} - \mathbf{F}\mathbf{F}')\mathbf{F} = 0$, which implies

$$\mathbf{R} = \mathbf{F}\mathbf{F}'. \tag{13.11}$$

We shall come back to this equation later. First a few other generalizations. The vector **g**, discussed so far as a single specimen, can be generalized to a set of m vectors **G**, defined by

$$\mathbf{G} = \mathbf{F}\Lambda^{-1}, \tag{13.12}$$

which is the generalization of (13.8). Also, we have a set of m y-vectors, collected in an $n \times m$ matrix **Y** defined by

$$\mathbf{Y} = \mathbf{XG} = \mathbf{XF}\Lambda^{-1}. \tag{13.13}$$

As a kind of check, we may note that the y-vectors are indeed independent (orthogonal) and have unit length:

$$\begin{aligned}\mathbf{Y}'\mathbf{Y}/n &= \mathbf{G}'\mathbf{X}'\mathbf{XG}/n = \mathbf{G}'\mathbf{RG} = \Lambda^{-1}\mathbf{F}'\mathbf{RF}\Lambda^{-1} = \Lambda^{-1}\mathbf{F}'(\mathbf{F}\Lambda)\Lambda^{-1} \\ &= \Lambda^{-1}\Lambda\Lambda\Lambda^{-1} = \mathbf{I},\end{aligned}$$

and that **F** gives the projections of the x-vectors on the y-vectors:

$$\mathbf{X}'\mathbf{Y}/n = \mathbf{X}'\mathbf{XG}/n = \mathbf{RG} = \mathbf{RF}\Lambda^{-1} = \mathbf{F}\Lambda\Lambda^{-1} = \mathbf{F}.$$

Let us summarize the results in words. We begin with a bundle of x-vectors representing the observed variables. This bundle is embedded in a V_m, which by itself is a subspace of an S_n. Within V_m we have located additional vectors **Y**. They are related to **X** by the relation **XG** = **Y**, and therefore are also expressed in terms of the n coordinates of the original S_n. The x-vectors have projections on the y-vectors, and these projections are given in a matrix **F**, associated with which is a diagonal matrix Λ that gives sums of squared projections. The system as a whole is set up in such a way that: the y-vectors are orthogonal; the first y-vector has the largest sum of squared projections on it; the second vector $\mathbf{y}_2$ is orthogonal to $\mathbf{y}_1$, and under this condition has the largest sum of squared projections on it; $\mathbf{y}_3$ is orthogonal to the plane of $\mathbf{y}_1$ and $\mathbf{y}_2$, and under this condition has the largest sum of squared projections on it; etc.

Let us now go back to equation (13.11). First, as mentioned in section 7.4, if **R** has rank p, with p smaller than m, then there are only p eigenvalues different from zero. This implies that we will be able to find a set of $m - p$ orthogonal vectors in V_m on which the projections of the x-vectors are all zero. More precisely: there is a V_{m-p} with zero projections on it, and within this V_{m-p} one can put up a set of $m - p$ orthogonal vectors in an infinite number of ways. The x-vectors then could be perfectly described within a V_p, and, in addition, if $\mathbf{F}_p$ contains the first p eigenvectors with non-zero eigenvalues, it remains true that $\mathbf{R} = \mathbf{F}_p\mathbf{F}_p'$.

In practice, however, the result is rather improbable. Most of the time $\mathbf{R}$ will have full rank m. Then we might be prepared to neglect directions in a V_m with small projections on them. Formally this means that we omit from $\mathbf{F}$ the final eigenvectors with small eigenvalues, and that we retain only, say, the first p eigenvectors. Let this remaining matrix be called $\mathbf{F}_p$ again. Then it is no longer true that $\mathbf{R} = \mathbf{F}_p\mathbf{F}_p'$.

However, the Eckhart-Young theorem discussed in section 7.4.6 shows that $\mathbf{F}_p\mathbf{F}_p'$ is the best approximation of $\mathbf{R}$ by a matrix of rank p. The matrix $\mathbf{R} - \mathbf{F}_p\mathbf{F}_p'$ gives the deviations of the retained solution from the observed $\mathbf{R}$, and the solution is defined by the fact that the sum of the squared deviations has a minimum. This is another, though not really different, way to interpret the result.

Still another interpretation emerges if we remember that there are two analogous representations of the $n \times m$ matrix $\mathbf{X}$. So far we concentrated on the vector picture, where m vectors are pictured in an S_n. The analogous picture takes the variables as dimensions and plots n points in an S_m. With two variables we then obtain the well-known scatter plot, in which points fall into an elliptical region if the variables are correlated. With m variables, we would have an m-dimensional scatter plot with points in a hyperelliptical region.

The equivalent of factor analysis in this picture is to search for vectors in the S_m with maximum spread of the projections on them. These vectors correspond to the axes of the hyperellipse: the largest amount of spread is found along the longest axis of this hyperellipse. To find this vector, assume that its direction cosines are $\mathbf{k}_1$, where $\mathbf{k}_1$ obeys the usual condition $\mathbf{k}_1'\mathbf{k}_1 = 1$. Projections of points on this vector then are given by the vector $\hat{\mathbf{y}}_1 = \mathbf{X}\mathbf{k}_1$. The variance of these projections is given by

$$\hat{\mathbf{y}}_1'\hat{\mathbf{y}}_1/n = \mathbf{k}_1'\mathbf{X}'\mathbf{X}\mathbf{k}_1/n = \mathbf{k}_1'\mathbf{R}\mathbf{k}_1.$$

Obviously, we should find a maximum for the latter expression under the condition $\mathbf{k}_1'\mathbf{k}_1 = 1$. Using undetermined multipliers, we can readily show that $\mathbf{k}_1$ can be found from the equation

$$\mathbf{R}\mathbf{k}_1 = \lambda_1\mathbf{k}_1,$$

so that $\mathbf{k}_1$ is an eigenvector of $\mathbf{R}$ normalized to unity. If we premultiply the last equation by $\mathbf{k}_1'$, we also see that the eigenvalue λ_1 represents the variance of the projections; for maximum variance, we obviously should take the largest eigenvalue. The second eigenvector $\mathbf{k}_2$ specifies the second longest axis of the hyperellipse; i.e., the vector orthogonal to $\mathbf{k}_1$, and with the largest variance of projections on it under this condition.

It can be shown that the vector $\mathbf{y}_1$, which we discussed earlier in the vector model, is just the standardized version of $\hat{\mathbf{y}}_1$. The variance of $\hat{\mathbf{y}}_1$ equals λ_1; therefore if we take $\mathbf{y}_1 = \tilde{\mathbf{y}}_1\lambda_1^{-\frac{1}{2}}$, then $\mathbf{y}_1$ is the standardized version of $\hat{\mathbf{y}}_1$. This defines $\mathbf{y}_1$ as $\mathbf{y}_1 = \mathbf{X}\mathbf{k}_1\lambda_1^{-\frac{1}{2}}$, which is the same as the earlier definition of $\mathbf{y}_1$ as $\mathbf{y}_1 = \mathbf{X}\mathbf{g}_1$, because $\mathbf{g}_1$ is an eigenvector of $\mathbf{R}$ normalized to the inverse of the eigenvalue.

What does the operation of standardizing the vectors $\hat{\mathbf{y}}$ mean geometrically? These vectors represent axes of the hyperellipse; they are already orthogonal, but they have different "lengths," so to speak. If we standardize the vectors, the geometrical counterpart is to let the hyperellipse shrink along its long axes and expand along its short axes, until it becomes a hypersphere with equal spread in all directions.

What does it mean if there are zero eigenvalues? Then there are vectors in an S_n on which the individual points have zero projections, i.e., vectors that do not discriminate between individual points. Similarly, if there are small eigenvalues, the geometric counterpart is that there are vectors in S_n with small variance between individual points. If we neglect those vectors, not too much is lost, because they hardly discriminate between individual measurements.

13.3. Definitions of Terms

The matrix of projections $\mathbf{F}$, or its reduced part $\mathbf{F}_p$, is called the *factor matrix*. Elements of $\mathbf{F}$ are called *factor loadings*; f_{ij} is the loading of variable i on factor j. The directions of the y-vectors in a V_m may be called *reference directions*; they form an optimal orthogonal coordinate system for the description of $\mathbf{R}$. The vectors $\mathbf{y}$ themselves are called *factors*. Remember that $\mathbf{y}$ is a vector in the V_m but also in the S_n; in fact, any $\mathbf{y}$ is an $n \times 1$ vector. A vector $\mathbf{y}$ therefore can be interpreted as an *unobserved variable*, a hypothetical, "latent" variable for which we have no direct observations. Its values can be calculated as linear compounds of the observed x-variables (vector $\mathbf{g}$ gives the appropriate weights). Elements of a vector $\mathbf{y}$ therefore can be interpreted as individual scores in the hypothetical variable; they are called *factor scores.* In this connection, it is useful to have another look at the elements of **F**. Since x-vectors and y-vectors have unit length in the V_m, it follows that the projection f_{ij} of $\mathbf{x}_i$ on $\mathbf{y}_j$ also represents the cosine of the angle between these vectors. But cosines of angles between vectors in an V_m stand for correlations; this implies that f_{ij} can also be interpreted as the correlation between the observed variable $\mathbf{x}_i$ and the hypothetical variables $\mathbf{y}_j$. Factor loadings are correlations between variables and factors.

It is an elementary matrix operation that

$$\mathbf{FF}' = \mathbf{f}_1\mathbf{f}_1' + \mathbf{f}_2\mathbf{f}_2' + \cdots + \mathbf{f}_m\mathbf{f}_m'.$$

Since $\mathbf{FF}' = \mathbf{R}$, we have

$$\mathbf{R} = \mathbf{f}_1\mathbf{f}_1' + \mathbf{f}_2\mathbf{f}_2' + \cdots + \mathbf{f}_m\mathbf{f}_m', \tag{13.14}$$

and it appears that **R** can be decomposed into additive $m \times m$ matrices, each of which is determined by one factor only and is therefore of rank one. The matrix $\mathbf{f}_i\mathbf{f}_i'$ is often called the *contribution* of the ith factor to the correlation matrix **R**; it is the portion of the correlation matrix that is "explained" by the ith factor. The first factor has the largest contribution, because the sum of its squared elements is largest. If this contribution is subtracted from **R**, we obtain the *first residual correlation matrix*

$$\mathbf{R}_1 = \mathbf{R} - \mathbf{f}_1\mathbf{f}_1'.$$

The second residual matrix $\mathbf{R}_2$ is then obtained as $\mathbf{R}_2 = \mathbf{R}_1 - \mathbf{f}_2\mathbf{f}_2'$. In this way, successive residual matrices can be calculated until the final residual matrix $\mathbf{R}_m$ has zero elements only.

An expression for **X** as a linear compound of **Y** can be obtained from (13.13). If this equation is postmultiplied by Λ, we have

$$\mathbf{XF} = \mathbf{Y}\Lambda = \mathbf{YF}'\mathbf{F},$$

and therefore

$$(\mathbf{X} - \mathbf{YF}')\mathbf{F} = \mathbf{0},$$

so that $(\mathbf{X} - \mathbf{YF}') = \mathbf{0}$ and therefore

$$\mathbf{X} = \mathbf{YF}'. \tag{13.15}$$

Written out for one specific variable x_i, we obtain the following vector equation:

$$\mathbf{x}_i = f_{i1}\mathbf{y}_1 + f_{i2}\mathbf{y}_2 + \cdots + f_{im}\mathbf{y}_m, \qquad (13.16)$$

in which a row of the factor matrix **F** reveals itself as a vector of weights for the linear compound. It follows that the variance of variable x_i (which is unity, as we agreed) can be expressed in terms of additive factor contributions. For, taking the inner product of vector $\mathbf{x}_i$ with itself, and using (13.16), we obtain

$$\mathbf{x}_i'\mathbf{x}_i/n = f_{i1}^2\mathbf{y}_1'\mathbf{y}_1/n + f_{i2}^2\mathbf{y}_2'\mathbf{y}_2/n + \cdots + f_{im}^2\mathbf{y}_m'\mathbf{y}_m/n, \qquad (13.17)$$

where all vector products of the type $\mathbf{y}_g'\mathbf{y}_h/n$ vanish since the y-vectors are independent. Also, all expressions $\mathbf{x}_i'\mathbf{x}_i/n$, $\mathbf{y}_1'\mathbf{y}_1/n, \ldots, \mathbf{y}_m'\mathbf{y}_m/n$ are unity, since all these variables are standardized, and equation (13.17) simplifies to

$$1 = f_{i1}^2 + f_{i2}^2 + \cdots + f_{im}^2,$$

which shows that the squared factor loadings in the ith row of **F** are components of the unit variance of variable x_i, which result also follows from equation (13.14). In other words, the squared factor loading indicates the amount of variance in the standardized observed variable "explained" by the factor. Therefore, if we sum squared elements over columns of **F** (which results in the eigenvalue, as you will remember), we have the total contribution of the factor to the total variance of all variables. This contribution is a portion of m, since m is the total variance of all observed variables.

Therefore λ_i is often expressed as a proportion of m or a percentage of m. It is the percentage of the total variance "explained" by the ith factor, or, if one hesitates to use the word "explained" here, the percentage indicates the relative variance that can be attributed to, or accounted for, by the ith factor. This is often called the *relative factor contribution.*

If **F** is a complete factor matrix that includes all factors with non-zero eigenvalues, then squares of row elements should add up to unity. This follows from (13.17), but also from the equation $\mathbf{R} = \mathbf{FF}'$, which implies that the inner product of a row vector of **F** with itself results in a diagonal element of **R** that is unity. However, if the factor matrix is not complete but is an incomplete version $\mathbf{F}_p$ with $m - p$ factors "neglected," then of course the portion of the variance of a variable that would be accounted for by the neglected factors will be missing. As a result, the sum of the squared elements in a row of $\mathbf{F}_p$ may be less than unity; this sum is often called the *communality* of the variable (a name that will be explained later).

This concludes our first approach to factor analysis. As we shall see in chapter 15, the procedure discussed above is only one of the many varieties of factor analysis; it is often called *principal-factors analysis* (the principal factors are the eigenvectors in the columns of **F**). There is also a term, *principal-components analysis*, which refers to the same type of analysis applied to the variance-covariance matrix rather than to the correlation matrix. In this sense principal-factors analysis is a special case of principal-components analysis: namely, principal-components analysis applied to standardized variables. Do *not* conclude from these remarks that the difference between principal factors and principal components is just a trivial matter of scaling, and that the two solutions are essentially identical. The truth of the matter is that eigenvectors are not invariant under rescaling of variables, so that the two solutions differ in other than trivial aspects. We will come back to this problem in section 15.2.1.

Some authors want to reserve the term factor analysis for the variety we shall discuss in section 13.5, in which diagonal elements of **R** are set lower than unity. In my opinion this distinction is not formally essential; so we can also use the term principal factor for eigenvalues obtained from **R** with unit diagonal elements.

13.4. Numerical Examples

13.4.1. The first illustration is artificial, but it has the advantage that the calculations are simple and can be repeated by a motivated reader on a desk calculator.

We are given the 6×4 matrix **X** of standardized measurements in table 13.1. The corresponding correlation matrix $\mathbf{R} = \mathbf{X}'\mathbf{X}/n$ is given in table 13.2.

Table 13.1. Hypothetical standardized data matrix.

x_1	x_2	x_3	x_4
1.5	−1.7	−1.0	−.80
.3	.4	.5	.58
−.9	1.3	−1.6	.76
1.0	.9	1.3	1.50
−.8	−.3	.7	−.78
−1.1	−.6	.1	−1.26

Table 13.2. Correlation matrix derived from table 13.1.

1.00	−.30	.12	.30
−.30	1.00	.12	.82
.12	.12	1.00	.192
.30	.82	.192	1.00

First we want to point out that **R** is of rank 3, since we have made x_4 linearly dependent on the first two variables, x_1 and x_2. The relation is $x_4 = .6x_1 + x_2$, as you may check in table 13.1. Therefore the vector $(.6 \quad 1 \quad 0 \quad -1)$ is an eigenvector of **R** with zero eigenvalue.

The other three eigenvectors, normalized to their eigenvalue, are given in table 13.3 in the format of the factor matrix **F**. You may verify that:

(a) **F** indeed gives eigenvectors: $\mathbf{RF} = \mathbf{F}\Lambda$;

(b) $\mathbf{F}'\mathbf{F} = \Lambda$;

(c) $\mathbf{FF}' = \mathbf{R}$;

(d) the eigenvalues add up to 4.

Table 13.4 gives the matrix $\mathbf{G} = \mathbf{F}\Lambda^{-1}$. Note that columns of **G** are obtained by dividing each column of **F** by its corresponding eigenvalue. Factor scores can be calculated from $\mathbf{Y} = \mathbf{XG}$, and are given in table 13.5. You might verify, at least for some items, that:

(a) $\mathbf{Y}'\mathbf{Y}/n = \mathbf{I}$;

(b) $\mathbf{X}'\mathbf{Y}/n = \mathbf{F}$;

(c) $\mathbf{X} = \mathbf{YF}'$.

Table 13.3. Factor matrix derived from table 13.2.

	f_1	f_2	f_3
x_1	.059	.936	−.347
x_2	.919	−.392	−.037
x_3	.343	.419	.841
x_4	.955	.169	−.245
eigenvalue	1.877	1.234	.889

Table 13.4. Transformation matrix G derived from table 13.3.

.031	.758	−.390
.489	−.317	−.042
.183	.339	.946
.509	.137	−.277

Table 13.5. Factor scores **Y** derived from tables 13.1 and 13.4 by $\mathbf{Y} = \mathbf{XG}$.

y_1	y_2	y_3
−1.375	1.227	−1.238
.592	.350	.179
.702	−1.533	−1.428
1.473	1.119	.386
−.440	−.381	1.203
−.951	−.782	.898

13.4.2. For a second example, we take up the correlation matrix for five sociological variables given in table 12.1. Table 13.6 gives the eigenvector matrix **F** and the eigenvalues. Just for illustrative purposes, we might agree that the last three eigenvectors could be neglected, so that the factor matrix would be reduced to the 5×2 matrix given in table 13.7. A column for communalities h^2 is added to this table (h^2 = sum of squared elements in same row).

Table 13.6. Factor matrix for correlation matrix given in table 12.1.

	f_1	f_2	f_3	f_4	f_5
x_1	.6806	.5896	−.3095	.2122	.2192
x_2	.7301	.4152	.4520	−.2703	−.1301
x_3	.8165	−.1758	−.3366	−.0394	−.4329
x_4	.7573	−.3378	.3022	.4677	.0453
x_5	.7707	−.3958	−.0952	−.3491	.3437
eigenvalue	2.8310	.8219	.5140	.4604	.3727

Table 13.7. First two factors, and communalities, for correlation matrix given in table 12.1.

	f_1	f_2	h^2
x_1	.6806	.5896	.8108
x_2	.7301	.4152	.7054
x_3	.8165	−.1758	.6976
x_4	.7573	−.3378	.6876
x_5	.7707	−.3958	.7506

The factor matrix in table 13.7 explains only part of **R**, and the residual matrix $\mathbf{R}_2 = \mathbf{R} - \mathbf{f}_1\mathbf{f}_1' - \mathbf{f}_2\mathbf{f}_2'$ is given in table 13.8. Note that the residual correlations are far from all being equal to zero, which shows that the two-factor solution is not very satisfactory.

A graph of the two-factor solution is given in figure 13.1. For its construction, table 13.7 is used as a matrix of coordinates of the five variable vectors in the two-factor space. Note that in this graph vectors have length h (square root of communality).

Finally, another picture (figure 13.2) of the two-factor solution is given, this one in the spirit of path analysis. It shows the two factors taken as two "ultimate" unobserved variables that are determiners of the observed x-variables. It follows immediately from equation (13.15) that factor loadings can be inserted in the figure as path coefficients, since (13.15) specifies "structural equations" in the sense of section 12.2.1.

Table 13.8. Second residual correlation matrix obtained from the matrix in table 12.1 after subtraction of contribution of first two factors given in table 13.6.

.1892	−.2257	.0009	.0157	.0308
	.2946	−.0851	.0043	.0066
		.3024	−.1397	−.1029
			.3124	−.1764
				.2494

13.5. Variables with Error

Our next step is an amendment to the procedure sketched thus far. As we saw in equation (13.15), the procedure resulted in a set of linear equations,

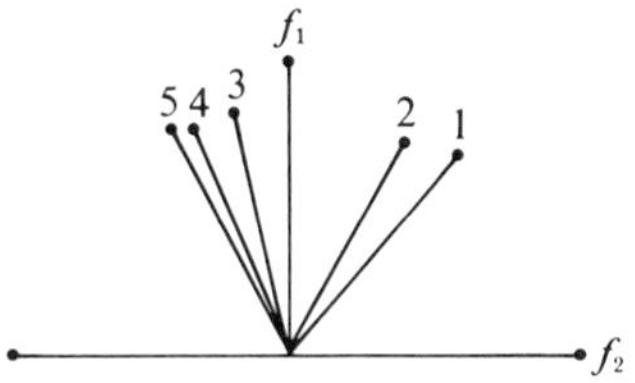

Figure 13.1 Graph of factor loadings first two factors, five sociological variables.

$\mathbf{X} = \mathbf{YF}'$, in which the observed variables are expressed as weighted compounds of the unobserved factors. For one specific variable x_i, we had equation (13.16):

$$\mathbf{x}_i = f_{i1}\mathbf{y}_1 + f_{i2}\mathbf{y}_2 + \cdots + f_{im}\mathbf{y}_m,$$

which does not contain an error term: $\mathbf{x}_i$ is fully determined by the factors. We now take a somewhat different model, in which an error term is added to the equation. Instead of equation (13.16) we then have

$$\mathbf{x}_i = f_{i1}\mathbf{y}_1 + f_{i2}\mathbf{y}_2 + \cdots + f_{im}\mathbf{y}_m + \mathbf{e}_i,$$

where $\mathbf{e}_i$ is the error term for $\mathbf{x}_i$. In generalized matrix notation,

$$\mathbf{X} = \mathbf{YF}' + \mathbf{E}, \tag{13.18}$$

where $\mathbf{E}$ is a $n \times m$ matrix of values in m error variables. We shall assume

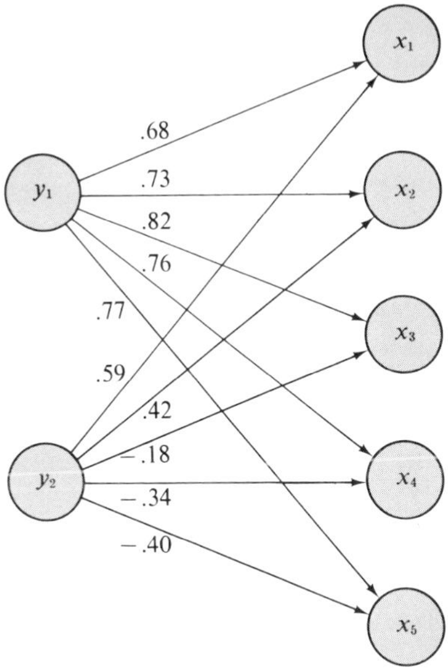

Figure 13.2 Path diagram for first two factors, five sociological variables.

that the error variables are uncorrelated:

$$\mathbf{E'E}/n = \mathbf{U}^2, \tag{13.19}$$

with $\mathbf{U}^2$ a diagonal matrix; an element u_i^2 represents the contribution of the error variable associated with x_i to the variance of x_i. Also, we assume that error variables are uncorrelated with factors:

$$\mathbf{E'Y} = \mathbf{0}. \tag{13.20}$$

The two assumptions (13.19) and (13.20) are parts of the mathematical model. It follows that we can write

$$\mathbf{R} = \mathbf{X'X}/n = (\mathbf{FY'} + \mathbf{E'})(\mathbf{YF'} + \mathbf{E})/n = \mathbf{FY'YF'}/n + \mathbf{U}^2. \tag{13.21}$$

If we also retain the assumption that the factors should be independent, then $\mathbf{Y'Y}/n = \mathbf{I}$, and equation (13.21) reads:

$$\mathbf{R} = \mathbf{FF'} + \mathbf{U}^2. \tag{13.22}$$

Let us suppose now, for the time being, that the error variances u_i^2 are known, so that $\mathbf{U}^2$ is given. We then can write

$$\mathbf{R}^* = \mathbf{R} - \mathbf{U}^2, \tag{13.23}$$

where $\mathbf{R}^*$ is called the *reduced correlation matrix*. It differs from the actual correlation matrix $\mathbf{R}$ only in the diagonal elements, which are unity in $\mathbf{R}$, but which equal $1 - u_i^2 = h_i^2$ in $\mathbf{R}^*$. The portion h_i^2 represents the variance of x_i that depends on factors the variables have in common; hence the name *communality* for h_i^2.

For the vector model, the implication is that $\mathbf{R}^*$ still represents cosines of angles in its off-diagonal elements, but that its diagonal elements now stand for vector lengths equal to h_i instead of to unity.

We could apply principal-components analysis to $\mathbf{R}^*$ rather than to $\mathbf{R}$, to find a factor matrix $\mathbf{F}$. We would have $\mathbf{R^*F} = \mathbf{F}\Lambda$ as the equivalent of (13.9), where $\mathbf{F'F} = \Lambda$ as in (13.10), and $\mathbf{FF'} = \mathbf{R}^*$ as the alternative for (13.11). In this way the equations (13.18) could be identified.*

In practice, however, values of u_i^2 will rarely be known; so all we can do is estimate them. One possible estimation procedure emerges if we look

* But notice that (13.13) no longer has a simple alternative; we will come back to that in section 13.10.

Table 13.9. First two principal components for matrix **R*** obtained from correlation matrix **R** in table 12.1 after diagonal elements are replaced by communalities as given in table 13.7.

	f_1	f_2	h^2
x_1	.6723	.5441	.7480
x_2	.6887	.2735	.5491
x_3	.7665	−.1440	.6083
x_4	.7067	−.2612	.5677
x_5	.7382	−.3510	.6681

back at tables 13.6 and 13.7. We found there that if only two factors are retained, values of h_i^2 can be computed. We might use these values for the diagonal elements of **R***, and subsequently analyze **R*** as indicated above. The result is a new 5×2 factor matrix, given in table 13.9 for the illustrative example. The revised factor matrix, in its turn, produces new estimates of communalities h_i^2. If they were the same as in the first factor matrix, we would have succeeded in our objective to find factors that reproduce the communalities inserted in **R***. However, it is most unlikely that this will happen after a first trial, and the procedure is to repeat the analysis by using the communalities in table 13.9 for a further trial. In this way we can go on, until the resulting values of h_i^2 are finally found to be equal to the values inserted at the beginning of the last trial. In general, this iterative procedure is found to converge. Table 13.11 gives the result after the

Table 13.10. Sixth iteration of first two factors for reduced correlation matrix obtained from table 12.1 after diagonal elements are replaced by communalities obtained in fifth iteration.

	f_1	f_2	h^2
x_1	.6709	.5258	.7290
x_2	.6360	.1687	.4308
x_3	.7567	−.1273	.5903
x_4	.6749	−.2057	.4969
x_5	.7232	−.3109	.6178

Table 13.11. Seventh iteration of first two factors for reduced correlation matrix obtained from table 12.1 by substituting diagonal elements by communalities in table 13.10.

	f_1	f_2	h^2
x_1	.6716	.5272	.7290
x_2	.6349	.1664	.4308
x_3	.7575	−.1285	.5903
x_4	.6743	−.2055	.4969
x_5	.7224	−.3097	.6178

seventh trial; it shows that the output values are no longer very much different from the input.

However, this convergence is only partly an indication of success, since the factors finally obtained might leave a residual correlation matrix with sizeable values in it. If so, we should repeat the analysis with one more factor, and see whether this gives a more satisfactory result. In our example, however, the result with only two factors is not too bad, since the residual matrix, shown in table 13.12, has values not very much different from zero.

The procedure described here shows that we use the revised factor model not only to account for an error term, but with a different objective as well: to insert values for h_i^2 in $\mathbf{R}^*$ in such a way that the number of factors retained can be kept small. In formal terms, this might be phrased as follows: matrix $\mathbf{R}$ in general will have rank m, equal to the number of variables, and therefore $\mathbf{R}$ will have m eigenvectors and require m factors. However, we can replace diagonal elements by h_i^2 in such a manner that the rank of $\mathbf{R}$ is reduced to less than m. What we try to do is estimate such

Table 13.12. Residual correlation matrix obtained from reduced correlation matrix after subtraction of contribution of first two factors as given in table 13.11.

−.0001	.0019	.0120	−.0125	.0001
	.0000	−.0216	.0230	−.0021
		−.0000	.0008	.0090
			−.0000	−.0098
				.0000

values for h_i^2 on the basis of the iterative procedure. The result is not that $\mathbf{R}^*$ really has small rank, but that it can be better approximated by a small-rank matrix than $\mathbf{R}$ itself could be.

In figure 13.3, the result obtained in table 13.11 is pictured as a path diagram. It might be compared with figure 13.2. In figure 13.2 we have two

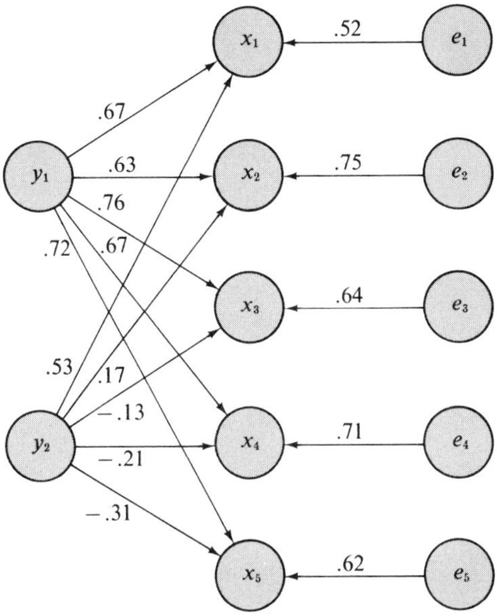

Figure 13.3 Path diagram, first two factors for reduced correlation matrix, five sociological variables.

ultimate unobserved factors that explain five variables; in figure 13.3 we have seven unobserved ultimate factors for the same five variables. This seems hardly a gain in parsimony; on the other hand, figure 13.3 requires only two "common factors," i.e., factors that determine more than one variable, whereas the other five ultimate factors are, presumably, sources of error variance. However, keep in mind that the picture might change considerably if we add more observed variables; portions of what now is called "error" might then appear to be tied to common factors that at first remained hidden because the data were too restricted to reveal their role. This situation is not peculiar to this variety of factor analysis, of course; new data can always reveal new things.

13.6. Interpretation of Factors

Given a factor matrix **F**, the question always arises of what the factors really mean. It is most unsatisfying, of course, to adopt a purely nominalistic form of operationalism, and say that one factor is factor *A* and another factor *B*. Actual research tends to be involved in the real world, and one wants to give factors names that stand for something we can talk about.

The only way open is to make a guess, on the basis of the observed variables the factor is correlated with (it also helps to see which variables

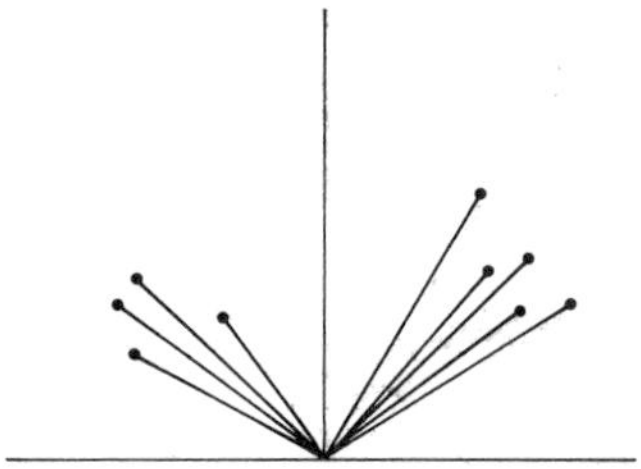

Figure 13.4
Factor loadings in two clusters.

the factor is not correlated with). For instance, in table 13.11, the first factor has positive loadings for all five variables; apparently the factor stands for what they all have in common. Perhaps it is reasonable to say that the factor reveals the process of occupational stratification as such, or that it stands for "family status." Such an interpretation is meaningful to the extent we can look on the observed variables as "indicators" of the underlying unobserved variable. The second factor, which has positive loadings for all variables relating to the father, and negative ones for the son's variables, might be labeled the father-versus-son factor.

Awkward as it is, this seems the only way to proceed. Often, however, the interpretation is obvious. Take the structure pictured in figure 13.4, and suppose that all vectors represent intelligence tests; they fall apart in two clusters, as we see. Suppose the vectors in one cluster are verbal intelligence tests, and the other cluster contains numerical intelligence tests. Then one factor, with positive loadings for all variables, obviously may be called "intelligence" (but what is intelligence?); the other factor distinguishes between verbal and numerical tests, and so is the "verbal-versus-numerical intelligence" factor. (In section 13.8 another approach will be discussed.)

13.7. How Many Factors?

In section 13.1 we stated that the objective of factor analysis is to reduce the dimensionality of the vector space, so that vectors can be described in terms of relatively few reference factors. Vectors in a V_m with zero projections on them can be neglected, and we agreed that we might also neglect factors with small eigenvalues, since their contribution to the variance of the individual measurements is small. However, what is "small"? This, of course, is a statistical problem, and we should use some kind of statistical test to decide whether a factor may be neglected or not.

Although we are not dealing with statistical matters, in this book, I can at least indicate a coarse rule of thumb. I do so with hesitation, because the rule is, in fact, very coarse, and should certainly not be applied as a final criterion. But in intermediate stages, or if one wants to make a rough judgment about the relevance of factors in publications where the author seems somewhat unclear about the question, the rule might have some use.

Remember that we agreed to take eigenvalues in decreasing order, so that λ_1 is the largest, λ_2 the second largest, etc. If we plot eigenvalues against their rank number, therefore, the result is a decreasing graph. Generally, the graph has a rounded L-shape: it starts with high values at the left, and the tail at the right tends to be rather flat. The rule says, first, that eigenvalues smaller than 1 may be neglected (note that the corresponding factor explains less variance than a single observed variable). Furthermore, if the graph shows a sharp drop after several initial high eigenvalues, whereas the eigenvalues to the right of the sharp drop decrease very smoothly and gradually, then these eigenvalues in the tail may be neglected. This is the rule, but I feel urged to repeat that, obviously, visual inspection of a graph is no real substitute for statistical testing of a hypothesis.

13.8. Rotation of Factors

Again, remember that the objective of factor analysis is to reduce the dimensionality of the vector space. Suppose now that, on the basis of a principal-components analysis, we have decided which dimensions can be neglected, and suppose that we retain p vectors ($p < m$) for an acceptable description of the vector bundle. Principal-components analysis then also indicates the positions of the p vectors in this reduced vector space.

However, given the V_p, there is no logical reason to adopt the eigenvectors as reference directions: any other (orthogonal) set of reference directions would describe the bundle as well. Eigenvectors are nice for a first identification of the reduced vector space, but once we have found the V_p, we need not prefer the eigenvectors as a reference system to any other set of coordinates.

For instance, look back at figure 13.4, and suppose that the V_2 pictured in it is acceptable as a reduced vector space; the two coordinates represent the first and second eigenvectors. However, the bundle of test vectors could be identified as well if we rotate the coordinates through about 45°,

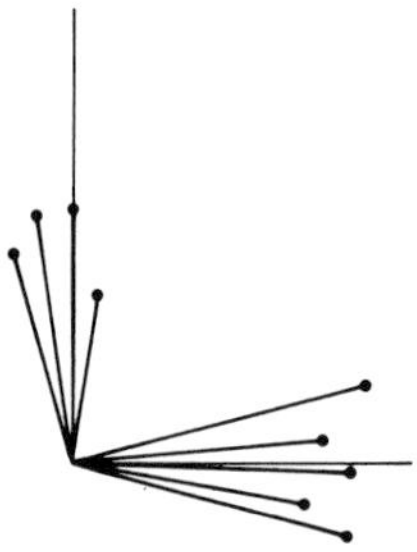

Figure 13.5
Rotated version of figure 13.4.

which results in figure 13.5. The two pictures are completely equivalent, of course, but we see that the interpretation of the factors changes; we no longer have one common intelligence factor and another factor for verbal-versus-numerical intelligence, but one factor for verbal intelligence (unrelated to numerical tests), and a second factor for numerical intelligence (unrelated to verbal tests). We might agree that the latter set of factors is much more meaningful than the first and therefore should be preferred. In other words, rotation of factors can be used to find the most meaningful set of factors.

To develop this idea further, we should specify criteria for rotation. In fact, there are two somewhat different principles that apply here. One principle is that a factor is meaningful if it identifies a cluster of vectors: a cluster of vectors indicates that the variables in the cluster have things in common and are distinguished from variables outside the cluster—the clustered variables are "indicators," so to speak, of a common underlying factor. This principle may be called the *primary-factor* principle. It implies that the factor can be easily interpreted, since it will have high loadings for all variables in the cluster.

The other principle starts at the other end, as it were, by trying to find factors that are unrelated to as many observed variables as possible. The interpretation problem is then simplified, because we only have to take into consideration those remaining variables that have loadings on the factor. Geometrically, this second principle implies that a factor is selected perpendicular to a V_{p-1} in the vector space (which has dimensionality p, with p the number of retained factors) in such a way that this V_{p-1} contains as many vectors as possible; these vectors then are orthogonal to the factor.

In figures 13.4 and 13.5, the two principles lead to the same result. The two new reference directions in figure 13.5 actually are centered in two separate clusters, and in this sense are a primary-factor solution. But since the two clusters are "orthogonal," roughly speaking, the two new factors are also orthogonal to a V_{p-1} (a line, here) on which many vectors have small projections.

The principles can be translated into more quantitative criteria. First of all, if **F** is a factor matrix obtained from a principal-components analysis, then rotation of factors means that we apply a transformation matrix **T** to **F** to obtain

$$\mathbf{G} = \mathbf{FT}, \tag{13.24}$$

where $\mathbf{T}'\mathbf{T} = \mathbf{I}$ (for an orthogonal rotation), and where **G** is the new, rotated factor matrix (of order $m \times p$). If we are looking for primary factors, the objective is to obtain a matrix **G** in which each column contains as many elements as possible that are different from zero (the corresponding variables form the cluster). Interpretation of the factor can then be based on inspection of what these variables have in common. For the other principle, however, we want the columns of **G** to contain as many zero or nearly zero loadings as possible. The corresponding observed variables are then contained in a V_{p-1} perpendicular to the factor represented by the column.

Since we want to use these criteria for all columns of **G** simultaneously, we are faced with a difficult problem, certainly not one that can be solved easily by mere inspection of the original matrix **F**. In the precomputer era, however, there was no other way: rotation was guided by drawing graphs of vector spaces, and trying to find suitable rotations by reasoned trial and error. Fast computers have made it possible to set up analytical criteria for optimizing, some of which we shall discuss briefly.

We start with the simple picture in figure 13.6, where one x-vector is

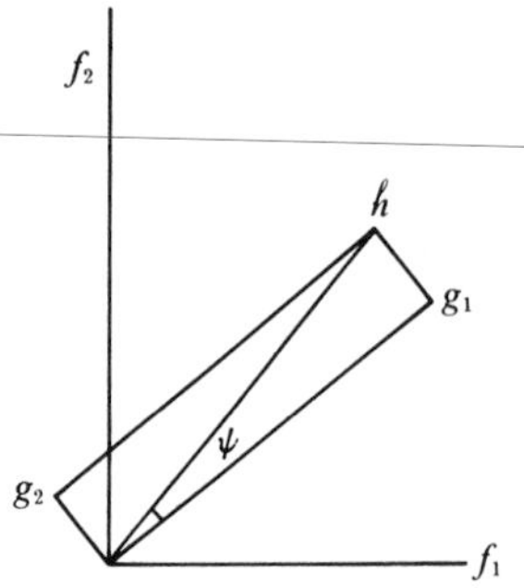

Figure 13.6
Principle of rotation in analytical solution.

pictured in a plane of two factors. We want the rotation to give the x-vector maximum projection on one factor and minimum projection on the other, for *all* pairs of factors. Such an optimum position is approximated if the angle between the x-vector and one of the factors is small; this angle is called ψ in figure 13.6. We then have

$$\begin{aligned} g_1 &= h \cos \psi, \\ g_2 &= h \sin \psi, \end{aligned} \tag{13.25}$$

where h is the length of the vector and g_1 and g_2 are new projections. It follows from (13.25) that

$$g_1 g_2 = h^2 \sin \psi \cos \psi = \tfrac{1}{2} h^2 \sin 2\psi. \tag{13.26}$$

If ψ is small, then sin 2ψ also will be small, but since h^2 is constant, it follows from (13.26) that a small value of sin 2ψ goes together with a small value of $g_1 g_2$. On the other hand, Pythagoras' theorem says that

$$g_1^2 + g_2^2 = h^2.$$

Therefore,

$$\begin{aligned} h^4 &= (g_1^2 + g_2^2)^2 = g_1^4 + g_2^4 + 2g_1^2 g_2^2 \\ &= g_1^4 + g_2^4 + \tfrac{1}{2} h^4 \sin^2 2\psi. \end{aligned} \tag{13.27}$$

But h^4 is a constant; therefore, if sin 2ψ decreases, the value of $g_1^4 + g_2^4$ must increase. Now we can apply the latter criterion to all tests and all factors simultaneously, and find a good compromise if we see to it that the rotation maximizes $\sum g^4$. In words, we try to find a transformation matrix **T** in such a way that the sum of the fourth powers of all elements in the resulting matrix **G** has a maximum.

This recipe can be interpreted in another way, as meaning that loadings should be either clearly different from zero, or nearly equal to zero, or that values of g^2 must be either large or small. The latter clause implies, however, that there will be a large variance for the values of g^2. Take a specific column of **G**; then the variance of the values g^2 for that column is

$$\sum_m (g^2 - \bar{g}^2)^2/m = \sum_m (g^4 - m\bar{g}^2)/m, \tag{13.28}$$

where $\bar{g}^2$ stands for the mean value of g^2 in the column:

$$\bar{g}^2 = \sum_m g^2/m. \tag{13.29}$$

If we compute the sum of the variances for all columns simultaneously, the appropriate form is

$$\sum_p \sum_m (g^4 - m\bar{g}^2)/m = \left(\sum_p \sum_m g^4/m\right) - \sum_p m\bar{g}^2. \tag{13.30}$$

The second term on the right of (13.30) reduces to

$$\sum_p m\bar{g}^2 = \sum_p \sum_m g^2. \tag{13.31}$$

Note that $\sum_p \sum_m g^2$ is the total sum of all squared loadings. However, for each row of **G**, the sum of the squared loadings equals h^2, and therefore the total sum of squared loadings adds up to $\sum h^2$, which is a constant. It follows that if we want to maximize (13.30), we need only maximize $\sum \sum g^4$, which is the same solution as we reached before.

Again, the idea that we maximize the sum of the fourth powers of g can be explained in still another way, by interpreting it as maximum kurtosis. The kurtosis of a distribution is defined as the sum of the fourth powers of deviations from the mean, divided by the squared sum of the squared deviations:

$$\sum x^4/(\sum x^2)^2.$$

It is a dimensionless measure for flatness versus peakedness of a distribution. The criterion for factor rotation can be phrased as that we want maximum flatness, i.e., high kurtosis, since factor loadings are then chased as much as possible away from the center of their distribution and become more easily interpretable. But since in a factor matrix the sum of all squared elements is constant, as we saw, high kurtosis is obtained by maximizing the fourth powers.

A disadvantage of the method is that variables with large communality contribute more to the ultimate solution than variables with smaller communality. If we do not like this idea, we had better divide elements in each row of **F** by the value of h in the same row, with the result that squared elements in a row will add up to unity. Or, in geometrical terms, vectors are extended until they all have unit length. It follows that we now should look for a maximum of $\sum (g/h)^4$. This solution, called the *varimax solution*, was developed by Kaiser (1958) and is a popular one among psychologists.

13.9. Oblique Reference System

Once we take the liberty of rotating factors, we may wonder whether the requirement that factors are independent must always be met. In fact, the answer is no; we might also use a reference system to describe V_p where the reference directions have oblique angles between them. To do so, we drop the condition on formula (13.24) that $\mathbf{T}'\mathbf{T} = \mathbf{I}$. (That is, diagonal elements of $\mathbf{T}'\mathbf{T}$ must still be unity, since columns of $\mathbf{T}$ give direction cosines, but the off-diagonal elements of $\mathbf{T}'\mathbf{T}$ need no longer equal zero. In fact, $\mathbf{T}'\mathbf{T}$ gives cosines of angles between the new reference directions.)

However, we first have to agree on what is meant by projections and coordinates in a nonorthogonal coordinate system. The simplest case is illustrated in figure 13.7, where two oblique coordinates are given under an angle ϕ. A point P located in their plane can be specified in two different ways:

(1) in terms of its perpendicular *projections* on the axes: the projection is the signed distance from the origin to the point where the perpendicular from P meets the coordinate. Projections will be symbolized by s.

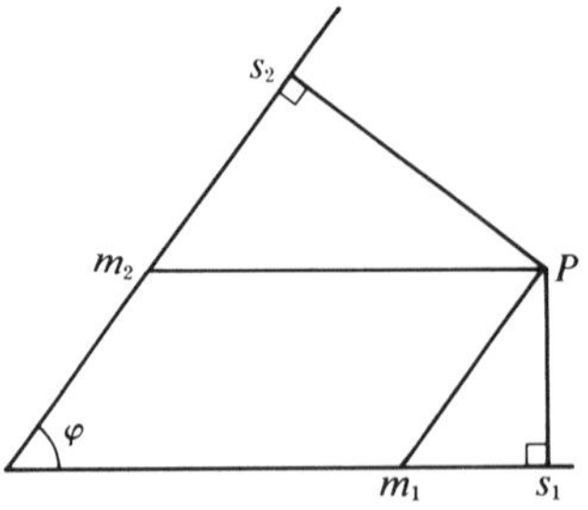

Figure 13.7
Projections s and coordinates m in an oblique system.

(2) in terms of oblique *coordinates.* Lines are drawn through P and parallel to the axes: coordinates are the signed distances from the origin to the points where these lines meet the axes. Coordinates are symbolized by m.

From figure 13.7, we can see that the relation between m and s is

$$\begin{aligned} s_1 &= m_1 + m_2 \cos \phi, \\ s_2 &= m_1 \cos \phi + m_2. \end{aligned} \tag{13.32}$$

In matrix notation,

$$\mathbf{s} = \boldsymbol{\phi}\mathbf{m}, \quad \text{or} \quad \mathbf{s}' = \mathbf{m}'\boldsymbol{\phi}, \tag{13.33}$$

where $\boldsymbol{\phi}$ is the 2×2 matrix of cosines between axes. Formula (13.33) remains valid for a space of higher dimensionality, in which a projection is defined as the signed distance from the origin to the point where a hyperplane through P and perpendicular to the axis meets the axis; a coordinate is the signed distance from the origin to the point on an axis where a hyperplane through P and parallel to all other axes meets this axis. Matrix $\boldsymbol{\phi}$ then becomes a $p \times p$ matrix, and $\mathbf{S}$ and $\mathbf{M}$ for a collection of n points becomes $n \times p$ matrices that satisfy

$$\mathbf{S} = \mathbf{M}\boldsymbol{\phi}. \tag{13.34}$$

For an orthogonal system, $\boldsymbol{\phi} = \mathbf{I}$, and $\mathbf{S} = \mathbf{M}$; in this special case, there is no difference between projections and coordinates.

Suppose now that $\mathbf{F}$ gives x-vectors in the orthogonal factor space defined by eigenvectors. We then may introduce a set of new, non-orthogonal coordinate axes, given by their direction cosines $\mathbf{T}$. It follows that angles between the new directions are

$$\mathbf{T}'\mathbf{T} = \boldsymbol{\phi}. \tag{13.35}$$

The new factor matrix will be the matrix of projections

$$\mathbf{S} = \mathbf{FT}, \tag{13.36}$$

and the matrix of corresponding coordinates is

$$\mathbf{M} = \mathbf{S}\boldsymbol{\phi} = \mathbf{FT}\boldsymbol{\phi}. \tag{13.37}$$

$\mathbf{S}$ is called the *factor structure*, and $\mathbf{M}$ the *factor pattern*. Again, for orthogonal axes, pattern and structure are identical, but for oblique axes they are not; in publications it would not be sufficient to give **S** without ϕ or **M**. Elements of the structure can be interpreted as correlations between variables and factors, and ϕ might be looked on as the matrix of correlations between unobserved factors.

Oblique factors may be used as a more natural approach to the primary-factor principle. The principle was characterized by the fact that factors are chosen through the center of clusters; however, clusters will in general not be orthogonal. The second principle required that hyperplanes be found such that many variables make small angles with them; again, such hyperplanes will not be orthogonal, in general. Analytic rotation procedures have been developed; they use criteria that are similar to those for orthogonal axes, but which account for the liberation from orthogonality. The oblique version of varimax is called *oblimax*. In practice, the oblimax procedure tends to produce solutions with small angles between factors. This is an unfortunate result, because it tends to confuse factors and runs counter to parsimony. Correlations between factors up to .30 might be acceptable, but larger ones seem not to serve economy of description.

In fact, a final criterion for a factor solution should be the empirical one. The fundamental disadvantage of the principal-components solution is that it is highly dependent on the specific variables chosen for the study. If one adds or omits variables, the solution is changed completely. Therefore, eigenvectors cannot be expected ever to attain the status of a proper scientific concept, since they will be different for each study and therefore cannot be identified across different studies. Some authors, like Cattell (1957), claim that the second principle for rotation (factors orthogonal to certain hyperplanes) results in factors that stand the empirical test better. This is why Cattell has given these factors the name of *source factors*, as contrasted to *surface factors* that follow the primary-factor principle.

13.10. Estimation of Factor Scores

Matrix **Y**, introduced in section 13.2, was interpreted as a matrix of unobserved factor scores. The remaining problem is how we can estimate such factor scores. For the "errorless" principal-factors solution, doing so is no problem, since the answer is given by formula (13.13): $\mathbf{Y} = \mathbf{XF}\Lambda^{-1}$. The formula remains valid even if **F** is not complete because some factors with low eigenvalues are neglected.

However, if the principal-components solution is applied to the reduced correlation matrix **R**, we run into problems. Here the basic equation is (13.18), which expresses the observed variables as a linear compound of the factor scores and an error term. But the error terms are unknown, and therefore we cannot simply reverse (13.18) to find an expression for **Y** in terms of **X**. The solution can lie in our recognizing that we are dealing with a multiple-correlation problem. In fact, for each factor, we know from the corresponding column of **F** (or **G**, if the factor matrix is rotated) what the correlations are between the factor and the observed variables. Estimation of factor scores therefore can be found from a multiple-regression formula, derived from equations (11.14) and (11.15). Instead of $\mathbf{b} = \mathbf{R}^{-1}\mathbf{r}$, the notation now becomes

$$\mathbf{w}_i = \mathbf{R}^{-1}\mathbf{f}_i, \qquad (13.38)$$

where $\mathbf{w}_i$ is a vector of regression weights to be applied in

$$\hat{\mathbf{y}}_i = \mathbf{X}\mathbf{w}_i, \qquad (13.39)$$

with $\hat{\mathbf{y}}_i$ a vector of estimated factor scores, with variance

$$\mathbf{w}_i'\mathbf{X}'\mathbf{X}\mathbf{w}_i/n = \mathbf{f}_i'\mathbf{R}^{-1}\mathbf{f}_i. \qquad (13.40)$$

In other words, the estimated factor scores are not in standard form, but have variance equal to the equivalent of a squared multiple correlation. To obtain standardized estimates, we still should divide $\hat{\mathbf{y}}_i$ by the square root of $\mathbf{f}_i'\mathbf{R}^{-1}\mathbf{f}_i$. The generalized version of (13.39) can be written as

$$\hat{\mathbf{Y}} = \mathbf{X}\mathbf{W} = \mathbf{X}\mathbf{R}^{-1}\mathbf{F}. \qquad (13.41)$$

Note that for a principal-components analysis of **R**, we have

$$\mathbf{R}^{-1}\mathbf{F} = \mathbf{F}\Lambda^{-1},$$

since **F** contains eigenvectors of **R**, which are also eigenvectors of $\mathbf{R}^{-1}$ (section 7.4.2); the corresponding eigenvalues are the reciprocals of the eigenvalues of **R**. However, if the analysis was applied to $\mathbf{R}^*$, then **F** no longer contains eigenvectors of $\mathbf{R}^{-1}$ and the simplification is impossible.

14

Canonical Correlation Analysis

14.1. General Introduction

Canonical correlation analysis can be looked on as a generalization of multiple correlation. In a multiple-correlation problem, we have a set of m variables x_i and one variable y; the objective is to find a linear compound of the x-variables that has maximum correlation with y. In canonical correlation, there is more than one y-variable, and the objective is to find a linear compound of the x-variables that has maximum correlation with a linear compound of the y-variables. The most suitable class of examples that comes to mind are those where the x-variables are from a different domain than the y-variables. For instance, the x-variables could be background variables referring to biographic data about psychiatric patients, and the y-variables descriptive variables about the patient's current behavior. The problem would be to find out whether there is some combination of background variables (a background "pattern") that has a high correlation with a combination of y-variables (a behavior pattern). Another example: the x-variables could be general sociological and economic variables measured for several cities, and the y-variables various criminality indices for those cities; again the problem would be whether cities can be characterized by a certain combination of socioeconomic measurements that has a high correlation with a combination of criminality indices.

Formally, the problem is the following. Let $\mathbf{X}$ be an $n \times m_1$ matrix of n observations on m_1 variables, and $\mathbf{Y}$ a related matrix of order $n \times m_2$ of n observations on m_2 variables. Let $\boldsymbol{\chi}$ be a linear compound of the x-variables; i.e., $\boldsymbol{\chi}$ is an $n \times 1$ vector defined by $\boldsymbol{\chi} = \mathbf{Xc}$, with $\mathbf{c}$ an $m_1 \times 1$ vector of weights. Similarly, define a linear compound of the y-variables, $\boldsymbol{\eta} = \mathbf{Yd}$, where $\mathbf{d}$ is an $m_2 \times 1$ vector of weights. The objective of canonical correlation analysis is to find $\mathbf{c}$ and $\mathbf{d}$ such that the correlation between $\boldsymbol{\chi}$ and $\boldsymbol{\eta}$ has a maximum. This correlation is called a *canonical correlation coefficient*.

We shall first give a formal treatment of the problem and thereafter, in section 14.3, a numerical example that follows the formal treatment very closely. (You might find it helpful to inspect this example while reading the formal exposition.)

14.2. Formal Treatment

14.2.1. We shall assume that the variables for which $\mathbf{X}$ and $\mathbf{Y}$ give measurements are standardized. We then have the following correlation matrices: $\mathbf{R}_{xx} = \mathbf{X}'\mathbf{X}/n$ (of order $m_1 \times m_1$), $\mathbf{R}_{xy} = \mathbf{X}'\mathbf{Y}/n$ (of order $m_1 \times m_2$), and $\mathbf{R}_{yy} = \mathbf{Y}'\mathbf{Y}/n$ (of order $m_2 \times m_2$) We shall write $\mathbf{R}_{yx} = \mathbf{R}_{xy}'$. These matrices could also be expressed as partitions of the general correlation matrix $\mathbf{R} = (\mathbf{X} \mid \mathbf{Y})'(\mathbf{X} \mid \mathbf{Y})/n$.

Further, we shall assume that $\boldsymbol{\chi} = \mathbf{Xc}$ and $\boldsymbol{\eta} = \mathbf{Yd}$ are also standardized. These assumptions are by no means essential; standardization simplifies the arguments, but does not affect the final solution.

The assumption that $\boldsymbol{\chi}$ and $\boldsymbol{\eta}$ are standardized can be translated into

$$\begin{aligned} \boldsymbol{\chi}'\boldsymbol{\chi}/n &= \mathbf{c}'\mathbf{X}'\mathbf{Xc}/n = \mathbf{c}'\mathbf{R}_{xx}\mathbf{c} = 1, \\ \boldsymbol{\eta}'\boldsymbol{\eta}/n &= \mathbf{d}'\mathbf{Y}'\mathbf{Yd}/n = \mathbf{d}'\mathbf{R}_{yy}\mathbf{d} = 1. \end{aligned} \tag{14.1}$$

The correlation between χ and η then becomes

$$\chi'\eta/n = \mathbf{c}'\mathbf{X}'\mathbf{Y}\mathbf{d}/n = \mathbf{c}'\mathbf{R}_{xy}\mathbf{d} = \mathbf{d}'\mathbf{R}_{yx}\mathbf{c} = \rho. \tag{14.2}$$

The objective is to find $\mathbf{c}$ and $\mathbf{d}$ such that ρ has a maximum value. Using the method of undetermined multipliers, we define a function

$$F = \mathbf{c}'\mathbf{R}_{xy}\mathbf{d} - \tfrac{1}{2}\mu(\mathbf{c}'\mathbf{R}_{xx}\mathbf{c} - 1) - \tfrac{1}{2}\lambda(\mathbf{d}'\mathbf{R}_{yy}\mathbf{d} - 1). \tag{14.3}$$

To find the maximum, the partial derivatives of F with respect to $\mathbf{c}$ and to $\mathbf{d}$ must be set equal to zero. The coefficient $\frac{1}{2}$ before the undetermined multipliers μ and λ simplify the resulting equations; since μ and λ are undetermined anyhow, we are free to use such a trick.

The result is

$$\begin{aligned} \partial F/\partial \mathbf{c}' &= \mathbf{R}_{xy}\mathbf{d} - \mu\mathbf{R}_{xx}\mathbf{c} = \mathbf{0}, \\ \partial F/\partial \mathbf{d}' &= \mathbf{R}_{yx}\mathbf{c} - \lambda\mathbf{R}_{yy}\mathbf{d} = \mathbf{0}, \end{aligned} \tag{14.4}$$

or

$$\begin{aligned} \mathbf{R}_{xy}\mathbf{d} &= \mu\mathbf{R}_{xx}\mathbf{c}, \\ \mathbf{R}_{yx}\mathbf{c} &= \lambda\mathbf{R}_{yy}\mathbf{d}. \end{aligned} \tag{14.5}$$

If we multiply the first equation of (14.5) by $\mathbf{c}'$ and the second by $\mathbf{d}'$ we see that

$$\begin{aligned} \mathbf{c}'\mathbf{R}_{xy}\mathbf{d} &= \mu\mathbf{c}'\mathbf{R}_{xx}\mathbf{c} = \mu, \\ \mathbf{d}'\mathbf{R}_{yx}\mathbf{c} &= \lambda\mathbf{d}'\mathbf{R}_{yy}\mathbf{d} = \lambda. \end{aligned} \tag{14.6}$$

But since $\mathbf{c}'\mathbf{R}_{xy}\mathbf{d} = \mathbf{d}'\mathbf{R}_{yx}\mathbf{c} = \rho$, it follows that $\mu = \lambda = \rho$. So we may rewrite (14.4) as

$$\begin{aligned} -\rho\mathbf{R}_{xx}\mathbf{c} + \mathbf{R}_{xy}\mathbf{d} &= \mathbf{0}, \\ \mathbf{R}_{yx}\mathbf{c} - \rho\mathbf{R}_{yy}\mathbf{d} &= \mathbf{0}. \end{aligned} \tag{14.7}$$

The set (14.7) defines a set of $m_1 + m_2 = m$ homogeneous equations with m unknowns in $\mathbf{c}$ and $\mathbf{d}$, and in addition an unknown coefficient ρ. We know from section 5.3 that such equations can be solved only if the determinant of their coefficients equals zero. This can be written as

$$\begin{vmatrix} -\rho\mathbf{R}_{xx} & \mathbf{R}_{xy} \\ \mathbf{R}_{yx} & -\rho\mathbf{R}_{yy} \end{vmatrix} = 0. \tag{14.8}$$

Expansion of the determinant will result in an equation with ρ as the only unknown, which makes it possible to find suitable values for ρ. However, the determinant in (14.8) can be simplified by using some of the rules given in section 4.1.3. One of the rules says that if we multiply the elements in a row or column by a constant, then the value of the determinant is multiplied by that constant. But since the value of the determinant in (14.8) is zero, we may without further ado multiply rows or columns by any constant. Using this, we multiply the first m_1 rows in (14.8) by $-\rho$, and subsequently divide the last m_2 columns by $-\rho$. The result is

$$\begin{vmatrix} \rho^2\mathbf{R}_{xx} & \mathbf{R}_{xy} \\ \mathbf{R}_{yx} & \mathbf{R}_{yy} \end{vmatrix} = 0. \qquad (14.9)$$

If we now expand the determinant, we obtain a polynomial of ${m_1}^{\text{th}}$ degree in ρ^2. In general, therefore, equation (14.9) will have m_1 different solutions for ρ^2 (provided that $m_1 \leqslant m_2$; if $m_1 > m_2$, we shall find $m_1 - m_2$ zero solutions). The positive roots of these solutions are the *canonical correlation coefficients*. For now we are interested only in the largest of them, since we are looking for a maximum correlation. Let this largest correlation be called ρ_1. Its value can be substituted in the set of equations (14.7), which thereby reduce to a set of simple homogeneous equations that can be solved for $\mathbf{c}$ and $\mathbf{d}$. Since this particular solution corresponds to ρ_1, we shall use subscripts and write $\mathbf{c}_1$ and $\mathbf{d}_1$.

The solution, however, is defined only up to an arbitrary scalar. Therefore, the solution obtained from (14.7) is only proportional to the final solution that we want and that should also satisfy the equations (14.1). Now let $\tilde{\mathbf{c}}_1$ and $\tilde{\mathbf{d}}_1$ be a solution obtained from (14.7). Then it is simple to identify $\mathbf{c}_1$ itself: all we have to do is calculate $\tilde{\mathbf{c}}_1'\mathbf{R}_{xx}\tilde{\mathbf{c}}_1$ and divide $\tilde{\mathbf{c}}_1$ by the square root of $\tilde{\mathbf{c}}_1'\mathbf{R}_{xx}\tilde{\mathbf{c}}_1$:

$$\mathbf{c}_1 = \tilde{\mathbf{c}}_1/(\tilde{\mathbf{c}}_1'\mathbf{R}_{xx}\tilde{\mathbf{c}}_1)^{\frac{1}{2}}.$$

Similarly for $\mathbf{d}_1$. Of course, if we used "uncorrected" weights $\tilde{\mathbf{c}}_1$ and $\tilde{\mathbf{d}}_1$ for the linear compounds $\tilde{\boldsymbol{\chi}}_1 = \mathbf{X}\tilde{\mathbf{c}}_1$ and $\tilde{\boldsymbol{\eta}}_1 = \mathbf{Y}\tilde{\mathbf{d}}_1$, this would not affect the correlation ρ_1 between $\tilde{\boldsymbol{\chi}}_1$ and $\tilde{\boldsymbol{\eta}}_1$, since the correlation is invariant under scalar transformations. But it would no longer be true that $\tilde{\boldsymbol{\chi}}_1$ and $\tilde{\boldsymbol{\eta}}_1$ are standardized.

The same procedure could be followed to find vectors $\mathbf{c}_2$ and $\mathbf{d}_2$ that are associated with the second canonical correlation ρ_2. From now on we shall always use subscripts for the individual vectors $\mathbf{c}_i$ and $\mathbf{d}_i$ corresponding to the subscript of ρ_i, and also to the subscripts of the linear compounds $\boldsymbol{\chi}_i$

and $\boldsymbol{\eta}_i$. We shall assume that the subscript corresponds with the order of magnitude of ρ_i. The symbol **C** will be used for a matrix of which $\mathbf{c}_1$, $\mathbf{c}_2$, etc., are columns; similarly for the matrix **D** with columns $\mathbf{d}_i$. Also, the notation **P** will be used for a diagonal matrix with diagonal elements ρ_i. Finally, we shall from now on use the symbols $\boldsymbol{\chi}$ and $\boldsymbol{\eta}$ for matrices with columns $\boldsymbol{\chi}_1$, $\boldsymbol{\chi}_2$, etc. for $\boldsymbol{\chi}$, and columns $\boldsymbol{\eta}_1$, $\boldsymbol{\eta}_2$, etc. for $\boldsymbol{\eta}$. An interpretation of this more general approach will be given in the next section.

14.2.2. The procedure sketched in the preceding section becomes very inefficient if the number of variables is large. Expansion of the determinant (14.9) then results into a somewhat unmanageable polynomial. It would be attractive to have a quicker procedure, and such a procedure can be developed. Going back to the set of equations (14.5), we can derive from the second of them

$$\mathbf{d} = \rho^{-1}\mathbf{R}_{yy}{}^{-1}\mathbf{R}_{yx}\mathbf{c}. \tag{14.10}$$

Substituting this in the first, we obtain

$$\mathbf{R}_{xy}\mathbf{R}_{yy}{}^{-1}\mathbf{R}_{yx}\mathbf{c} = \rho^2\mathbf{R}_{xx}\mathbf{c},$$

or, after premultiplication by $\mathbf{R}_{xx}{}^{-1}$,

$$\mathbf{R}_{xx}{}^{-1}\mathbf{R}_{xy}\mathbf{R}_{yy}{}^{-1}\mathbf{R}_{yx}\mathbf{c} = \rho^2\mathbf{c}, \tag{14.11}$$

where we assume that the inverses $\mathbf{R}_{xx}{}^{-1}$ and $\mathbf{R}_{yy}{}^{-1}$ exist. Equation (14.11) shows that **c** can be calculated as a right eigenvector of the rather complicated nonsymmetric matrix at the left, with ρ^2 as the corresponding eigenvalue. Of course, for the maximum canonical correlation, we should take the eigenvector $\mathbf{c}_1$ with the largest eigenvalue associated to it. We meet the same problem discussed in the preceding section, in that (14.11) gives a solution determined up to an arbitrary scalar; to satisfy (14.1) we must normalize the eigenvector $\mathbf{c}_1$ so that $\mathbf{c}_1'\mathbf{R}_{xx}\mathbf{c}_1 = 1$.

The corresponding vector $\mathbf{d}_1$ can then be found directly from (14.10). An alternative would be to set up a similar eigenvector equation for **d** by interchanging **c** and **d**, and x and y, in (14.11). The equation becomes

$$\mathbf{R}_{yy}{}^{-1}\mathbf{R}_{yx}\mathbf{R}_{xx}{}^{-1}\mathbf{R}_{xy}\mathbf{d} = \rho^2\mathbf{d}. \tag{14.12}$$

Note that (14.11) and (14.12) produce identical eigenvalues. A disadvantage of (14.11) and (14.12) is that the matrices at the left are not symmetric. Some algorithms for determining eigenvectors do assume that the matrix is symmetric, and so cannot be used to solve (14.11) or (14.12). However, it is possible to find a symmetric alternative. The trick is to use an auxiliary vector

$$\mathbf{q} = \mathbf{R}_{xx}^{\frac{1}{2}}\mathbf{c}, \tag{14.13}$$

which implies $\mathbf{c} = \mathbf{R}_{xx}^{-\frac{1}{2}}\mathbf{q}$ (compare section 7.4.8 for the meaning of $\mathbf{R}_{xx}^{-\frac{1}{2}}$ and $\mathbf{R}_{xx}^{-\frac{1}{2}}$). Substitution of (14.13) in (14.11) gives:

$$\mathbf{R}_{xx}^{-1}\mathbf{R}_{xy}\mathbf{R}_{yy}^{-1}\mathbf{R}_{yx}\mathbf{R}_{xx}^{-\frac{1}{2}}\mathbf{q} = \rho^2\mathbf{R}_{xx}^{-\frac{1}{2}}\mathbf{q}$$

or, after premultiplication by $\mathbf{R}_{xx}^{\frac{1}{2}}$:

$$\mathbf{R}_{xx}^{-\frac{1}{2}}\mathbf{R}_{xy}\mathbf{R}_{yy}^{-1}\mathbf{R}_{yx}\mathbf{R}_{xx}^{-\frac{1}{2}}\mathbf{q} = \rho^2\mathbf{q}. \tag{14.14}$$

The matrix at the left is in fact symmetric, so that (14.14) is a simple eigenvector equation.

Obviously, since the matrix at the left of (14.14) is of order $m_1 \times m_1$, there are m_1 different solutions $\mathbf{q}_1$ for the eigenvector. We could collect them as columns of the eigenvector matrix $\mathbf{Q}$ for which it will be true that $\mathbf{Q}'\mathbf{Q}$ is a diagonal matrix. If we normalize the eigenvectors $\mathbf{q}_i$ to unity, we have $\mathbf{Q}'\mathbf{Q} = \mathbf{I}$.

But we also have, as a generalization of (14.13), $\mathbf{Q} = \mathbf{R}_{xx}^{\frac{1}{2}}\mathbf{C}$. Substituting, we find $\mathbf{Q}'\mathbf{Q} = \mathbf{C}'\mathbf{R}_{xx}\mathbf{C} = \mathbf{I}$. On the other hand, we have $\boldsymbol{\chi} = \mathbf{X}\mathbf{C}$, and therefore $\boldsymbol{\chi}'\boldsymbol{\chi}/n = \mathbf{C}'\mathbf{R}_{xx}\mathbf{C}$. So we have the important result that the application of the canonical weights $\mathbf{C}$ to the matrix $\mathbf{X}$ results in a set of independent and standardized compounds. Similarly, the set $\boldsymbol{\eta} = \mathbf{Y}\mathbf{D}$ is also orthogonal, and standardized.

For the correlations between these sets, we have $\boldsymbol{\chi}'\boldsymbol{\eta}/n = \mathbf{C}'\mathbf{R}_{xy}\mathbf{D}$. It can be shown that this correlation matrix is the diagonal matrix $\mathbf{P}$. The proof is easily derived from the first equation of (14.5). Its generalized version reads

$$\mathbf{R}_{xy}\mathbf{D} = \mathbf{R}_{xx}\mathbf{C}\mathbf{P}, \tag{14.15}$$

and premultiplication by $\mathbf{C}'$ shows that

$$\mathbf{C}'\mathbf{R}_{xy}\mathbf{D} = \mathbf{C}'\mathbf{R}_{xx}\mathbf{C}\mathbf{P} = \mathbf{P}. \tag{14.16}$$

It follows that if we take pairs of compound variables, one from the set $\boldsymbol{\chi}$ and one from the set $\boldsymbol{\eta}$, the correlation between them will be zero unless the variables have the same subscript, in which case the correlation equals ρ_i. Canonical correlation analysis now begins to appear very similar to factor analysis. In fact, the set χ can be interpreted as an orthogonal factor solution for **X**. The factor loadings would be given by the matrix of correlations between the sets **X** and χ:

$$\mathbf{X}'\chi/n = \mathbf{X}'\mathbf{X}\mathbf{C}/n = \mathbf{R}_{xx}\mathbf{C}.$$

Simultaneously, we have a factor solution for the set **Y** with $\boldsymbol{\eta}$ as the orthogonal factors, and factor loadings $\mathbf{R}_{yy}\mathbf{D}$. The two factor solutions are interrelated by the fact that the vector χ_i makes an acute angle with $\boldsymbol{\eta}_i$, whereas every pair $\boldsymbol{\eta}_i$, χ_j, with $i \neq j$, is orthogonal.

14.3. Numerical Illustrations

14.3.1. As a working example, we take a simple problem with $m_1 = m_2 = 2$. Table 14.1 gives the matrices **X** and **Y**, with standardized variables, and table 14.2 the corresponding correlation matrix **R**. The equations (14.7), written out, are

$$\begin{pmatrix} -\rho & -.31\rho & .64 & .80 \\ -.31\rho & -\rho & .34 & .00 \\ .64 & .34 & -\rho & -.13\rho \\ .80 & .00 & -.13\rho & -\rho \end{pmatrix} \begin{pmatrix} \mathbf{c} \\ \cdots \\ \mathbf{d} \end{pmatrix} = 0. \tag{14.17}$$

The determinant (14.9) becomes:

$$\begin{vmatrix} \rho^2 & .31\rho^2 & .64 & .80 \\ .31\rho^2 & \rho^2 & .34 & .00 \\ .64 & .34 & 1.00 & .13 \\ .80 & .00 & .13 & 1.00 \end{vmatrix} = 0. \tag{14.18}$$

Expanding (14.18) we obtain the polynomial equation

$$.8886\rho^4 - .9190\rho^2 + .0740 = 0$$

with solutions

$$\begin{aligned} \rho_1{}^2 &= .946, \qquad \rho_1 = .97, \\ \rho_2{}^2 &= .088, \qquad \rho_2 = .30. \end{aligned} \tag{14.19}$$

Table 14.1. Working example for canonical correlation analysis.

x_1	x_2	y_1	y_2
−1.44	−1.64	−1.83	−.32
−.96	.55	.61	−1.94
.00	−.55	.61	.00
.96	.00	.00	.65
1.44	.00	1.22	1.29
.00	1.64	−.61	.32

The first solution ρ_1 is substituted in (14.17) to obtain

$$\begin{aligned} -.97c_{11} - .30c_{12} + .64d_{11} + .80d_{12} &= 0, \\ -.30c_{11} - .97c_{12} + .34d_{11} + .00d_{12} &= 0, \\ .64c_{11} + .34c_{12} - .97d_{11} - .12d_{12} &= 0, \\ .80c_{11} + .00c_{12} - .12d_{11} - .97d_{12} &= 0. \end{aligned} \tag{14.20}$$

To find a solution, we may set an arbitrary value on one of the unknowns, say $\tilde{\mathbf{d}}_{12} = 1$. The other unknowns then can be identified as

$$\tilde{\mathbf{c}}_{11} = 1.32, \qquad \tilde{\mathbf{c}}_{12} = -.173, \qquad \tilde{\mathbf{d}}_{11} = .680.$$

To find the vector $\mathbf{c}_1$ itself, we calculate $\tilde{\mathbf{c}}_1{}'\mathbf{R}_{xx}\tilde{\mathbf{c}}_1 = 1.63 = (1.28)^2$. The solution for $\mathbf{c}_1$ therefore becomes

$$\mathbf{c}_1 = \begin{pmatrix} 1.32 \\ -.173 \end{pmatrix} \bigg/ 1.28 = \begin{pmatrix} 1.034 \\ -.135 \end{pmatrix}. \tag{14.21a}$$

Table 14.2. Correlation matrix for data in table 14.1.

	x_1	x_2	y_1	y_2
x_1	1.00	.31	.64	.80
x_2	.31	1.00	.34	.00
y_1	.64	.34	1.00	.13
y_2	.80	.00	.13	1.00

A solution for $\mathbf{d}_1$ is found in the same way:

$$\mathbf{d}_1 = \begin{pmatrix} .680 \\ 1.000 \end{pmatrix} \Big/ 1.28 = \begin{pmatrix} .532 \\ .781 \end{pmatrix}. \tag{14.21b}$$

By substituting the second solution $\rho_2 = .30$ in the equations (14.17), we can find the vectors $\mathbf{c}_2$ and $\mathbf{d}_2$. They are

$$\mathbf{c}_2 = \begin{pmatrix} -.195 \\ 1.043 \end{pmatrix}, \quad \text{and} \quad \mathbf{d}_2 = \begin{pmatrix} .857 \\ -.638 \end{pmatrix}. \tag{14.22}$$

If we had preferred to use equation (14.11), we would have found, for the matrix at the left in this equation,

$$\begin{pmatrix} .9678 & .1647 \\ -.1147 & .0661 \end{pmatrix},$$

of which $\mathbf{c}_1$ and $\mathbf{c}_2$ are the two right eigenvectors, and .946 and .088 the two eigenvalues.

14.3.2. The second example is based upon a study by Duncan, Haller, and Portes (1968). The data refer to ten variables related to educational and occupational aspirations of 17-year-old boys. For all boys in the sample, five measurements were taken:

x_1 = parental aspiration ("the degree to which parents encouraged the respondent to have high levels of achievement");

x_2 = family socioeconomic status;

x_3 = intelligence;

y_1 = occupational aspiration;

y_2 = educational aspiration.

The other five variables have the same content as the first five, but they were measured for the boy mentioned by each respondent as his best

Table 14.3. Correlations between variables as described in text (Duncan, Haller, and Portes 1968).

x_1	x_2	x_3	x_4	x_5	x_6	y_1	y_2	y_3	y_4
1.0000	.1839	.0489	.0186	.0782	.1147	.2137	.2742	.1124	.0839
	1.0000	.2220	.1861	.3355	.1021	.4105	.4043	.2903	.2598
		1.000	.2707	.2302	.0931	.3240	.4047	.3054	.2786
			1.0000	.2950	−.0438	.2930	.2407	.4105	.3607
				1.0000	.2087	.2995	.2863	.5191	.5007
					1.0000	.0760	.0702	.2784	.1988
						1.0000	.6247	.3269	.4216
							1.0000	.3669	.3275
								1.0000	.6404
									1.0000

friend. The variables are here identified with the labels:

x_4 = intelligence (friend);

x_5 = socioeconomic status (friend);

x_6 = parental aspirations (friend);

y_3 = educational aspirations (friend);

y_4 = occupational aspirations (friend).

The matrix of correlations between all variables is given in table 14.3. For this example, we shall calculate canonical correlations between the six x-variables and the first two y-variables (later, in sections 15.5 and 16.4, we will come back to this example). **C** in this example will be a 6×2 matrix, and **D** a 2×2 matrix. They are shown in tables 14.4 and 14.5,

Table 14.4. Transformation weights C to be applied to x-variables to obtain canonical variables χ.

x_1	.3258	−.4107
x_2	.4808	.3401
x_3	.4559	−.7181
x_4	.2024	.6893
x_5	.1838	.1247
x_6	−.0266	.1737

Table 14.5. Transformation weights D to be applied to variables y_1 and y_2 in order to obtain canonical variables η.

	η_1	η_2
y_1	.4642	1.1934
y_2	.6419	−1.1080

respectively. The corresponding canonical correlations are given in table 14.6. Also, table 14.7 gives correlations between the observed variables **X** and the set $\boldsymbol{\chi}$, and correlations between **Y** and $\boldsymbol{\eta}$.

Figure 14.1 gives a diagram in terms of path analysis. It postulates that the **X** variables are ultimate variables that completely determine the latent variables $\boldsymbol{\chi}$ (with path coefficients as indicated in **C**), and that $\boldsymbol{\eta}$ forms another set of latent variables, determined partly by the variables $\boldsymbol{\chi}$ and partly by exogenous "error" (error comes in with a contribution equal to the square root of $1 - \rho^2$). The variables **Y**, finally, are interpreted as completely determined by the set $\boldsymbol{\eta}$. Note that the dependence of $\boldsymbol{\eta}_2$ on $\boldsymbol{\chi}_2$ is minimal, so small, in fact, that one might omit $\boldsymbol{\chi}_2$ and look on $\boldsymbol{\eta}_2$ as an exogenous variable.

The picture can be changed slightly by omitting the variables $\boldsymbol{\chi}$. We then draw the arrows directly from the x-variables to the two $\boldsymbol{\eta}$ variables, on the understanding that the coefficients on these paths now have to be multiplied by their corresponding canonical correlations. The result is that $\boldsymbol{\eta}$ is interpreted as dependent partly on the x-variables and partly on exogenous influence, and as determining the y-variables completely. One can easily think of η_1 as a latent "aspiration" variable, of which y_1 and y_2 are indicators.

Many variations of this picture are possible. In fact, in the publication by Duncan *et al.* canonical analysis was not used. Instead, a variable that

Table 14.6. Canonical correlations corresponding to tables 14.4 and 14.5.

	η_1	η_2
χ_1	.6092	—
χ_2	—	.1431

Table 14.7. Correlations between observed variables and canonical variables.

	χ_1	χ_2
x_1	.4517	−.3410
x_2	.7388	.2928
x_3	.6733	−.4312
x_4	.4769	.5800
x_5	.5299	.2809
x_6	.1318	.0901
	η_1	η_2
y_1	.8652	.5013
y_2	.9319	−.3625

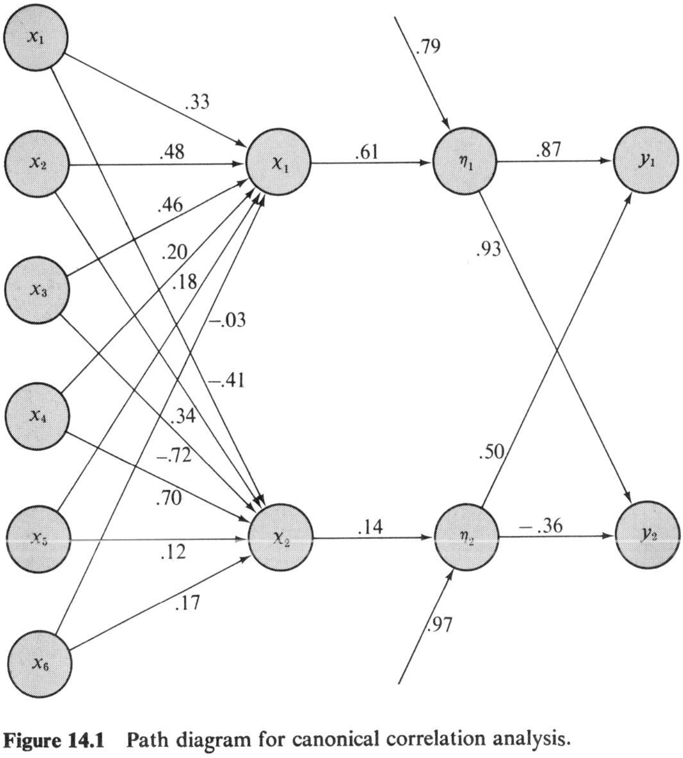

Figure 14.1 Path diagram for canonical correlation analysis.

we shall call g was postulated to serve the same function as η does in our diagram, in that it had to explain as much of the correlation between the y's as possible. (We will come back to this approach in section 15.5.)

It is also possible to draw pictures in terms of the vector space for canonical solutions. Actually, we may take $\boldsymbol{\chi}_1$ and $\boldsymbol{\chi}_2$ as reference vectors. The matrix $\mathbf{X}'\boldsymbol{\chi}/n = \mathbf{X}'\mathbf{X}\mathbf{C}/n = \mathbf{R}_{xx}\mathbf{C}$ then shows the projections of the x-vectors on the hyperplane spanned by $\boldsymbol{\chi}_1$ and $\boldsymbol{\chi}_2$. Canonical correlations show how long the projections of $\boldsymbol{\eta}_1$ and $\boldsymbol{\eta}_2$ are on this plane; the directions

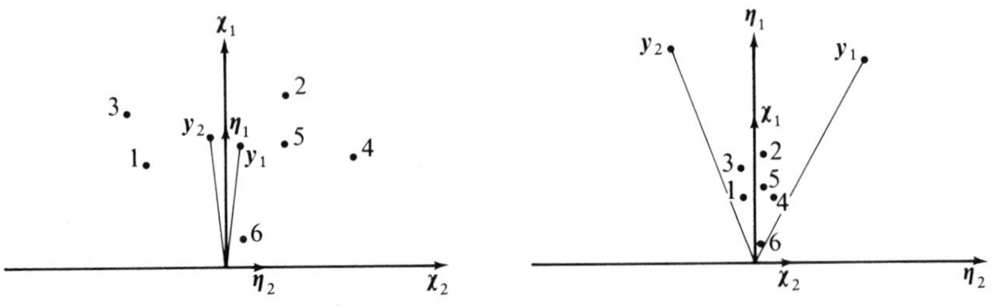

Figure 14.2 Projections of vectors on the (χ_1, χ_2)-plane (upper diagram) and on the (η_1, η_2)-plane (lower diagram).

of these vectors coincide with those of $\boldsymbol{\chi}_1$ and $\boldsymbol{\chi}_2$, respectively. We might also draw projections of the y-variables on this plane, by using

$$\mathbf{Y}'\boldsymbol{\chi}/n = \mathbf{Y}'\mathbf{X}\mathbf{C}/n = \mathbf{R}_{yx}\mathbf{C}.$$

A similar and corresponding picture can be drawn for the hyperplane spanned by $\boldsymbol{\eta}_1$ and $\boldsymbol{\eta}_2$. Such drawings are given in figure 14.2 for this example; note their similarity to factor solutions.

14.4. Canonical Analysis with Variables that are Subject to Error

One problem remains, which is that the equations $\boldsymbol{\chi} = \mathbf{X}\mathbf{C}$ and $\boldsymbol{\eta} = \mathbf{Y}\mathbf{D}$, fundamental for canonical analysis, do not contain error terms. This is a problem because, for instance, in the path diagram in figure 14.1 we do not permit an exogenous influence for $\boldsymbol{\chi}_1$; yet it seems more than reasonable that such an influence might be present. We then should use equations

$$\begin{aligned} \boldsymbol{\chi} &= \mathbf{X}\mathbf{C} + \mathbf{E}_x, \\ \boldsymbol{\eta} &= \mathbf{Y}\mathbf{D} + \mathbf{E}_y, \end{aligned} \qquad (14.23)$$

in which $\mathbf{E}_x$ and $\mathbf{E}_y$ stand for the error components (i.e., exogenous influences, which are called "error" just for convenience).

It can be easily shown that this change has only a minor effect on the techniques for calculating **C**, **D**, and canonical correlations. The only difference is that we should use reduced correlation matrices $\mathbf{R}_{xx}^*$ and $\mathbf{R}_{yy}^*$, to be obtained from $\mathbf{R}_{xx}$ and $\mathbf{R}_{yy}$ by subtracting error variance components from the diagonal elements. Such error components of variance will usually not be known, and will have to be estimated.

14.5. Other Uses of Canonical Analysis

To conclude this chapter, we want to mention a few other applications of canonical analysis. One interesting application is pattern matching. Imagine, for instance, a study on the analysis of speech sounds: we ask the two speakers *A* and *B* to utter, say, 15 different vowels. The produced speech sounds are analyzed according to some physical procedure (determination of three formants, say). We thus obtain a 15×3 matrix of physical measurements for speaker *A*, and a 15×3 matrix of measurements for speaker *B*. Let us call these matrices **X** and **Y**, respectively. The problem is to find out whether these two matrices are related. In geometrical terms, we can picture **X** as 15 points in an S_3, and **Y** as 15 points in another three-space, $S_{3'}$. The problem can be formulated as follows: can we find a vector in S_3 such that the projections of the vowel points on it are maximally correlated with the projections of the points on some vector in $S_{3'}$?

This question can be directly answered by canonical analysis. In fact, the vectors $\mathbf{c}_1$ and $\mathbf{d}_1$ as defined in section 14.2 indicate how these vectors must be chosen. More precisely, if we normalize $\mathbf{c}_1$ and $\mathbf{d}_1$ to unity, we have the two vectors of direction cosines (note that such a transformation does not affect the canonical correlation). In addition, $\mathbf{c}_2$ and $\mathbf{d}_2$ would give another pair of vectors of direction cosines (after normalization to unity) such that the projections on them correlate, with correlation ρ_2. In general, the matrices **C** and **D** indicate pairs of correlating vectors in both spaces.

A drawback of this procedure might be that the matrices **C** and **D** are not orthogonal. That is, the vectors indicated by them do not form orthogonal coordinate systems. We cannot superimpose one space on the other without stretching and shrinking them. If this is a serious disadvantage, we can use another variety of canonical analysis. This variety emerges if, instead of the side conditions $\mathbf{c}'\mathbf{R}_{xx}\mathbf{c} = 1$ and $\mathbf{d}'\mathbf{R}_{yy}\mathbf{d} = 1$, we impose the conditions $\mathbf{c}'\mathbf{c} = 1$ and $\mathbf{d}'\mathbf{d} = 1$, so that $\mathbf{c}$ and $\mathbf{d}$ are defined as direction cosines from the start. Furthermore, instead of trying to maximize the correlation between the compound variables (between the projections on the vectors defined by $\mathbf{c}$ and $\mathbf{d}$, respectively), we will maximize the covariance. In formal notation: in the first space, S_3 here, we want to introduce a vector with direction cosines $\mathbf{c}$ such that the projections $\boldsymbol{\chi} = \mathbf{Xc}$ on this vector have maximum covariance with the projections $\boldsymbol{\eta} = \mathbf{Yd}$ in the second space, $S_{3'}$, where $\mathbf{d}$ gives direction cosines for a vector in this second space.

Assuming that $\mathbf{X}$ and $\mathbf{Y}$ give deviations of the means, the covariance is $\boldsymbol{\chi}'\boldsymbol{\eta}/n = \mathbf{c}'\mathbf{X}'\mathbf{Yd}/n = \mathbf{c}'\mathbf{Vd}$, where $\mathbf{V}$ is the variance-covariance matrix $\mathbf{X}'\mathbf{Y}/n$. The objective is, therefore, to maximize $\mathbf{c}'\mathbf{Vd}$ under the conditions $\mathbf{c}'\mathbf{c} = 1$ and $\mathbf{d}'\mathbf{d} = 1$. Using undetermined multipliers it is easily shown that

$$\mathbf{Vd} = \mu\mathbf{c},$$
$$\mathbf{V}'\mathbf{c} = \mu\mathbf{d},$$

where μ is the covariance $\boldsymbol{\chi}'\boldsymbol{\eta}/n$. From the first equation we can solve $\mathbf{c} = \mu^{-1}\mathbf{Vd}$; substituting this in the second equation, we obtain

$$\mu^{-1}\mathbf{V}'\mathbf{Vd} = \mu\mathbf{d},$$

or $\mathbf{V}'\mathbf{Vd} = \mu^2\mathbf{d}$. Similarly, it can be shown that $\mathbf{VV}'\mathbf{c} = \mu^2\mathbf{c}$. The vectors $\mathbf{c}$ and $\mathbf{d}$ therefore can be found as eigenvectors of the symmetric matrices $\mathbf{VV}'$ and $\mathbf{V}'\mathbf{V}$. If we collect the eigenvector solutions in matrices $\mathbf{C}$ and $\mathbf{D}$, we will have $\mathbf{C}'\mathbf{C} = \mathbf{I}$ and $\mathbf{D}'\mathbf{D} = \mathbf{I}$, assuming that the eigenvectors are normalized to unity.

Sometimes one wants to rotate a pattern towards a certain "target." Formally we would then have the following problem: $\mathbf{X}$ is the data matrix, and $\mathbf{Y}$ is a target matrix. We want to find a transformation $\mathbf{XT} = \mathbf{U}$ such that $\mathbf{U}$ matches $\mathbf{Y}$. The point we want to make here is that there is no essential difference between canonical matching (where *both* data matrices are transformed) and target matching (where only *one* matrix is transformed and the other kept fixed), because if $\boldsymbol{\chi} = \mathbf{XC}$, $\boldsymbol{\eta} = \mathbf{YD}$, and $\boldsymbol{\chi}$ matches $\boldsymbol{\eta}$, then $\mathbf{XCD}^{-1}$ will match $\mathbf{Y}$. So the canonical solution can easily be changed into a target solution.

More difficult problems arise if we want to transform three or more data matrices in such a way that there is some match between all compounds simultaneously. A treatment of such problems can be found in Horst (1965) and in Van de Geer (1968).

15

Varieties of Factor Analysis

15.1. Introduction

There are many varieties of factor analysis. Principle-components analysis, sketched in Chapter 13, is only one of them. In this chapter we shall discuss some others, and relate them to path analysis and canonical analysis. One might say that each variety applies a different principle to the identifiability problem, which is that there is no unique solution for the basic equation (13.18),

$$\mathbf{X} = \mathbf{Y}\mathbf{F}' + \mathbf{E},$$

since there are infinitely many solutions for $\mathbf{F}'$. One solution is to take for $\mathbf{F}'$ the factor matrix that results from principal-components analysis. Or, since there is no objection to a rotation of the factor matrix $\mathbf{G} = \mathbf{F}\mathbf{T}$, we could also write the basic equation in terms of $\mathbf{G}$. But there are an infinite number of possible rotation matrices $\mathbf{T}$, orthogonal or oblique. They all result in a solution for the basic equation, all of them equally good.

In such a situation, the solution is said to be unidentifiable. The concept of *identifiability* is an important one in econometry, as we will show in more detail in Chapter 16.* For now, let us define the concept in formal terms. First of all, we need the concept of model. A *model* is a mathematical formulation of assumptions that are taken to be valid. For factor analysis, the model would include

$$\mathbf{X} = \mathbf{YF}' + \mathbf{E}$$

as the basic equation, together with assumptions $\mathbf{E}'\mathbf{E}/n = \mathbf{U}^2$, $\mathbf{E}'\mathbf{Y} = \mathbf{0}$, and, for the orthogonal case, $\mathbf{Y}'\mathbf{Y}/n =$ diagonal. If the model is applied to given data $\mathbf{X}$, the result we want is a *solution* for $\mathbf{F}'$. This matrix constitutes the *parameters* of the model; they have to be inferred from the empirical data. If we succeed in finding a solution, the model can be said to explain the data; that is, data generated by the model, with the parameters solved, would be undistinguishable from real data.

The model with $\mathbf{F}'$ solved, or, in general, the model with parameters solved, is sometimes called a *structure*. In our factor example, the solved equation $\mathbf{X} = \mathbf{YF}' + \mathbf{E}$ would then be called *structural equations*. The identifiability problem arises if there is no unique solution for the parameters, i.e., if there are many solutions that are equally good. The solution then is said to be unidentifiable. For our example, we see that the solution for $\mathbf{F}$ is not identifiable, since there are infinitely many solutions, equally good.

In such a situation we could impose further constraints and include them in the model. In other words, we add assumptions, and the result might be that the solution becomes identifiable. For instance, in factor analysis the optimization procedures implied by varimax can be interpreted as additional assumptions of the model, by which a unique solution becomes identified. Varimax, however, is only one variety; other varieties of factor analysis add different assumptions to identify a solution.

We see that the researcher has to make decisions in two stages, so to speak. The first decision stage is to choose a model, i.e., to specify the mathematical assumptions that he takes to be valid. The second stage is to determine whether the data are consistent with a solution of the model. But if a solution can be accepted, it does not follow that other models would not lead to equally good solutions. Very often in social research the researcher mixes the two decision stages; if he finds some model to be

* On identifiability, see Koopmans 1953, Koopmans and Bausch 1959, and Koopmans and Reiersøl 1950.

unsatisfactory, he amends his model, so that the first decision stage comes to depend on the second. This mixing creates much confusion in methodological discussions, because researchers sometimes tend to reject one model because another is found to work. Factor analysis is rejected because path analysis works, for instance.

It is extremely difficult in science to judge the merits of various models. There is no single criterion; and the only real criterion is to see how productive a model is in the long run. However, models tend to merge with scientific concepts as such. In personality research, for instance, factors obtained from factor analysis tend to become proper scientific concepts, and it is difficult (some methodologists might even say impossible, I among them) to separate the empirical evidence for those concepts from the mathematical assumptions in the model. In an advanced science one cannot tell what portion of a concept is empirical and what portion assumed. At first sight, models look like just tools, just techniques, just hired servants to do housecleaning jobs. However, before long, the relation develops into something more than a relation between maidservant and master: the master marries the girl.

15.2. Canonical Factor Analysis

15.2.1. Formal Treatment. The method of canonical factor analysis was developed by Rao (1952). One of its merits, compared to principal-components analysis and many other varieties of factor analysis, is that its result is invariant under rescaling of variables. If, for instance, principal-components analysis is applied to variables that are not standardized (e.g., the variance-covariance matrix is used instead of the correlation matrix), the result is different from that obtained for standardized variables. This is not so for canonical factor analysis.

The additional criteria that make the solution identifiable are borrowed here from canonical correlation analysis. The major difference is that the set of observed variables **Y** in canonical correlation analysis is replaced by the set of unobserved factor scores.

The basic features of the model for canonical factor analysis are the following. First, the observed variables are written as linear functions of unobserved values **Y**,

$$\mathbf{X} = \mathbf{YF}' + \mathbf{E}, \tag{15.1}$$

which may be replaced, writing **M** for **YF**′, by

$$\mathbf{X} = \mathbf{M} + \mathbf{E}. \tag{15.2}$$

For convenience, we shall assume that **X** is standardized (this does not influence the result—it merely simplifies some derivations). Then **M** and **E** will not be standardized, since doing so could not be consistent with equation (15.2). Further assumptions are: $\mathbf{M}'\mathbf{E}/n = \mathbf{0}$; $\mathbf{E}'\mathbf{E}/n = \mathbf{U}^2$ is diagonal; $\mathbf{Y}'\mathbf{Y}/n = \mathbf{I}$. Nothing new so far.

The objective of canonical factor analysis is to find transformations of the observed variables that make them optimally correlated with linear combinations of *m*-variables. This objective implies that we are looking for transformation vectors *k* and *c*,

$$\begin{aligned} \mathbf{Xk} &= \mathbf{v}, \\ \mathbf{Mc} &= \mathbf{w}, \end{aligned} \tag{15.3}$$

such that $\mathbf{v}'\mathbf{w}/n$ has a maximum if the side conditions $\mathbf{v}'\mathbf{v}/n = 1$, and $\mathbf{w}'\mathbf{w}/n = 1$, are imposed. Then $\mathbf{v}'\mathbf{w}/n$ is a correlation, and we maximize the correlation. To solve, we substitute

$$\begin{aligned} \mathbf{v}'\mathbf{v}/n &= \mathbf{k}'\mathbf{X}'\mathbf{Xk}/n = \mathbf{k}'\mathbf{Rk}, \\ \mathbf{w}'\mathbf{w}/n &= \mathbf{c}'\mathbf{M}'\mathbf{Mc}/n = \mathbf{c}'\mathbf{R}^*\mathbf{c}, \\ \mathbf{v}'\mathbf{w}/n &= \mathbf{k}'\mathbf{X}'\mathbf{Mc}/n = \mathbf{k}'\mathbf{R}^*\mathbf{c}, \end{aligned} \tag{15.4}$$

where $\mathbf{R}^*$ is the reduced correlation matrix $\mathbf{R} - \mathbf{U}^2$. The objective now can be reformulated as that we want a maximum for $\mathbf{k}'\mathbf{R}^*\mathbf{c}$ under the conditions $\mathbf{k}'\mathbf{Rk} = 1$ and $\mathbf{c}'\mathbf{R}^*\mathbf{c} = 1$.

We introduce an auxiliary function F in the usual way,

$$F = \mathbf{k}'\mathbf{R}^*\mathbf{c} - \tfrac{1}{2}\mu(\mathbf{k}'\mathbf{Rk} - 1) - \tfrac{1}{2}\lambda(\mathbf{c}'\mathbf{R}^*\mathbf{c} - 1), \tag{15.5}$$

and set its partial derivatives equal to zero:

$$\partial F/\partial \mathbf{k}' = \mathbf{R}^*\mathbf{c} - \mu\mathbf{Rk} = \mathbf{0}; \tag{15.6a}$$

$$\partial F/\partial \mathbf{c}' = \mathbf{R}^*\mathbf{k} - \lambda\mathbf{R}^*\mathbf{c} = \mathbf{0}. \tag{15.6b}$$

If we multiply (15.6a) by $\mathbf{k}'$ and (15.6b) by $\mathbf{c}'$, we see that $\mathbf{k}'\mathbf{R}^*\mathbf{c} = \mu\mathbf{k}'\mathbf{Rk} = \mu$, and $\mathbf{c}'\mathbf{R}^*\mathbf{k} = \lambda\mathbf{c}'\mathbf{R}^*\mathbf{c} = \lambda$, so that we may write $\mu = \lambda = \rho$, where ρ is a canonical correlation between **v** and **w**. Also, equation (15.6a) shows that $\mathbf{k} = \rho\mathbf{c}$ (if we exclude the possibility that both **k** and **c** are zero vectors, which in fact is excluded by the conditions). It follows that we may write $\mathbf{c} = \mathbf{k}\rho^{-1}$, and if we substitute this in equation (15.6a), we obtain

$$\mathbf{R}^*\mathbf{k}\rho^{-1} = \rho\mathbf{R}\mathbf{k},$$

or

$$\mathbf{R}^*\mathbf{k} = \mathbf{R}\mathbf{k}\rho^2. \tag{15.7}$$

The problem is how to find a solution for (15.7). It can be shown that (15.7) can be transformed into an ordinary eigenvector problem. First, substitute $\mathbf{R} = \mathbf{R}^* + \mathbf{U}^2$:

$$\mathbf{R}^*\mathbf{k} = (\mathbf{R}^* + \mathbf{U}^2)\mathbf{k}\rho^2, \tag{15.8}$$

or, rearranged,

$$\mathbf{R}^*\mathbf{k} = \mathbf{U}^2\mathbf{k}\rho^2/(1 - \rho^2). \tag{15.9}$$

For convenience, we shall write $\alpha = \rho^2/(1 - \rho^2)$. Equation (15.9) then becomes

$$\mathbf{R}^*\mathbf{k} = \alpha\mathbf{U}^2\mathbf{k}. \tag{15.10}$$

We shall now use the same trick as in section 14.2.2; i.e., we introduce an auxiliary vector

$$\mathbf{q} = \mathbf{U}\mathbf{k} \tag{15.11}$$

or $\mathbf{k} = \mathbf{U}^{-1}\mathbf{q}$. Substitute this in (15.10) to obtain

$$\mathbf{R}^*\mathbf{U}^{-1}\mathbf{q} = \alpha\mathbf{U}\mathbf{q},$$

which, both sides premultiplied by $\mathbf{U}^{-1}$, becomes

$$\mathbf{U}^{-1}\mathbf{R}^*\mathbf{U}^{-1}\mathbf{q} = \alpha\mathbf{q}, \tag{15.12}$$

and (15.12) is an ordinary eigenvector problem that we can solve. In general, we shall find m different eigenvectors **q**, which we can collect in an eigenvector matrix **Q** (of order $m \times m$). We shall assume that the eigenvectors are normalized to unity, so that $\mathbf{Q}'\mathbf{Q} = \mathbf{I}$. The corresponding eigenvalues will be collected in a diagonal matrix **A**.

Similarly, we obtain an $m \times m$ matrix **K** from the generalized version of equation (15.11):

$$\mathbf{Q} = \mathbf{U}\mathbf{K}, \qquad \mathbf{K} = \mathbf{U}^{-1}\mathbf{Q}. \tag{15.13}$$

Equation (15.12) can then be generalized to

$$\mathbf{U}^{-1}\mathbf{R}^*\mathbf{U}^{-1}\mathbf{Q} = \mathbf{QA}. \tag{15.14}$$

It follows that

$$\mathbf{U}^{-1}\mathbf{R}^*\mathbf{U}^{-1} = \mathbf{QAQ}',$$

and therefore

$$\mathbf{R}^* = \mathbf{UQAQ'U}. \tag{15.15}$$

A factor solution for $\mathbf{R}^*$ follows directly, for the essential idea of factor analysis is that $\mathbf{R}^*$ is decomposed as $\mathbf{FF}'$. Therefore, if we rewrite (15.15) as

$$\mathbf{R}^* = (\mathbf{UQA}^{\frac{1}{2}})(\mathbf{A}^{\frac{1}{2}}\mathbf{Q'U}), \tag{15.16}$$

it is clear that $\mathbf{F}$ can be identified as

$$\mathbf{F} = \mathbf{UQA}^{\frac{1}{2}}. \tag{15.17}$$

It is interesting to note that a factor solution for $\mathbf{R}$ can also be found very easily. Suppose this factor solution is $\mathbf{G}$, so that

$$\mathbf{R} = \mathbf{GG}'.$$

However,

$$\mathbf{R} = \mathbf{R}^* + \mathbf{U}^2 = \mathbf{UQAQ'U} + \mathbf{U}^2 = \mathbf{UQ(A + I)Q'U},$$

so obviously $\mathbf{G}$ can be identified as

$$\mathbf{G} = \mathbf{UQ(A + I)}^{\frac{1}{2}}. \tag{15.18}$$

Columns of $\mathbf{F}$ are then proportional to columns of $\mathbf{G}$.

$\mathbf{F}$ can be interpreted as the matrix of correlations between the observed variables $\mathbf{X}$ and the latent systematic variables $\mathbf{W}$. To see this, we have to remember the result $\mathbf{c} = \mathbf{k}\rho^{-1}$, or, generalized to the full matrix, $\mathbf{C} = \mathbf{KP}^{-1}$, where $\mathbf{P}$ is the diagonal matrix with elements ρ_i. For the covariance matrix $\mathbf{X'W}/n$, we find:

$$\begin{aligned}\mathbf{X'W}/n &= \mathbf{X'MC}/n = \mathbf{R}^*\mathbf{C} = \mathbf{R}^*\mathbf{KP}^{-1} = \mathbf{UQAQ'UKP}^{-1} && \text{[from (15.15)]}\\ &= \mathbf{UQAQ'UU}^{-1}\mathbf{QP}^{-1} && \text{[from (15.13)]}\\ &= \mathbf{UQAQ'QP}^{-1} = \mathbf{UQAP}^{-1}.\end{aligned} \tag{15.19}$$

$\mathbf{X'W}/n$ is a covariance matrix, not a correlation matrix, since the variables **W** are not standardized. In fact, their variance-covariance matrix $\mathbf{W}'\mathbf{W}/n$ is

$$\begin{aligned}\mathbf{W}'\mathbf{W}/n &= \mathbf{C}'\mathbf{M}'\mathbf{M}\mathbf{C}/n = \mathbf{C}'\mathbf{R}^*\mathbf{C} = \mathbf{P}^{-1}\mathbf{K}'\mathbf{R}^*\mathbf{K}\mathbf{P}^{-1} \\ &= \mathbf{P}^{-1}\mathbf{Q}'\mathbf{U}^{-1}\mathbf{R}^*\mathbf{U}^{-1}\mathbf{Q}\mathbf{P}^{-1} = \mathbf{P}^{-1}\mathbf{A}\mathbf{P}^{-1}, \qquad (15.20)\end{aligned}$$

and it is easily realized that $\mathbf{W}'\mathbf{W}/n$ is a diagonal matrix. Therefore, to find a correlation between the variables **X** and **W**, all we have to do is to divide columns of (15.19) by the corresponding square root of elements of (15.20). Since (15.20) gives a diagonal matrix, we can rewrite $\mathbf{P}^{-1}\mathbf{A}\mathbf{P}^{-1}$ as $\mathbf{P}^{-2}\mathbf{A}$, and the inverse of its square root is $\mathbf{P}\mathbf{A}^{-\frac{1}{2}}$. Therefore, the correlation matrix is found if we postmultiply (15.19) by the latter matrix:

$$\mathbf{U}\mathbf{Q}\mathbf{A}\mathbf{P}^{-1}(\mathbf{P}\mathbf{A}^{-\frac{1}{2}}) = \mathbf{U}\mathbf{Q}\mathbf{A}^{\frac{1}{2}} = \mathbf{F}. \qquad \text{Q.E.D.}$$

Also, it can be shown that **G** gives correlations between the variables in **X** and the variables in **V**. Derivations are similar to those shown above, and are left to the reader as an exercise.

In passing, and to avoid confusion, we should remark that previously we did set up the side condition $\mathbf{K}'\mathbf{R}\mathbf{K} = \mathbf{I}$. Obviously, when we agreed that $\mathbf{Q}'\mathbf{Q} = \mathbf{I}$, we violated this condition, since it can be shown that with the present version of **K**, we have $\mathbf{K}'\mathbf{R}\mathbf{K} = \mathbf{A} + \mathbf{I}$. In other words, with the present **K** the transformation $\mathbf{X}\mathbf{K} = \mathbf{V}$ results in variables **V** that are independent but not standardized. To standardize them, we should normalize **K** differently. However, this step is not essential, after all, and therefore we shall adhere to **K** in its present version.

It is interesting to look at the results from a somewhat different angle, which perhaps makes more clear what canonical factor analysis really does. To see this, let us look at the transformation $\mathbf{X}\mathbf{K} = \mathbf{V}$. Since $\mathbf{X} = \mathbf{M} + \mathbf{E}$, the matrix **V** can be decomposed into two additive portions, $\mathbf{V} = \mathbf{M}\mathbf{K} + \mathbf{E}\mathbf{K}$, and similarly for the variance-covariance matrix $\mathbf{V}'\mathbf{V}/n$, where we have

$$\begin{aligned}\mathbf{V}'\mathbf{V}/n &= \mathbf{K}'\mathbf{X}'\mathbf{X}\mathbf{K}/n = \mathbf{K}'(\mathbf{M}'\mathbf{M}/n)\mathbf{K} + \mathbf{K}'(\mathbf{E}'\mathbf{E}/n)\mathbf{K} \\ &= \mathbf{K}'\mathbf{R}^*\mathbf{K} + \mathbf{K}'\mathbf{U}^2\mathbf{K}. \qquad (15.21)\end{aligned}$$

This shows the two portions of the variance-covariance matrix. The first portion, $\mathbf{K}'\mathbf{R}^*\mathbf{K}$, represents (co)variance due to systematic components, and the second portion, $\mathbf{K}'\mathbf{U}^2\mathbf{K}$, represents (co)variance due to error components. For the latter portion, we can write

$$\mathbf{K}'\mathbf{U}^2\mathbf{K} = \mathbf{Q}'\mathbf{U}^{-1}\mathbf{U}^2\mathbf{U}^{-1}\mathbf{Q} = \mathbf{Q}'\mathbf{Q} = \mathbf{I}, \tag{15.22}$$

where we have used equation (15.13). Equation (15.22) shows that the variance-covariance portion in $\mathbf{V}'\mathbf{V}/n$ due to error is the unit matrix. Therefore, components $\mathbf{EK}$ in $\mathbf{V}$ can be said to be independent and to have unit variance. In other words, the space of the error components is made "spherical."

If we now look at equation (15.14) again, we discover another aspect in it. Imagine that we rescale the observed variables by applying $\mathbf{U}^{-1}$ as a diagonal matrix of scalars: $\mathbf{X}_s = \mathbf{X}\mathbf{U}^{-1}$. Then $\mathbf{X}_s'\mathbf{X}_s/n$ becomes the variance-covariance matrix for these rescaled variables, and we have $\mathbf{X}_s'\mathbf{X}_s/n = \mathbf{U}^{-1}\mathbf{R}\mathbf{U}^{-1}$. Again, since $\mathbf{R} = \mathbf{R}^* + \mathbf{U}^2$, this matrix can be decomposed into two additive portions:

$$\mathbf{U}^{-1}\mathbf{R}\mathbf{U}^{-1} = \mathbf{U}^{-1}\mathbf{R}^*\mathbf{U}^{-1} + \mathbf{U}^{-1}\mathbf{U}^2\mathbf{U}^{-1} = \mathbf{U}^{-1}\mathbf{R}^*\mathbf{U}^{-1} + \mathbf{I}.$$

Here the portion $\mathbf{U}^{-1}\mathbf{R}^*\mathbf{U}^{-1}$ stands for (co)variance due to systematic components, and the portion due to error components is the unit matrix. So this also means that the error space has been made spherical. In addition, equation (15.14) now shows that $\mathbf{Q}$ just gives a principal-components solution for the systematic portion of the variance-covariance matrix. A geometrical interpretation is that the observed measurements are plotted in a space with the variables in $\mathbf{M}$ and $\mathbf{E}$ as coordinates. What we do first is transform the space (by the postmultiplication by $\mathbf{U}^{-1}$) so that the cloud of points becomes spherical with respect to the $\mathbf{E}$-coordinates. Subsequently, we apply a principal-components analysis to the (elliptic) cloud of points in the remaining part of the space defined by the $\mathbf{M}$-coordinates.

We may repeat here that the canonical factor solution is invariant under any rescaling of the variables. In general, rescaling of variables can be expressed as

$$\mathbf{X}_s = \mathbf{XS}, \tag{15.23}$$

where $\mathbf{S}$ is a diagonal matrix of scalars. Obviously, the variance-covariance matrix $\mathbf{X}_s'\mathbf{X}_s/n$ can be split into two additive matrices,

$$\mathbf{X}_s'\mathbf{X}_s/n = \mathbf{M}_s'\mathbf{M}_s/n + \mathbf{E}_s'\mathbf{E}_s/n, \tag{15.24}$$

where $\mathbf{M}_s = \mathbf{MS}$ and $\mathbf{E}_s = \mathbf{ES}$. It is easily shown that

$$\mathbf{M}_s'\mathbf{M}_s/n = \mathbf{SR^*S} \tag{15.25}$$

and

$$\mathbf{E}_s'\mathbf{E}_s/n = \mathbf{SU^2S}. \tag{15.26}$$

Let $\mathbf{SU^2S} = \mathbf{U}_s^2$, so that we can write $\mathbf{U}_s = \mathbf{US} = \mathbf{SU}$. It follows that equation (15.14) will not be affected by the scalar transformation, since

$$\mathbf{U^{-1}R^*U^{-1}} = \mathbf{U^{-1}S^{-1}SR^*SS^{-1}U^{-1}} = \mathbf{U}_s^{-1}(\mathbf{M}_s'\mathbf{M}_s/n)\mathbf{U}_s^{-1}, \tag{15.27}$$

and it appears that for the rescaled variables, we shall find the same solution for $\mathbf{Q}$ and $\mathbf{A}$. The solution of the factor matrix was $\mathbf{F} = \mathbf{UQA}^{\frac{1}{2}}$, and for rescaled variables we would obtain a matrix

$$\mathbf{F}_s = \mathbf{U}_s\mathbf{QA}^{\frac{1}{2}} = \mathbf{SUQA}^{\frac{1}{2}} = \mathbf{SF}.$$

This means that $\mathbf{F}_s$ is obtained from $\mathbf{F}$ if we apply the scalars to the appropriate rows of $\mathbf{F}$. $\mathbf{F}_s$ should perhaps be called not a factor matrix, but a matrix of components of variance.

The point is that, to find a canonical solution for rescaled variables, all we need do is rescale the canonical solution for the original variables. At first sight, one might think this a trivial statement that is true for any factor solution. In general, a factor solution satisfies $\mathbf{R} = \mathbf{FF}'$. If we rescale the variables, we will obtain a variance-covariance matrix $\mathbf{SRS}$, and since $\mathbf{SRS} = \mathbf{SFF'S}$, it follows that $\mathbf{SF}$ is a matrix of components of variance. Now, if $\mathbf{F}$ is a canonical factor solution, $\mathbf{SF}$ is also a canonical factor solution. In contrast, if $\mathbf{F}$ is a principal-factor solution of $\mathbf{R}$, then $\mathbf{FS}$ is *not* a principal-components solution of $\mathbf{SRS}$. In this sense, the canonical solution is invariant under rescaling of the variables, whereas the principal-factor solution is not.

15.2.2. Maximum Likelihood Solution. Thus far we have assumed that $\mathbf{U}^2$ is a known matrix, and on that assumption have developed techniques for canonical factor analysis. In actual studies, however, the assumption will not be met, and $\mathbf{U}^2$ should then be estimated. One possible procedure might be to estimate h_i^2 from several iterations of the principal-factor method, as described in section 13.5. Then u_i^2 will be estimated as $1 - h_i^2$. This method would result in a canonical factor solution that is nothing but a rotated version of the principal-factor solution.

There is, however, a much more interesting approach to the problem of estimating $\mathbf{U}^2$ combined with canonical analysis. This approach is the *maximum likelihood solution.* Essentially, this solution is based on statistical considerations. It is assumed that the observations are sampled from a multinormal distribution with variance covariance matrix $\boldsymbol{\Sigma}$, and that $\boldsymbol{\Sigma}$ can be decomposed according to $\boldsymbol{\Sigma} = \boldsymbol{\Phi}\boldsymbol{\Phi}' + \boldsymbol{\Psi}^2$. Given that the number of variables is m, $\boldsymbol{\Psi}^2$ is a diagonal m x m matrix of specific components of variance (i.e., specific to one variable only), whereas $\boldsymbol{\Phi}$ represents an $m \times p$ matrix referring to p common components of variance. Here $\boldsymbol{\Phi}$ and $\boldsymbol{\Psi}$ are the population parameters that should be estimated from the sampled data.

The general reasoning behind the maximum likelihood solution is as follows. We begin by assuming that the nature of the population distribution of the variates is known; for instance, it is a multinormal distribution. The multinormal density distribution is given in expression (8.14). Parameters of this distribution are: the vector of population means $\bar{\mathbf{x}}$, and the population variance-covariance matrix $\mathbf{A}$. Suppose now that we select values for the population parameters, and also that we are given a vector of observations $\mathbf{x}_i$; then the density of $\mathbf{x}_i$ can be calculated as $g(\mathbf{x}_i)$ if we set $\mathbf{x} = \mathbf{x}_i$ on the right side of (8.14). For another vector of observations $\mathbf{x}_j$ we can do the same, and evaluate $g(\mathbf{x}_j)$. Then if the observations are independent samples, their joint density is given by $g(\mathbf{x}_i) \cdot g(\mathbf{x}_j)$. Generalizing, if we have n observation vectors $\mathbf{x}_k$, where $k = 1, \ldots, n$, the joint density is

$$L = \prod_k g(\mathbf{x}_k),$$

which is called the *likelihood* of the set of observations for the particular selection of population parameters. Of course, L will vary, depending on the choice of the population parameters. It follows that, in general, there will be some selection of population parameters that maximizes L. This choice is the maximum likelihood estimation of the population parameters. Loosely speaking, it is the set of population parameters under which the set of observations is the least surprising.

So far the general idea. In our application, we have to estimate $\boldsymbol{\Phi}$ and $\boldsymbol{\Psi}$. The details of the derivations are somewhat complex and will not be discussed here. But you may intuitively realize that the canonical factor solution must be related to the maximum likelihood solution, since canonical factors are selected on the basis of best correlations with systematic "error-free" components, M. Therefore, if $\mathbf{U}^2$ were known, or, in the terminology we are using here, if $\boldsymbol{\Psi}$ were known, canonical factor analysis would produce the best estimate of $\boldsymbol{\Phi}$. However, if $\mathbf{U}^2$ is not known, we have to proceed more carefully. We then have to apply an iterative procedure, which will be demonstrated now for the correlation matrix $\mathbf{R}$; it could also be applied to the variance-covariance matrix of rescaled variables, with invariant result.

The procedure starts with estimating a one-factor solution. As a first estimate of this single factor, we might take the first principal factor of $\mathbf{R}$, defined by $\mathbf{R}\mathbf{f}_{11} = \mathbf{f}_{11}\lambda_{11}$, where, of course, $\mathbf{f}_{11}$ is scaled so that $\mathbf{f}_{11}'\mathbf{f}_{11} = \lambda_{11}$. For a first estimate of $\mathbf{U}^2$, we take the diagonal elements of $\mathbf{R} - \mathbf{f}_{11}\mathbf{f}_{11}'$ (the first residual correlation matrix). Let us call this diagonal matrix $\mathbf{U}_{11}{}^2$. Then our first canonical factor solution will obey

$$\mathbf{U}_{11}{}^{-1}(\mathbf{R} - \mathbf{U}_{11}{}^2)\mathbf{U}_{11}{}^{-1}\mathbf{q}_{11} = \mathbf{q}_{11}\alpha_{11}, \tag{15.28}$$

which equation is an application of (15.14). Here α_{11} is the largest eigenvalue; $\mathbf{q}_{11}$ is normalized to one; and $\mathbf{U}_{11}\mathbf{q}_{11}\alpha_{11}{}^{\frac{1}{2}}$ is the subsequent estimate $\mathbf{f}_{12}$ of the first factor.

We then repeat the procedure on the basis of $\mathbf{f}_{12}$. Again we estimate $\mathbf{U}_{12}{}^2 =$ diagonal $(\mathbf{R} - \mathbf{f}_{12}\mathbf{f}_{12}')$, and we find a new estimate $\mathbf{f}_{13}$ from

$$\mathbf{U}_{12}{}^{-1}(\mathbf{R} - \mathbf{U}_{12}{}^2)\mathbf{U}_{12}{}^{-1}\mathbf{q}_{12} = \mathbf{q}_{12}\alpha_{12} \tag{15.29}$$

and

$$\mathbf{f}_{13} = \mathbf{U}_{12}\mathbf{q}_{12}\alpha_{12}{}^{\frac{1}{2}}. \tag{15.30}$$

In this way we continue until finally the difference between the ith estimate $\mathbf{f}_{1i}$ and the $(i+1)^{\text{th}}$ estimate $\mathbf{f}_{1,i+1}$ is negligible.

Then the process is widened to a two-factor solution. First, we calculate the residual matrix $\mathbf{R}_1 = \mathbf{R} - \mathbf{f}_{1i}\mathbf{f}_{1i}'$. As a first estimate of the second factor, we take the eigenvector of this residual matrix that has the largest eigenvalue:

$$\mathbf{R}_1\mathbf{f}_{21} = \mathbf{f}_{21}\lambda_{21}.$$

We then form a provisional factor matrix $\mathbf{F}_{21}$ with $\mathbf{f}_{1i}$ and $\mathbf{f}_{21}$ as its columns. A new estimate of $\mathbf{U}^2$ is found from

$$\mathbf{U}_{21}{}^2 = \text{diagonal } (\mathbf{R} - \mathbf{F}_{21}\mathbf{F}_{21}'),$$

and the canonical equation can be written as

$$\mathbf{U}_{21}{}^{-1}(\mathbf{R} - \mathbf{U}_{21}{}^2)\mathbf{U}_{21}{}^{-1}\mathbf{Q}_{21} = \mathbf{Q}_{21}\mathbf{A}_{21}.$$

A next estimate of the factor matrix becomes

$$\mathbf{F}_{22} = \mathbf{U}_{21}\mathbf{Q}_{21}\mathbf{A}_{21}^{\frac{1}{2}}. \tag{15.31}$$

The subsequent estimate of $\mathbf{U}^2$ is then

$$\mathbf{U}_{22}^2 = \text{diagonal } (\mathbf{R} - \mathbf{F}_{22}\mathbf{F}_{22}'),$$

and we derive a new estimate $\mathbf{F}_{23}$ from the canonical equation. Again, we continue until the estimate of the two-factor solution has converged sufficiently.

The process is then continued by extending to a three-factor solution (if we want to go that far). Details are similar to the extension to the two-factor solution, and can be generated by the reader. The analysis would continue in this manner until p factors are drawn.

This number p is either determined beforehand, or the iterative process is stopped because it fails to converge if it is extended to more factors. One main advantage of the maximum likelihood method is that, if p is fixed beforehand, the solution can be tested for goodness of fit by a chi-square test, for details of which, see the literature (e.g., Lawley and Maxwell 1963, Morrison 1967).

15.2.3. Numerical Illustration of Canonical Factor Analysis. For an illustration, we take a matrix of correlations between five sociological variables, the same variables that were used in section 12.3. However, the correlation matrix here is slightly different from that in section 12.3. The latter gave the actual values of correlations as found by Blau and Duncan; here the correlations have been changed a little, in order to give us an opportunity later to demonstrate an interesting point.

Table 15.1. Modified correlation matrix for five sociological variables (to be compared to table 12.1).

	x_1	x_2	x_3	x_4	x_5
x_1	.810	.516	.453	.328	.339
x_2		.968	.438	.417	.405
x_3			.928	.538	.595
x_4				.837	.541
x_5					.950

Table 15.2. Canonical factor matrix for reduced correlation matrix given in table 15.10.

	I	II	III	IV
x_1	.611	.122	−.321	.564
x_2	.942	−.280	.004	−.044
x_3	.660	.676	−.131	−.134
x_4	.567	.393	.566	.205
x_5	.543	.380	.144	.012

The matrix is given in table 15.1. Note that diagonal elements are not unity, but estimates of communalities (how they were arrived at will be explained later in a different context; see section 17.2). The canonical factor matrix is given in table 15.2, the matrix of eigenvectors $\mathbf{Q}$ (normalized to one) and corresponding eigenvalues in table 15.3. Figure 15.1 gives a diagram in the spirit of path analysis. An interpretation of the factors might be as follows. The first one indicates that all variables measure the same underlying variable ("family status"? "social stratification"?), with the father's occupational status being the best indicator. The second factor reflects something like "son's status." The third is somewhat specific for the son's first job, and the fourth for the father's education.

There is a special problem about the fact that $\mathbf{q}_5$ in table 15.3 has eigenvalue 1, which is why the factor solution has four factors only; in formula (15.17) the corresponding column vanishes. We will come back to this point in section 17.1.

Table 15.3. Eigenvectors normalized to one, and eigenvalues of matrix $\mathbf{U}^{-1}\mathbf{R}\mathbf{U}^{-1}$. To obtain eigenvalues of matrix $\mathbf{U}^{-1}\mathbf{R} * \mathbf{U}^{-1}$, subtract 1.000 from given values.

	q_1	q_2	q_3	q_4	q_5
x_1	.226	.088	−.442	.864	.001
x_2	.852	−.495	.013	−.166	−.021
x_3	.396	.792	−.291	−.330	−.142
x_4	.226	.306	.840	.340	−.184
x_5	.119	.163	.118	.011	.972
eigenvalue	39.573	11.120	3.789	3.240	1.000

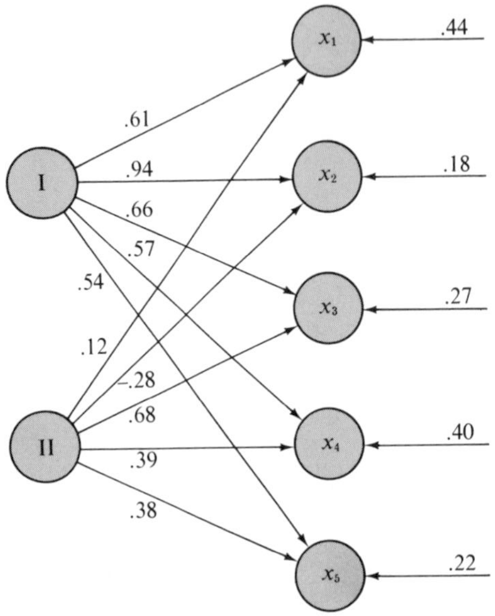

Figure 15.1 Path diagram for canonical factor analysis, first two factors, five sociological variables.

15.2.4. Formal Treatment of Canonical Discriminant Factor Analysis. A special application of canonical factor analysis obtains if the observations can be partitioned by rows into "blocks," where observations in the same block share some common characteristic. To illustrate the idea, imagine a study in the physics of speech: 15 different vowels are recorded for each of 10 different speakers. Each recorded sound is then analyzed in 18 measurements (e.g., the total energy for each of 18 bands of the frequency spectrum). The result in a 150 × 18 matrix of observations that can be partitioned into 15 submatrices, one for each vowel. Geometrically, we have 150 points in an S_{18}. These points are grouped into 15 subgroups, each of 10 points for the same vowel.

For an example like this, the main interest will be in how the vowels are differentiated by subgroup means. Differences between speakers for the same vowel then can be treated as "error."

In the following, I will demonstrate that this situation can be handled in the same way as in the canonical factor problem. For each observation x_{ijk} (the kth observation in subgroup j on variable i; x_{ijk} is measured as a deviation from the over-all mean for variable i), we can write:

$$x_{ijk} = m_{ij} + e_{ijk},$$

where m_{ij} is the mean of variable i within subgroup j, and where e_{ijk} is an error component. This implies, for the total matrix **X**,

$$\mathbf{X} = \mathbf{M} + \mathbf{E}, \tag{15.32}$$

where the rows in **M** are identical within subgroups. It follows that the matrix of sums of squares and cross-products $\mathbf{T} = \mathbf{X}'\mathbf{X}$ can be decomposed according to

$$\mathbf{T} = \mathbf{X}'\mathbf{X} = (\mathbf{M} + \mathbf{E})'(\mathbf{M} + \mathbf{E}) = \mathbf{M}'\mathbf{M} + \mathbf{E}'\mathbf{E} + \mathbf{M}'\mathbf{E} + \mathbf{E}'\mathbf{M}. \tag{15.33}$$

We shall assume that $\mathbf{M}'\mathbf{E} = \mathbf{0}$, i.e., that error components are uncorrelated with subgroup means. Equation (15.33) then reduces to

$$\mathbf{T} = \mathbf{M}'\mathbf{M} + \mathbf{E}'\mathbf{E} = \mathbf{B} + \mathbf{W}, \tag{15.34}$$

where we use the symbol **B** for the portion dependent on differences *between* groups, and the symbol **W** for the portion dependent on differences *within* groups. $\mathbf{M}'\mathbf{M}$ could be simplified to $\sum n_j \bar{\mathbf{x}}_j \bar{\mathbf{x}}_j{}'$, where n_j is the number of observations in block j, and $\bar{\mathbf{x}}_j$ the vector of means for this block.

The objective is to find weights **c** and **d** such that the compounds **Xc** and **Md** have maximum correlation. But, as we saw in section 15.2.1, it is an equivalent procedure to transform **X** in such a way that the error portion becomes "spherical," and then to apply a principal-components analysis to the systematic portion. The obvious transformation to spherical error is $\mathbf{X}\mathbf{W}^{-\frac{1}{2}}$. We then have, for the matrix of sums of squares and cross products,

$$\begin{aligned}(\mathbf{X}\mathbf{W}^{-\frac{1}{2}})'(\mathbf{X}\mathbf{W}^{-\frac{1}{2}}) &= \mathbf{W}^{-\frac{1}{2}}\mathbf{X}'\mathbf{X}\mathbf{W}^{-\frac{1}{2}} = \mathbf{W}^{-\frac{1}{2}}\mathbf{M}'\mathbf{M}\mathbf{W}^{-\frac{1}{2}} + \mathbf{W}^{-\frac{1}{2}}\mathbf{E}'\mathbf{E}\mathbf{W}^{-\frac{1}{2}} \\ &= \mathbf{W}^{-\frac{1}{2}}\mathbf{B}\mathbf{W}^{-\frac{1}{2}} + \mathbf{I},\end{aligned} \tag{15.35}$$

so that, in fact, the portion of the matrix dependent on error becomes the unit matrix. For the next step, we use the equivalent of equation (15.14),

$$\mathbf{W}^{-\frac{1}{2}}\mathbf{B}\mathbf{W}^{-\frac{1}{2}}\mathbf{Q} = \mathbf{Q}\Lambda, \tag{15.36}$$

with **Q** normalized to unity, so that $\mathbf{Q}'\mathbf{Q} = \mathbf{I}$ and $\mathbf{Q}\mathbf{Q}' = \mathbf{I}$. It follows that

$$\mathbf{B} = \mathbf{W}^{\frac{1}{2}}\mathbf{Q}\Lambda\mathbf{Q}'\mathbf{W}^{\frac{1}{2}}, \tag{15.37}$$

and that

$$\mathbf{F}_B = \mathbf{W}^{\frac{1}{2}}\mathbf{Q}\Lambda^{\frac{1}{2}} \tag{15.38}$$

is a matrix of loadings on the canonical components. To complete the picture, we note that we can also give a matrix of loadings for **T**. Since

$$\mathbf{T} = \mathbf{B} + \mathbf{W} = \mathbf{W}^{\frac{1}{2}}\mathbf{Q}\Lambda\mathbf{Q}'\mathbf{W}^{\frac{1}{2}} + \mathbf{W} = \mathbf{W}^{\frac{1}{2}}(\mathbf{Q}\Lambda\mathbf{Q}' + \mathbf{I})\mathbf{W}^{\frac{1}{2}}$$

$$= \mathbf{W}^{\frac{1}{2}}(\mathbf{Q}\Lambda\mathbf{Q}' + \mathbf{Q}\mathbf{Q}')\mathbf{W}^{\frac{1}{2}} = \mathbf{W}^{\frac{1}{2}}\mathbf{Q}(\Lambda + \mathbf{I})\mathbf{Q}'\mathbf{W}^{\frac{1}{2}}, \quad (15.39)$$

it follows directly that

$$\mathbf{F}_T = \mathbf{W}^{\frac{1}{2}}\mathbf{Q}(\Lambda + \mathbf{I})^{\frac{1}{2}} \quad (15.40)$$

is a matrix of loadings corresponding to **T**.

If we want factor matrices, we should use the fact that the total correlation matrix $\mathbf{R}_T$ can be written as

$$\mathbf{R}_T = \mathbf{D}_T^{-\frac{1}{2}}\mathbf{T}\mathbf{D}_T^{-\frac{1}{2}}, \quad (15.41)$$

where $\mathbf{D}_T$ is a diagonal matrix in which the diagonal elements are identical to the diagonal elements of **T** itself. It follows that

$$\mathbf{G}_T = \mathbf{D}_T^{-\frac{1}{2}}\mathbf{F}_T = \mathbf{D}_T^{-\frac{1}{2}}\mathbf{W}^{\frac{1}{2}}\mathbf{Q}(\Lambda + \mathbf{I})^{\frac{1}{2}} \quad (15.42)$$

is a factor matrix for $\mathbf{R}_T$. In the same manner we could easily obtain a factor matrix for

$$\mathbf{R}_B = \mathbf{D}_B^{-\frac{1}{2}}\mathbf{B}\mathbf{D}_B^{-\frac{1}{2}};$$

it is

$$\mathbf{G}_B = \mathbf{D}_B^{-\frac{1}{2}}\mathbf{F}_B = \mathbf{D}_B^{-\frac{1}{2}}\mathbf{W}^{\frac{1}{2}}\mathbf{Q}\Lambda^{\frac{1}{2}}. \quad (15.43)$$

15.2.5. Numerical Illustration. For a numerical illustration, we use artificial data, where it may be imagined that they refer to the analysis of five vowels, each spoken by three different speakers, and measured for four spectral bands. The corresponding 15 × 4 matrix is given in table 15.4, from which we calculate the matrix **T** of sums of squares and cross-products of deviations from the means. We find

$$\mathbf{T} = \begin{pmatrix} 3{,}246 & 716 & 728 & 772 \\ 716 & 2{,}538 & 1{,}739 & 55 \\ 728 & 1{,}739 & 2{,}850 & -1{,}185 \\ 772 & 55 & -1{,}185 & 2{,}060 \end{pmatrix}.$$

Table 15.4. Hypothetical measurements on five different vowels in four different variables.

x_1	x_2	x_3	x_4
26	9	4	19
10	14	−5	3
11	26	11	8
15	−10	−12	13
8	−12	−14	9
−2	5	−7	4
−7	−23	−21	3
−14	−2	−1	16
−23	−5	−12	2
−4	−9	−6	−8
−11	−21	−10	−13
−29	11	8	−9
12	5	17	−11
12	3	30	−16
−4	9	18	−20
Block totals:			
47	49	10	30
21	−17	−33	26
−44	−30	−34	21
−44	−19	−8	−30
20	17	65	−47

The portion **B** can be calculated from $\mathbf{B} = \sum n_j \bar{\mathbf{x}}_j \bar{\mathbf{x}}_j'$, and is found to be

$$\mathbf{B} = \begin{pmatrix} 2{,}307.33 & 1{,}480.67 & 975 & 470.67 \\ 1{,}480.67 & 1{,}413.33 & 1{,}109.33 & 56.33 \\ 975 & 1{,}109.33 & 2{,}211.33 & -1{,}362.33 \\ 470.67 & 56.33 & -1{,}362.33 & 1{,}708.67 \end{pmatrix}.$$

Finally, **W** can be found indirectly by subtracting **B** from **T**, or directly by adding matrices $\mathbf{X}_j'\mathbf{X}_j - n_j \bar{\mathbf{x}}_j \bar{\mathbf{x}}_j'$, where $\mathbf{X}_j$ is the partition of **X** corresponding to block j. The result is

$$\mathbf{W} = \begin{pmatrix} 938.67 & -764.67 & -247 & 301.33 \\ -764.67 & 1{,}124.67 & 629.67 & -1.33 \\ -247 & 629.67 & 638.67 & 177.33 \\ 301.33 & -1.33 & 177.33 & 351.33 \end{pmatrix}.$$

We then calculate

$$\mathbf{W}^{-\frac{1}{2}} = \begin{pmatrix} .0647 & .0339 & -.0037 & -.0316 \\ .0339 & .0639 & -.0282 & -.0116 \\ -.0037 & -.0282 & .0647 & -.0157 \\ -.0316 & -.0116 & -.0157 & .0815 \end{pmatrix}$$

and

$$\mathbf{W}^{-\frac{1}{2}}\mathbf{T}\mathbf{W}^{-\frac{1}{2}} = \begin{pmatrix} 17.408 & 11.316 & 4.474 & -10.803 \\ 11.316 & 10.621 & -.466 & -3.475 \\ 4.475 & -.466 & 10.503 & -12.150 \\ -10.803 & -3.475 & -12.150 & 18.815 \end{pmatrix}.$$

The eigenvectors of the latter matrix are, normalized to one,

$$\mathbf{Q} = \begin{pmatrix} -.595 & .451 & .311 & -.588 \\ -.322 & .595 & -.102 & .729 \\ -.374 & -.513 & .689 & .350 \\ .635 & .423 & .646 & .025 \end{pmatrix},$$

with eigenvalues $(\Lambda + \mathbf{I})$ equal to 37.860, 17.129, 1.199, and 1.159, respectively.

A next step might be to calculate the between-groups correlation matrix. It can be found that

$$\mathbf{R_B} = \begin{pmatrix} 1.00 & .82 & .43 & .24 \\ .82 & 1.00 & .63 & .04 \\ .43 & .63 & 1.00 & -.70 \\ .24 & .04 & -.70 & 1.00 \end{pmatrix}.$$

The corresponding factor matrix $\mathbf{G} = \mathbf{D_B}^{-\frac{1}{2}}\mathbf{W}^{\frac{1}{2}}\mathbf{Q}\Lambda^{\frac{1}{2}}$ is given in table 15.5, which shows that the first two factors explain most of the variance, and that we might neglect the last two factors.

Finally, for this example, we have a look at the transformed matrix **XC**. First, the matrix of weights **C** can be found from $\mathbf{C} = \mathbf{W}^{-\frac{1}{2}}\mathbf{Q}$:

$$\mathbf{C} = \begin{pmatrix} -.068 & .038 & -.006 & -.015 \\ -.038 & .063 & -.023 & .017 \\ -.023 & -.058 & .036 & .004 \\ .080 & .021 & .033 & .007 \end{pmatrix}.$$

Table 15.5. Factor matrix for between-groups correlations, with amount of total variance explained by each factor.

	I	II	III	IV
x_1	−.677	.693	.118	−.216
x_2	−.765	.541	.020	.347
x_3	−.937	−.265	.152	.151
x_4	.553	.827	.175	−.016
	2.206	1.527	.068	.190

Note that **C** is not simply a rotation matrix, but also implies a differential change of scale in order to make error components spherical. Since only the first two factors were acceptable, we have calculated values for $\mathbf{v}_1 = \mathbf{X}\mathbf{c}_1$ and $\mathbf{v}_2 = \mathbf{X}\mathbf{c}_2$; they are given in table 15.6. The results are pictured in figure 15.2, in which note the circular error: individual points for a vowel scatter around the mean with equal spread in all directions.

Table 15.6. Transformed variables, obtained from table 15.4. Within-groups variance is spherical (within-groups sums of squares add up to one).

$\mathbf{v}_1$	$\mathbf{v}_2$
−.677	1.725
−.851	1.615
−1.336	1.583
.671	.916
.948	.556
.429	.732
2.061	−.425
2.333	−.256
2.188	−.444
.107	−.539
.725	−1.433
.656	−1.065
−2.275	−.456
−2.898	−1.447
−2.080	−1.062

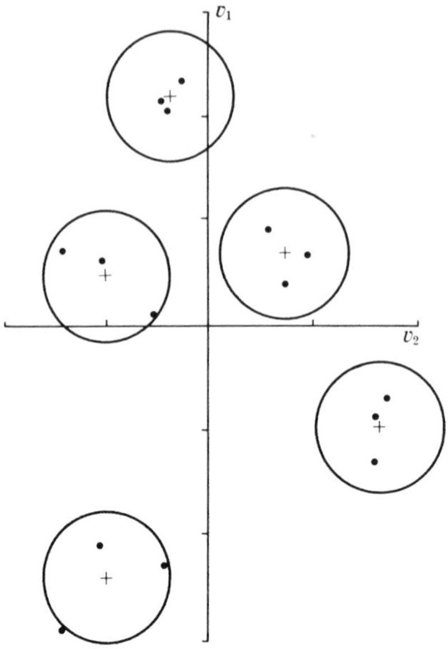

Figure 15.2 Graph of projections given in table 15.6. Group averages are indicated by +. Circles are drawn with radius equal to two standard deviations (standard deviation is the square root of within-groups variance).

15.2.6. Another Interpretation. There is another way of looking at the procedure. Suppose, in fact, that we apply a transformation vector **k** to **X**. The result is a vector of transformed unobserved scores **y**, say, **y** = **Xk**. The corresponding sum of squares **y′y** can be divided into a within and a between portion, for it is easily seen that

$$\mathbf{y}'\mathbf{y} = \mathbf{k}'\mathbf{X}'\mathbf{X}\mathbf{k} = \mathbf{k}'\mathbf{T}\mathbf{k} = \mathbf{k}'(\mathbf{B} + \mathbf{W})\mathbf{k} = \mathbf{k}'\mathbf{B}\mathbf{k} + \mathbf{k}'\mathbf{W}\mathbf{k}. \quad (15.44)$$

Now **k′Bk** represents variations in **y** that can be attributed to differences between means, and **k′Wk** is "error" variation. Suppose we want to maximize the first portion relative to the last; we write

$$\alpha = \mathbf{k}'\mathbf{B}\mathbf{k}/\mathbf{k}'\mathbf{W}\mathbf{k}, \quad (15.45)$$

so that we want a maximum for α. The result will be the same as when we take a maximum for

$$\log \alpha = \log (\mathbf{k}'\mathbf{B}\mathbf{k}) - \log (\mathbf{k}'\mathbf{W}\mathbf{k}).$$

Partial derivatives with respect to **k** are

$$\partial \log \alpha/\partial \mathbf{k} = 2\mathbf{B}\mathbf{k}/\mathbf{k}'\mathbf{B}\mathbf{k} - 2\mathbf{W}\mathbf{k}/\mathbf{k}'\mathbf{W}\mathbf{k}, \quad (15.46)$$

where we use the rule from calculus that the derivative of log $(f(x))$ equals $f'(x)/f(x)$.

If we set the partial derivatives equal to zero, and multiply by $\mathbf{k'Bk}$, we find

$$\mathbf{2Bk - 2Wk(k'Bk)/k'Wk = 0}, \tag{15.47}$$

but $\mathbf{k'Bk/k'Wk} = \alpha$; it follows that

$$\mathbf{Bk} - \alpha\mathbf{Wk} = \mathbf{0}. \tag{15.48}$$

Note, however, that this last equation is essentially the same as (15.10), and therefore implies the equivalent of (15.14), which is equation (15.36). This result demonstrates that canonical factor analysis might as well have been developed from the principle that we maximize variation between blocks with respect to variation within blocks. This explains why the technique is called *canonical "discriminant" factor analysis:* it is a technique for identifying factors that discriminate optimally between block averages relative to within-blocks scatter. This interpretation of the result essentially says that we normalize error variance to identical value in all directions (i.e., we make error variance "spherical"), and then look for the directions with successive maxima for between-blocks variance. The solution is nothing but a principal-components solution applied after transformation to spherical-error variance.

15.3. Alpha-Factor Analysis

15.3.1. Formal Treatment. Alpha-factor analysis is inspired by considerations about the generalizability of results (Kaiser and Caffrey 1965). It looks on the observed variables as if they were a sample from a universe of variables, and the criterion for the factor solution is to achieve maximum correspondence between factors derived from the sample and factors that would be obtained from the universe. The easiest way to start is to look on the problem as a special regression problem. Sample factors are then variables subject to error, so that their variance can be divided into two portions: one systematic portion explained by universe factor variance, and another portion that must be attributed to error. Formula (10.10) then becomes relevant. There the variance $\sigma_2{}^2$ is divided into a portion $\sigma^2_{2.1}$ for scatter about regression or error variance, and a portion $(\text{cov})^2/\sigma_1{}^2$ for variance due to regression. For a relative measure of variance due to regression, we have to divide both sides in formula (10.10) by σ_2^2; the measure for relative variance due to regression becomes $(\text{cov})^2/\sigma_1^2 \cdot \sigma_2^2$. This value is called α in alpha-factor analysis, and the objective is to select factors for which α is maximized.

Now let

$$\mathbf{y} = \mathbf{X}\mathbf{w} \tag{15.49}$$

express factor scores as estimated from observed variables, with $\mathbf{X}$ for the matrix of observed values (order $n \times m$), and $\mathbf{w}$ a vector of weights (order $m \times 1$). Corresponding to (15.49), is an equation

$$\boldsymbol{\eta} = \chi\omega, \tag{15.50}$$

which gives the same for the universe of variables. Here χ has order $n \times \infty$, and ω is of order $\infty \times 1$. We want to maximize α for the pair $\mathbf{y}$, $\boldsymbol{\eta}$. The expression for α is

$$\alpha = \frac{(\mathbf{y}'\boldsymbol{\eta}/)/n^2}{(\mathbf{y}'\mathbf{y}) \cdot (\boldsymbol{\eta}'\boldsymbol{\eta})/n^2} = \frac{(\mathbf{w}'\mathbf{X}'\chi\omega)^2}{(\mathbf{w}'\mathbf{X}'\mathbf{X}\mathbf{w}) \cdot (\omega'\chi'\chi\omega)}\,. \tag{15.51}$$

Here $\mathbf{w}'\mathbf{X}'\mathbf{X}\mathbf{w}/n$ can be estimated from $\mathbf{w}'\mathbf{R}^*\mathbf{w}$, where $\mathbf{R}^*$ is the reduced correlation matrix $\mathbf{R}^* = \mathbf{R} - \mathbf{U}^2$. However, we do not have a similar expression for terms with the universe variables in them. All we can do is to approximate then somehow, on the basis of reasonable assumptions. First, let us investigate these assumptions.

The term $\mathbf{w}'\mathbf{X}'\chi\omega/n$ can be regarded as a sum of $m \times \infty$ terms, each of which is a covariance (an element of $\mathbf{X}'\chi/n$) multiplied by appropriate weights. Let $\overline{\mathbf{w}'\mathbf{X}\chi\omega}$ stand for the average of these terms. A similar expression $\overline{\omega'\chi'\chi\omega}$ stands for the average of such elements in the matrix $\mathbf{w}'\chi'\chi\omega$. If we now look at $\mathbf{w}'\mathbf{X}'\mathbf{X}\mathbf{w}$, this form also contains terms with weighted covariances in them, insofar as we look at off-diagonal elements. However, diagonal elements in $\mathbf{X}'\mathbf{X}/n$ refer to sample variances. Now let $\overline{\mathbf{w}'\mathbf{X}'\mathbf{X}\mathbf{w}}$ stand for the average taken over only the $m(m-1)$ covariance terms. The assumption is then

$$\left(\overline{\mathbf{w}'\mathbf{X}'\chi\omega}\right)^2 = \left(\overline{\omega'\chi'\chi\omega}\right) \cdot \left(\overline{\mathbf{w}'\mathbf{X}'\mathbf{X}\mathbf{w}}\right), \tag{15.52}$$

which says, in words, that the covariance between a sample variable and a universe variable is, on the average, equal to the harmonic mean of sample covariance and universe covariance. We then may write

$$\frac{(\mathbf{w}'\mathbf{X}'\chi\omega)^2}{\omega'\chi'\chi\omega} = \frac{(m \times \infty)^2\left(\overline{\mathbf{w}'\mathbf{X}'\chi\omega}\right)^2}{(\infty)^2\left(\overline{\omega'\chi'\chi\omega}\right)} = m^2 \cdot \frac{\left(\overline{\mathbf{w}'\mathbf{X}'\chi\omega}\right)^2}{\left(\overline{\omega'\chi'\chi\omega}\right)} = m^2 \cdot \left(\overline{\mathbf{w}'\mathbf{X}'\mathbf{X}\mathbf{w}}\right). \tag{15.53}$$

This can be substituted in (15.51), with the result that

$$\alpha = m^2 \frac{\left(\overline{\mathbf{w}'\mathbf{X}'\mathbf{X}\mathbf{w}}\right)}{\mathbf{w}'\mathbf{X}'\mathbf{X}\mathbf{w}}. \tag{15.54}$$

However, since $\overline{\mathbf{w}'\mathbf{X}'\mathbf{X}\mathbf{w}}$ averages over the $m(m-1)$ off-diagonal cells only, we may write

$$\overline{\mathbf{w}'\mathbf{X}'\mathbf{X}\mathbf{w}} = (\mathbf{w}'\mathbf{R}^*\mathbf{w} - \mathbf{w}'\mathbf{H}^2\mathbf{w})/m(m-1), \tag{15.55}$$

where $\mathbf{H}^2$ stands for the diagonal matrix of communalities $h_i^2 = 1 - u_i^2$. Note, therefore, that the form in the numerator in the right side of (15.55) represents a matrix with zero elements on the diagonal. Substituting (15.55) in (15.54), we have

$$\alpha = \frac{m}{m-1} \cdot \frac{\mathbf{w}'\mathbf{R}^*\mathbf{w} - \mathbf{w}'\mathbf{H}^2\mathbf{w}}{\mathbf{w}'\mathbf{R}^*\mathbf{w}} = \frac{m}{m-1}\left(1 - \frac{\mathbf{w}'\mathbf{H}^2\mathbf{w}}{\mathbf{w}'\mathbf{R}^*\mathbf{w}}\right). \tag{15.56}$$

A maximum for α will be found if we minimize $\mathbf{w}'\mathbf{H}^2\mathbf{w}/\mathbf{w}'\mathbf{R}^*\mathbf{w}$, or if we determine a maximum for the inverse of the latter expression, which is $\mathbf{w}'\mathbf{R}^*\mathbf{w}/\mathbf{w}'\mathbf{H}^2\mathbf{w}$. Call this γ; the objective is to find a maximum for γ, which will also be a maximum for $\log \gamma$. We can therefore say that the objective is to maximize

$$\log \gamma = \log (\mathbf{w}'\mathbf{R}^*\mathbf{w}) - \log (\mathbf{w}'\mathbf{H}^2\mathbf{w}). \tag{15.57}$$

Partial derivatives are

$$\partial(\log \gamma)/\partial \mathbf{w}' = 2\mathbf{R}^*\mathbf{w}/\mathbf{w}'\mathbf{R}^*\mathbf{w} - 2\mathbf{H}^2\mathbf{w}/\mathbf{w}'\mathbf{H}^2\mathbf{w}. \tag{15.58}$$

The expression on the right side of (15.58) must be set equal to zero. It follows that

$$\mathbf{R}^*\mathbf{w}/\mathbf{w}'\mathbf{R}^*\mathbf{w} = \mathbf{H}^2\mathbf{w}/\mathbf{w}'\mathbf{H}^2\mathbf{w} \tag{15.59}$$

or, if we multiply both sides by $\mathbf{w}'\mathbf{R}^*\mathbf{w}$,

$$\mathbf{R}^*\mathbf{w} = \frac{\mathbf{w}'\mathbf{R}^*\mathbf{w}}{\mathbf{w}'\mathbf{H}^2\mathbf{w}} \cdot \mathbf{H}^2\mathbf{w} = \gamma\mathbf{H}^2\mathbf{w}. \tag{15.60}$$

Equation (15.60) can be transformed into an ordinary eigenvector equation by using the same trick as for equation (14.13). Let $\mathbf{q} = \mathbf{H}\mathbf{w}$. We then have $\mathbf{R}^*\mathbf{H}^{-1}\mathbf{q} = \gamma\mathbf{H}\mathbf{q}$, or

$$\mathbf{H}^{-1}\mathbf{R}^*\mathbf{H}^{-1}\mathbf{q} = \gamma\mathbf{q} \tag{15.61}$$

which can be solved for the eigenvectors $\mathbf{q}$ with corresponding eigenvalues γ. In general, we have m solutions, and if $\mathbf{Q}$ is the eigenvector matrix and $\mathbf{\Gamma}$ the corresponding diagonal matrix of eigenvalues, (15.61) can be generalized to

$$\mathbf{H}^{-1}\mathbf{R}^*\mathbf{H}^{-1}\mathbf{Q} = \mathbf{Q}\mathbf{\Gamma}, \tag{15.62}$$

where we assume that $\mathbf{Q}$ is normalized to one. It follows immediately that

$$\mathbf{R}^* = \mathbf{H}\mathbf{Q}\mathbf{\Gamma}\mathbf{Q}'\mathbf{H}, \tag{15.63}$$

and that a factor matrix $\mathbf{F}$, for which it must be true that

$$\mathbf{R}^* = \mathbf{F}\mathbf{F}',$$

can be identified as

$$\mathbf{F} = \mathbf{H}\mathbf{Q}\mathbf{\Gamma}^{\frac{1}{2}}. \tag{15.64}$$

Corresponding values for α are found from

$$\alpha = \frac{m}{m-1}\left(1 - \frac{1}{\gamma}\right). \tag{15.65}$$

15.3.2. Relation Between Alpha-Factor Analysis and Canonical Factor Analysis. We can understand better what alpha-factor analysis does by taking a fresh look at the matrix $\mathbf{H}^{-1}\mathbf{R}^*\mathbf{H}^{-1}$ in equation (15.48). The effect of premultiplying and postmultiplying by $\mathbf{H}^{-1}$ is that diagonal elements become equal to unity. Elements of $\mathbf{R}^*$ can be interpreted to represent $h_ih_j \cos(i, j)$, as we saw in section 3.1. So the elements in $\mathbf{H}^{-1}\mathbf{R}^*\mathbf{H}^{-1}$ just represent cosines of angles between vectors in the vector model; they are equal to $\cos(i, j)$, which is unity for $i = j$ on the diagonal. Or, in the vector model, the operation has the effect of extending vectors to unit length. It is easily recognized, then, that these vectors now represent the systematic components of the original variables after they have been standardized.

Their original variance was h_i^2, which has now been made equal to one. Therefore, elements of $\mathbf{H}^{-1}\mathbf{R}^*\mathbf{H}^{-1}$ can be interpreted as correlations between systematic components of the variables, after error has been eliminated from them.

To see this in detail, take the basic equation

$$\mathbf{X} = \mathbf{M} + \mathbf{E},$$

which was equation (15.2). Here **M** stands for the systematic components in **X** and **E** for the error components. For some correlation r_{ij}, we have

$$r_{ij} = \mathbf{x}_i'\mathbf{x}_j/n \qquad (i \neq j),$$

but $\mathbf{x}_i'\mathbf{x}_j = \mathbf{m}_i'\mathbf{m}_j + \mathbf{e}_i'\mathbf{e}_j = \mathbf{m}_i'\mathbf{m}_j$, since error variables are independent. The term $\mathbf{m}_i'\mathbf{m}_j/n$ represents the covariance between the systematic components m_i and m_j; their variances are $\mathbf{m}_i'\mathbf{m}_i/n = h_i^2$ and $\mathbf{m}_j'\mathbf{m}_j/n = h_j^2$. So, to obtain a correlation between systematic parts, we have to divide $\mathbf{m}_i'\mathbf{m}_j/n$ by the product of the two standard deviations, that is by h_ih_j. This operation is generalized by writing $\mathbf{H}^{-1}\mathbf{R}^*\mathbf{H}^{-1}$. Alpha-factor analysis comes down to factorizing this matrix.

A close analogy between alpha analysis and canonical factor analysis now appears. In canonical factor analysis, as we saw, the error components were made spherical by applying a transformation $\mathbf{U}^{-1}\mathbf{R}^*\mathbf{U}^{-1}$. Alpha analysis, on the other hand, uses a transformation by which the systematic components are made spherical. The effect, of course, is different. For one thing, the result of alpha analysis depends on m, the number of variables. If this number increases, then communalities h^2 will tend to increase, and the difference between **R** and **R*** becomes smaller. The longer the vectors are, the less it matters whether we standardize them; it follows that alpha analysis depends on the size of m. This is not true for canonical analysis, since the transformation $\mathbf{U}^{-1}\mathbf{R}^*\mathbf{U}^{-1}$, which makes the error parts spherical, does not depend so much on the absolute sizes of error variances, but on their relative sizes, and relative sizes may not at all be affected by m. On the other hand, relative sizes will be very much dependent on the sample size n. Therefore, where alpha analysis depends on m, canonical factor analysis depends on n.

Both analyses result in a factor matrix that is invariant under rescaling of the variables. We saw that this was one of the advantages of canonical factor analysis; the advantage is shared by alpha analysis. Also, alpha analysis assumes that values of u^2 are known or can be estimated. They might be estimated from a first ordinary principal-components analysis, in which case alpha analysis will be just another method for obtaining a rotated version of the principal-components factor matrix that satisfies certain criteria. We shall not give a numerical illustration.

15.4. Square-Root Factor Method

The square-root factor method was rather popular in days when fast computers were not yet available. Its distinctive feature is that the numerical calculations are relatively easy and straightforward, and do not require the solution of an eigenvector problem, say, which is a troublesome and time-consuming process if it has to be done with desk calculators. Recently, therefore, the square-root method has more or less lost its attractiveness, since hand calculation is no longer required. However, as we shall see, the method has certain similarities to path analysis, and the objective of this section is to show these similarities in detail.

15.4.1. Formal Treatment. The formal treatment does not require heavy mathematical machinery. Actually, the square-root solution of factor analysis imposes the additional criterion that the factor matrix **F** should be triangular; i.e., it has zero elements above the diagonal. Since elements of a factor matrix can be interpreted as correlations between observed variables and unobserved latent variables called factors, the criterion implies that the first observed variable has zero correlation with all factors except the first, that the second observed variable has zero correlations with all factors except the first two, etc. This, of course, assumes that we select a certain order for the observed variables, since otherwise the factor solution is still not identified (we would have as many solutions as there are permutations for the m variables).

The technique is very simple. We shall illustrate how it can be applied to **R**, where it is assumed that the observed variables in **R** have the prescribed order. For the illustration, we take the five sociological variables used earlier for an illustration of path analysis (section 12.3). The matrix of correlations was

$$\mathbf{R} = \begin{pmatrix} 1.000 & .516 & .453 & .332 & .322 \\ .516 & 1.000 & .438 & .417 & .405 \\ .453 & .438 & 1.000 & .538 & .596 \\ .332 & .417 & .538 & 1.000 & .541 \\ .322 & .405 & .596 & .541 & 1.000 \end{pmatrix}.$$

For the first column of the factor matrix $\mathbf{f}_1$, take the first column of $\mathbf{R}$:

$$\mathbf{f}_1 = \begin{pmatrix} 1.000 \\ .516 \\ .453 \\ .332 \\ .322 \end{pmatrix}.$$

It follows that $\mathbf{f}_1\mathbf{f}_1'$ gives the contribution of the first factor. We have

$$\mathbf{f}_1\mathbf{f}_1' = \begin{pmatrix} 1.000 & .516 & .453 & .332 & .322 \\ .516 & .266 & .234 & .171 & .166 \\ .453 & .234 & .205 & .150 & .146 \\ .332 & .171 & .150 & .110 & .107 \\ .322 & .166 & .146 & .107 & .104 \end{pmatrix},$$

and the first residual matrix is

$$\mathbf{R}_1 = \mathbf{R} - \mathbf{f}_1\mathbf{f}_1' = \begin{pmatrix} .000 & .000 & .000 & .000 & .000 \\ .000 & .734 & .204 & .246 & .239 \\ .000 & .204 & .795 & .388 & .450 \\ .000 & .246 & .388 & .890 & .434 \\ .000 & .239 & .450 & .434 & .896 \end{pmatrix},$$

where it should be noted that the first row (and column) of $\mathbf{R}_1$ has zero elements only. In fact, $\mathbf{f}_1$ was chosen to give this result. It follows that remaining factors are related to the last four variables only, since all correlations between x_1 and the others have been explained already.

The next step is to identify $\mathbf{f}_2$. The trick is $\mathbf{f}_2 = \mathbf{r}_{2;1}/(r_{22;1})^{\frac{1}{2}}$ where $\mathbf{r}_{2;1}$ is the second column of $\mathbf{R}_1$ and $r_{22;1}$ is the diagonal element in that column. The result is

$$\mathbf{f}_2 = \begin{pmatrix} .000 \\ .857 \\ .238 \\ .287 \\ .279 \end{pmatrix}.$$

The second residual matrix $\mathbf{R}_2$ now becomes

$$\mathbf{R}_2 = \mathbf{R}_1 - \mathbf{f}_2\mathbf{f}_2' = \begin{pmatrix} .000 & .000 & .000 & .000 & .000 \\ .000 & .000 & .000 & .000 & .000 \\ .000 & .000 & .738 & .319 & .384 \\ .000 & .000 & .319 & .807 & .354 \\ .000 & .000 & .384 & .354 & .819 \end{pmatrix},$$

with the first two rows (and columns) of the residual matrix equal to zero. It follows that further factors are related to the last three variables only. To find them, just repeat the process: take the next column of $\mathbf{R}_2$, and divide by the square root of the diagonal element to find the next column of the factor matrix. The result is

$$\mathbf{f}_3 = \begin{pmatrix} .000 \\ .000 \\ .859 \\ .372 \\ .447 \end{pmatrix}, \quad \mathbf{f}_4 = \begin{pmatrix} .000 \\ .000 \\ .000 \\ .818 \\ .229 \end{pmatrix}, \quad \mathbf{f}_5 = \begin{pmatrix} .000 \\ .000 \\ .000 \\ .000 \\ .752 \end{pmatrix},$$

and taken together, the five columns $\mathbf{f}$ can be collected in a triangular factor matrix $\mathbf{F}$.

Before going on, we might note some peculiarities of the process. First, elements of $\mathbf{R}_1$ can be shown to be equal to

$$r_{ij} - r_{1i} \cdot r_{1j},$$

but this is just the expression in the numerator of the partial-correlation formula (11.13). To find the partial correlation itself, we should still divide by the square root of $1 - r_{1i}^2$ and the square root of $1 - r_{1j}^2$. However, these two quantities are given in the corresponding diagonal elements of $\mathbf{R}_1$. So the result is: if we divide all elements in $\mathbf{R}_1$ by the square root of the diagonal element in the same row and in the same column, then the resulting quantities are partial correlations. Note that diagonal elements become one.

This also applies to the subsequent residual matrices. In fact, if in $\mathbf{R}_2$ we divide by the square root diagonal elements in the same column and same row, the result is the partial correlation $r_{ij \cdot 12}$. Therefore, the process is also very useful if we want to calculate such higher-order partial correlations.

Let us now look back at the factor solution. The basic equation associated with it is

$$\mathbf{X} = \mathbf{YF}', \tag{15.66}$$

where $\mathbf{F}'$ is used as a matrix of weights to be applied to factor scores in order to express the observed variables as linear functions of unobserved factors. The model assumes no error. It follows, if (15.67) is written out, that

$$\begin{aligned} \mathbf{x}_1 &= \mathbf{y}_1, \\ \mathbf{x}_2 &= f_{21}\mathbf{y}_1 + f_{22}\mathbf{y}_2, \\ \mathbf{x}_3 &= f_{31}\mathbf{y}_1 + f_{32}\mathbf{y}_2 + f_{33}\mathbf{y}_3, \\ &\text{etc.} \end{aligned} \tag{15.67}$$

This, in fact, is a solution in terms of path analysis, since each observed variable is expressed as a linear function of the unobserved "ultimate" variables y, which are independent among themselves. Note, however, that the set (15.67) can be rewritten in terms of observed variables. All we have to do is to substitute y_i from the expressions in earlier equations. For instance, in the second equation, replace $\mathbf{y}_1$ by $\mathbf{x}_1$ (as given in the first equation), and the result is

$$\mathbf{x}_2 = f_{21}\mathbf{x}_1 + f_{22}\mathbf{y}_2$$

Similarly, from the first two equations we can solve

$$\mathbf{y}_2 = (\mathbf{x}_2 - f_{21}\mathbf{y}_1)/f_{22} = (\mathbf{x}_2 - f_{21}\mathbf{x}_1)/f_{22},$$

and substitute in the third:

$$\begin{aligned} \mathbf{x}_3 &= f_{31}\mathbf{y}_1 + f_{32}\mathbf{y}_2 + f_{33}\mathbf{y}_3 = f_{31}\mathbf{x}_1 + f_{32}(\mathbf{x}_2 - f_{21}\mathbf{x}_1)/f_{22} + f_{33}\mathbf{y}_3 \\ &= (f_{31} - f_{32}f_{21}/f_{22})\mathbf{x}_1 + (f_{32}/f_{22})\mathbf{x}_2 + f_{33}\mathbf{y}_3, \end{aligned}$$

and $\mathbf{x}_3$ is expressed as a linear function of the preceding observed variables, with, in addition, a component $f_{33}\mathbf{y}_3$, where $\mathbf{y}_3$ is independent of the first two observed variables. This could be reformulated to read

$$\mathbf{x}_3 = b_{31}\mathbf{x}_1 + b_{32}\mathbf{x}_2 + f_{33}\mathbf{e}_3,$$

where we use the notation b for the new coefficients, and where we write $\mathbf{e}_3$ instead of $\mathbf{y}_3$. The equation is then of the same type as in set (12.9) for path analysis.

In fact, in the illustration, we have

$$x_2 = .516x_1 + .857e_2,$$
$$x_3 = .310x_1 + .278x_2 + .859e_3,$$

which is exactly the same as solution (12.15) for the path coefficients. This would also be true for the remaining equation for x_4 and x_5. The factors y that result from the square-root method are now seen to be nothing but the ultimate factors x_1 and the exogenous e-variables.

This could have been proven more directly. To see this, first note what the procedure described here implies. Given the relation

$$\mathbf{X} = \mathbf{YF}',$$

it follows that

$$\mathbf{X}_i = \mathbf{YF}_i', \tag{15.68}$$

$$\mathbf{x}_{i+1} = \mathbf{Yf}_{i+1}, \tag{15.69}$$

where $\mathbf{F}_i'$ is an $m \times i$ matrix obtained from $\mathbf{F}'$ by omitting the last $m - i$ columns, and $\mathbf{f}_{i+1}$ is the $(i + 1)^{\text{th}}$ column of $\mathbf{F}'$. However, the elements in the last $m - i$ rows of $\mathbf{F}_i'$ are all zero, since $\mathbf{F}$ is triangular. Let us write $\mathbf{F}_i{}^{*\prime}$ for the matrix of order $i \times i$ if we omit these zero rows. We then can replace (15.68) by

$$\mathbf{X}_i = \mathbf{Y}_i\mathbf{F}_i{}^{*\prime},$$

where $\mathbf{Y}_i$ has the first i columns of $\mathbf{Y}$. It follows that

$$\mathbf{Y}_i = \mathbf{X}_i\mathbf{F}_i^{*\prime}.^{-1} \tag{15.70}$$

Now for (15.69), we may write

$$\mathbf{x}_{i+1} = \mathbf{Y}_i\mathbf{f}_{i+1}^* + \mathbf{y}_{i+1}f_{i+1,i+1}, \tag{15.71}$$

where $\mathbf{f}_{i+1}^*$ stands for an $i \times i$ vector obtained from $\mathbf{f}_{i+1}$ by omitting the last $m - i$ elements. It follows that

$$\mathbf{x}_{i+1} = \mathbf{X}_i\mathbf{F}_i^{*\prime-1}\mathbf{f}_{i+1}^* + \mathbf{y}_{i+1}f_{i+1,i+1}, \tag{15.72}$$

where $\mathbf{x}_{i+1}$ now is expressed as a linear compound of the preceding i x-variables and its specific component $\mathbf{y}_{i+1}$.

We have to prove that the coefficients in (15.72) are the same as those for path analysis. In section 12.4 it was already proved that these coefficients are multiple-regression weights,

$$\mathbf{b}_{i+1} = \mathbf{R}_i^{-1}\mathbf{r}_{i+1}, \tag{15.73}$$

where $\mathbf{R}_i$ is the matrix of correlations for the first i x-variables, and $\mathbf{r}_{i+1}$ is the vector of correlations between variable x_{i+1} and the preceding i x-variables. So our task is to identify (15.73) with the first term on the right of (15.72). For this purpose, we use the fundamental theorem of factor analysis $\mathbf{R} = \mathbf{FF}'$, from which it follows that

$$\mathbf{R}_i = \mathbf{F}_i\mathbf{F}_i'$$

and

$$\mathbf{r}_{i+1} = \mathbf{F}_i\mathbf{f}_{i+1}.$$

However, given that in our special case $\mathbf{F}$ is triangular, so that the last $m - i$ columns of $\mathbf{F}_i$ contain zero's, this can be rewritten as

$$\mathbf{R}_i = \mathbf{F}_i^*\mathbf{F}_i^{*\prime},$$

$$\mathbf{r}_{i+1} = \mathbf{F}_i^*\mathbf{f}_{i+1}^*.$$

Substituting this in (15.73), we obtain

$$\mathbf{b}_{i+1} = (\mathbf{F}_i^*\mathbf{F}_i^{*\prime})^{-1}\mathbf{F}_i^*\mathbf{f}_{i+1}^* = \mathbf{F}_i^{*\prime-1}\mathbf{f}_{i+1}^*, \tag{15.74}$$

which completes the proof.

However, it is more insightful to look at the vector space. What the square-root solution does, actually, is to select the position of the reference factors in a special way. First, $\mathbf{y}_1$ is chosen to coincide with $\mathbf{x}_1$. The result is that the projections of the other x-vectors on $\mathbf{y}_1$ are equal to the correlations r_{1i}. Then the second factor $\mathbf{y}_2$ must be orthogonal to $\mathbf{x}_1$, and is therefore located in the hyperplane perpendicular to $\mathbf{x}_1$. It is selected to coincide with the projection of $\mathbf{x}_2$ on this hyperplane. (This selection explains, by the way, why the residual matrix $\mathbf{R}_1$ is related to partial correlations $r_{ij\cdot 1}$, since, as we saw in section 11.1.4, partial correlations are mapped as cosines of angles between projections in a plane orthogonal to $\mathbf{x}_1$.) It follows also that $\mathbf{x}_2$ is located in the plane through $\mathbf{y}_1$ and $\mathbf{y}_2$ and is therefore linearly dependent on $\mathbf{y}_1$ and $\mathbf{y}_2$; this makes it possible to express $\mathbf{x}_2$ as a linear function of $\mathbf{x}_1$ and $\mathbf{y}_2$. The reasoning must then be generalized to subsequent factors: $\mathbf{y}_3$ is chosen in the hyperplane perpendicular to both

$\mathbf{x}_1$ and $\mathbf{y}_2$, and in this hyperplane coincides with the projection of $\mathbf{x}_3$; reversely, $\mathbf{x}_3$ is located in the hyperplane through $\mathbf{x}_1$, $\mathbf{y}_2$, and $\mathbf{y}_3$; therefore, $\mathbf{x}_3$ can be expressed as a linear function of these three variables, or as a linear function of $\mathbf{x}_1$, $\mathbf{x}_2$, and $\mathbf{y}_3$; and so on.

It follows immediately that, for instance, the projection of $\mathbf{x}_3$ on the plane defined by $\mathbf{x}_1$ and $\mathbf{y}_2$ is equal to the multiple correlation between $\mathbf{x}_3$ and the combination $\mathbf{x}_1$, $\mathbf{x}_2$, since, as we saw, the multiple correlation is precisely mapped as this projection. Therefore the coefficient of $\mathbf{y}_3$ in the equation for $\mathbf{x}_3$ must be equal to the square root of the variance about regression; and similarly for subsequent equations. In fact, the square-root factor analysis is just this specific application of multiple-correlation analysis in order to identify factors.

15.4.2. Square-root Analysis of Variables Subject to Error. We will now take a short look at what should be done if we want to factorize $\mathbf{R}^*$ instead of $\mathbf{R}$, which is the appropriate procedure if estimates of error variance in each variable are available. Square-root analysis then proceeds in almost exactly the same way, the only difference being that, for the first factor $\mathbf{f}_1$ of the factor matrix $\mathbf{F}$, we take $\mathbf{r}_1/h_1$ instead of $\mathbf{r}_1$ itself; i.e., we take the first column of $\mathbf{R}^*$ and divide the elements by the square root of the diagonal element in it. This agrees completely with the rule mentioned earlier for subsequent factor columns. For instance, with

$$\mathbf{R}^* = \begin{pmatrix} .64 & .32 & .40 \\ .32 & .65 & .48 \\ .40 & .48 & .77 \end{pmatrix},$$

the factor matrix becomes

$$\mathbf{F} = \begin{pmatrix} .8 & 0 & 0 \\ .4 & .7 & 0 \\ .5 & .4 & .6 \end{pmatrix}.$$

In terms of path analysis, the difference from analysis of $\mathbf{R}$ is that in each equation we have, apart from the y's, also a genuine error term, e_i:

$$\mathbf{x}_1 = f_{11}\mathbf{y}_1 + \mathbf{e}_1,$$
$$\mathbf{x}_2 = f_{21}\mathbf{y}_1 + f_{22}\mathbf{y}_2 + \mathbf{e}_2,$$
$$\mathbf{x}_3 = f_{31}\mathbf{y}_1 + f_{32}\mathbf{y}_2 + f_{33}\mathbf{y}_3 + \mathbf{e}_3,$$

etc.

In general, the model would read

$$\mathbf{X} = \mathbf{Y}\mathbf{F}' + \mathbf{E},$$

with the y's independent, the e's also, and the y's uncorrelated with the e's. Again, we could eliminate, e.g., $\mathbf{y}_1$ from the second equation by substituting for it $\mathbf{y}_1 = (\mathbf{x}_1 - \mathbf{e}_1)/f_{11}$; this procedure seems much less attractive than in the errorless case, but the interpretation remains very much similar. Instead of taking $\mathbf{x}_1$ itself as an ultimate factor, we now take $\mathbf{y}_1$, its systematic component, only; similarly, $\mathbf{y}_2$ is the systematic component of $\mathbf{x}_2$ that is independent from $\mathbf{y}_1$; and so on. The additional components $\mathbf{e}_1$ form a set of new ultimate factors, which cannot be given an interpretation and therefore might be taken to stand for genuine error variance. However, the warning must be added that components $\mathbf{e}_i$ might be shown to contain systematic variance, too, if we add new variables to the study. For instance, in the example with five sociological variables, one of the variables is "son's education," which might be found to have a certain error component: imagine now that data for a new variable, "son's intelligence," become available and could be brought into the study; then it might appear that what first looked like error can be partly identified with the son's intelligence.

For the vector model, there is not much difference when compared to the analysis of **R**. Of course, in analyzing $\mathbf{R}^*$, we assume that vectors representing the variables have length h_i instead of unity. Also, residual matrices now cannot be so easily related to partial correlations. But, on the whole, the model is very much similar.

You should realize that the factor solution given by a square-root analysis is just a rotation of the principal-components solution. As we saw, a factor matrix may be rotated in infinitely many ways, and there is one rotation that makes a factor matrix triangular for a given order of the variables. Square-root analysis gives this solution directly, but one could as well start with the principal-components solution and rotate it to triangular form.

15.4.3. Square-root Factor Analysis and Path Analysis. In section 15.4.1 it was shown that the result of square-root analysis is identical to that of path analysis. In fact, the factors obtained from square-root analysis are the same as the ultimate variables assumed in path analysis.

This result might seem surprising, since in the literature, path analysis is often explicitly contrasted to factor analysis. It is then said that path analysis is a useful technique for identifying a causal model, but factor analysis is not. Apparently the close relation between path analysis and factor analysis is seldom recognized.

Duncan (1966, p. 16), for instance, concludes his introductory paper on path analysis with the remark: "An indoctrination in path analysis makes one skeptical of the claim that 'modern factor analysis' allows one to leave all the work to the computer." In a general sense, one could not agree more; blind and mindless application of factor analysis can never be recommended. But the citation at least suggests that factor analysis has undesirable features that are absent from path analysis. As we saw, path analysis is just a special variety of factor analysis, if one wishes to see it that way, and so has no special virtues of its own. One might argue perhaps, since square-root analysis requires a specific temporal order for the variables, that such a requirement is alien to factor analysis as such. In my opinion, however, the decision about the temporal order of the variables is trivial from the formal point of view, regardless of its importance to the researcher who wants to identify causal relations.

Blalock (1961, p. 169) notes:

"We should thus be explicitly aware of the fact that factor analysis techniques and the interpretation given to the factors extracted ordinarily presuppose certain limited kinds of causal models . . . . The mere fact that factor analysis provides a technique for substituting a small number of unmeasured variables for a large number of measured variables does not mean that it should be used indiscriminately whenever the number of variables becomes too large for adequate conceptualization."

Apparently, Blalock has in mind only a specific kind of application of factor analysis, such as in research on mental tests, where one is looking for a reduced set of underlying factors of which the tests are indicators. This type of analysis would be inappropriate for sociological research where one uses census data "or other types of analysis in which some of the measured variables are caused by other measured variables" (p. 183). This seems to suggest that in sociology there is no need for latent variables, since all observed variables have a clear and distinct interpretation, whereas in test research, the tests by themselves have only an approximate interpretation to start with, and the ultimate interpretation emerges from the pattern of correlations among different tests and between tests and criteria. However, it is certainly questionable whether we have here a real difference between sociology and psychology. In sociology, as well, variables can be approximate indicators, or mixed indicators. In fact, the case might be argued that sociology much too often tends to adhere to observed variables as if they were basic entities and as if there were no need for a higher level of abstraction.

15.5. Special Case

15.5.1. Introduction. As a special exercise more than as a basic issue, in this section we shall treat a problem mentioned by Duncan *et al.* (1968) in the study referred to in section 14.3. As was remarked in the final paragraph of that section, canonical analysis can be interpreted as a kind of factor analysis: it introduces two orthogonal sets of reference factors, $\boldsymbol{\chi}$ and $\boldsymbol{\eta}$, for the space of the x-variables and of the y-variables, respectively. In the example used in section 14.3, for instance, we might take the transformed variables $\boldsymbol{\eta}_1$ and $\boldsymbol{\eta}_2$ as factors; we would have the following factor matrix for the y-variables:

	η_1	η_2
y_1	.8652	.5013
y_2	.9319	−.3625

To find this matrix, remember that factor loadings represent correlations between observed variables and factors. Table 14.7 therefore, can be thought of as a factor matrix.

We shall now complete the matrix above by adding the χ's as row variables. In other words, we also insert projections (correlations) of $\boldsymbol{\chi}_1$ and $\boldsymbol{\chi}_2$ on the factors $\boldsymbol{\eta}_1$ and $\boldsymbol{\eta}_2$; these projections are nothing but the canonical correlations. This explains the first two columns of the factor matrix given in table 15.7. However, we need more columns in order to explain the variance of χ_1 and χ_2 fully, since the sums of the squared row elements are not equal to unity. This is remedied by adding the columns u_1 and u_2, where the loading of χ_1 on u_1 is equal to $\sqrt{1 - \rho_1^2}$, and the loading of χ_2 on u_2 equal to $\sqrt{1 - \rho_2^2}$.

Note that χ_1 and χ_2, if we look on them as factors for the space of x-variables, are not sufficient to explain $\mathbf{R}_{xx}$, and that we would need additional factors χ_3 to χ_6. However, these additional factors are all

Table 15.7. Factor matrix for χ_1, χ_2, y_1, and y_2.

	η_1	η_2	u_1	u_2
χ_1	.6092	—	.7930	—
χ_2	—	.1431	—	.9897
y_1	.8652	.5013	—	—
y_2	.9319	−.3625	—	—

uncorrelated with the η's and with the y-variables. Therefore, they would have zero loadings in the four factors of table 15.7, and we are not interested in them, because all the relation between the set of x-variables and the set of y-variables is covered by χ_1 and χ_2, so to speak.

Now, as we saw in section 14.3, η_1 could be interpreted as an "aspiration" factor, with y_1 and y_2 as indicators. However, η_1 does not explain the correlation between the two y's, since this correlation also depends on η_2. The problem now is to identify an aspiration factor that all by itself accounts for the observed correlation r_{y1y2}. Then, as a second requirement, we also want this factor to be maximally dependent on the x-variables, or, since it comes to the same thing, on the two canonical variables χ_1 and χ_2.

15.5.2. Formal Treatment. First, we note that table 15.7 gives a complete description of the relations between all variables given in the rows. It follows that the only freedom we have to find a different solution of the factors is to rotate the factor matrix in table 15.7. Let this matrix be $\mathbf{F}$; then another solution is $\mathbf{G}$, with $\mathbf{G} = \mathbf{FT}$, and $\mathbf{T}$ an orthogonal rotation matrix.

Suppose now that $\mathbf{t}$ is the first column of $\mathbf{T}$, so that $\mathbf{Ft} = \mathbf{g}_1$ gives the first rotated factor. Let this be the aspiration factor we want to find; then the requirements are that $\mathbf{g}_1$ should account completely for the correlation between y_1 and y_2. Or, in formal notation,

$$g_{31} \cdot g_{41} = r_{y1y2} \tag{15.75}$$

with g_{31} and g_{41} being the third and fourth elements of $\mathbf{g}_1$. The second requirement is that $\mathbf{g}_1$ should be maximally related to χ_1 and χ_2. But since these two χ's are independent, their joint contribution to the variance of the factor is given by the sum of their squared loadings on the factor. So the requirement becomes that we want a maximum for

$$g_{11}^2 + g_{21}^2. \tag{15.76}$$

First, to simplify notation, we will give special symbols for partitions of the factor matrix **F** in table 15.7. The table can be partitioned into four 2 × 2 matrices. The partition at the upper left then is **P**, the matrix of canonical correlations. The upper right matrix will be called **U** (with $\mathbf{U}^2 = \mathbf{I} - \mathbf{P}^2$). The lower left matrix will be called $\mathbf{F}_y$; it is the factor matrix for the two *y*-variables. Its rows will be indicated as $\mathbf{f}_1'$ and $\mathbf{f}_2'$. Finally, the lower right partition is a zero matrix. Then we shall also partition the transformation vector **t** into two 2 × 1 vectors, to be called $\mathbf{t}_1$ and $\mathbf{t}_2$; their elements are t_{11} and t_{21} for $\mathbf{t}_1$, and t_{12} and t_{22} for $\mathbf{t}_2$.

Using this notation, we now reformulate the two requirements given in (15.75) and (15.76). Instead of (15.75), we now have

$$(\mathbf{f}_1'\mathbf{t}_1) \cdot (\mathbf{f}_2'\mathbf{t}_1) = r_{y_1 y_2}. \tag{15.77}$$

The second requirement is that we want a maximum for $g_{11}^2 + g_{21}^2$. But, since

$$\begin{pmatrix} g_{11} \\ g_{21} \end{pmatrix} = \mathbf{P}\mathbf{t}_1 + \mathbf{U}\mathbf{t}_2,$$

we have to find a maximum for

$$(\mathbf{P}\mathbf{t}_1 + \mathbf{U}\mathbf{t}_2)'(\mathbf{P}\mathbf{t}_1 + \mathbf{U}\mathbf{t}_2). \tag{15.78}$$

In addition, since **T** is an orthogonal transformation, there is the condition that $\mathbf{t}'\mathbf{t} = 1$, or

$$\mathbf{t}_1'\mathbf{t}_1 + \mathbf{t}_2'\mathbf{t}_2 = 1. \tag{15.79}$$

To solve, one's first idea might be to use the method of undetermined multipliers. However, this method fails here, since it results in equations that cannot be solved analytically. Therefore we have to use a different approach, specifically, an iterative trial-and-error approach: we just select values that satisfy the conditions (15.77) and (15.79), and calculate (15.78); we then try different values and again calculate (15.78), and so on, until we find a maximum result for (15.78).

This process can be set up very systematically. First, note that (15.77) defines a quadratic equation in t_{11} and t_{21}. In our example this equation is

$$(.8653t_{11} + .5013t_{21})(.9319t_{11} - .3625t_{21}) = .6246 \tag{15.80}$$

Table 15.8. Rotated version of matrix in table 15.7.

	g_1	g_2	g_3	g_4
χ_1	.9142	−.2972	−.2747	−.0197
χ_2	.0180	.0922	−.1103	.9894
y_1	.7801	.6258	—	—
y_2	.8008	—	.5989	—

and it defines a hyperbola in the plane with t_{11} and t_{21} as coordinates. Permissible values for t_{11} and t_{21} therefore are restricted to points on this hyperbola. In addition, however, they are restricted by condition (15.79), from which it follows that $\mathbf{t}_1'\mathbf{t}_1 \leqslant 1$. This means that permissible points are restricted to the region inside a circle with radius one. Now if we look at the part of the hyperbola inside this circle, we shall in general find that either t_{11} is a univalued function of t_{21} or the other way round; in our example, the first alternative happens to be true. So the easiest approach is to take values for t_{21} throughout the permitted range, and calculate the corresponding value of t_{11}.

Given that part, it follows that $\mathbf{t}_2'\mathbf{t}_2 = 1 - \mathbf{t}_1'\mathbf{t}_1$, or that permissible values of $\mathbf{t}_2$ are restricted to points on a circle with radius equal to the square root of $1 - \mathbf{t}_1'\mathbf{t}_1$. This also can be inspected systematically, by following the circle.

Using this approach, it is found that the solution is $\mathbf{t}_1' = (.8763 \quad .0437)$, and $\mathbf{t}_2' = (.4796 \quad .0119)$. The resulting factor column $\mathbf{g}_1$ is given in table 15.8.

However, table 15.8 gives a complete factor matrix, and what remains to be shown is how the other columns were arrived at. They were found by setting the criterion that y_1 should have zero loadings on g_3 and g_4, whereas y_2 has zero loadings on g_2 and g_4. Together with the condition that $\mathbf{T}$ is orthogonal, this is sufficient to identify $\mathbf{T}$, and therefore the complete factor matrix $\mathbf{G}$. The solution for $\mathbf{T}$ is given in table 15.9.

Table 15.9. Rotation matrix **T** applied to matrix in table 15.7 to obtain table 15.8.

.8763	.2905	.3845	—
.0437	.7469	−.6636	—
.4796	−.5979	−.6418	−.0248
.0119	−.0148	−.0155	.9997

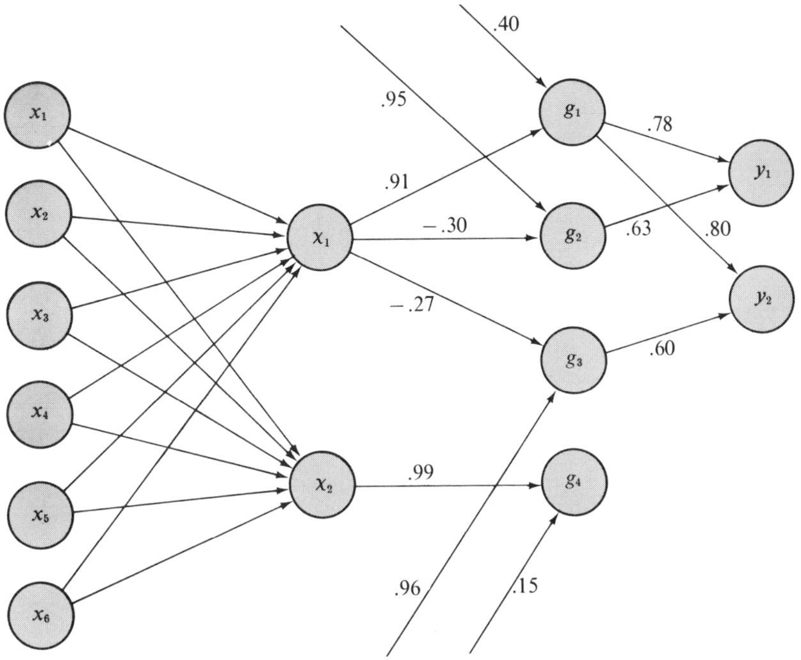

Figure 15.3 Path diagram with aspiration factor g_1 that accounts for all the correlation between the indicators y_1 and y_2. Path coefficients for paths going from x-variables are the same as in figure 14.1.

Factor scores for g_1 now could be estimated from $\boldsymbol{\chi}_1$ and $\boldsymbol{\chi}_2$ alone, by using the values in the first column of G: $.9142\boldsymbol{\chi}_1 + .0180\boldsymbol{\chi}_2$. This expression, in turn, can be expressed in the original variables, since we know how χ_1 and χ_2 are linear functions of the x-variables (from table 14.7). The result is

$$\mathbf{g}_1 = .18\mathbf{x}_1 + .27\mathbf{x}_2 + .25\mathbf{x}_3 + .11\mathbf{x}_4 + .10\mathbf{x}_5 - .01\mathbf{x}_6.$$

The result is used for a path diagram in figure 15.3. Note that g_2 and g_3 are specific factors for y_1 and y_2, respectively, and they also might have been expressed in terms of the x's. Note also that we have to add exogenous error terms, since the g's are not completely determined by the χ's. E.g., the exogenous term for g_1 is found as a residual variance by subtracting the contribution ($.9142^2 + .0180^2$) from unity; the path coefficient is the square root of this quantity.

So much for this special exercise. We have inserted it to show that ideas from canonical correlation analysis, factor analysis, and path analysis can be combined and integrated for some special purpose.

16

Nonrecursive Linear Models

16.1. Introduction

When we discussed path analysis in chapter 12, our interest was restricted to recursive models, i.e., models in which it never occurs that variable *a* influences variable *b*, either directly or indirectly, while at the same time *b* is a direct or indirect determiner of *a*. In terms of the path diagram, in a recursive models arrows are never allowed to form closed loops. In this chapter we will show how this restriction can be dropped, to allow us to draw path diagrams with arrows going back and forth. To give an example, in the sociological study of aspirations, described in section 14.3, an aspiration factor was postulated for respondents; it was assumed that this aspiration factor depended on background variables referring to the respondent himself, and also on background factors related to the respondent's best friend. However, one could as well postulate a similar aspiration factor for the best friend, with y_3 and y_4 as indicators, that also depended on such background factors. The resulting path diagram would be very similar to figure 14.1, would be its mirror image. It then seems only reasonable to assume that the respondent's aspiration is influenced by the aspiration of his best friend, and vice versa. In the path diagram we would then have an arrow going from the respondent's aspiration to the aspiration of his best friend, and at the same time an arrow going in the opposite direction. So far we have not been able to specify such models.

Nonrecursive models have been developed mainly for economic applications. A classic example is the Keynesian system that relates

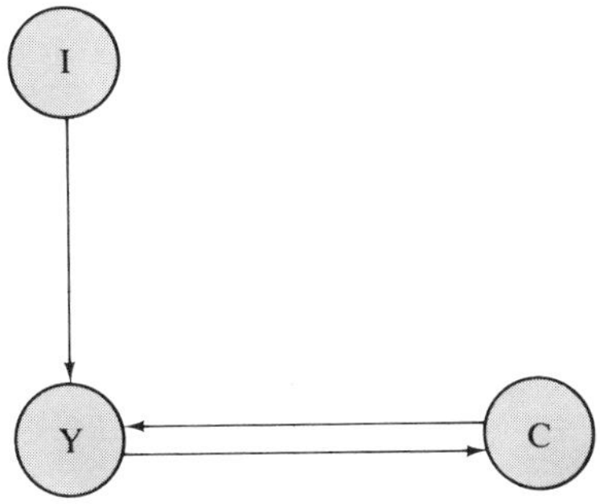

Figure 16.1 Keynesian model of relations between investment (I), income (Y), and consumption (C).

income, consumption, and investment. The model states that consumption C is a (linear) function of income Y, and further, that income Y is a linear function of consumption C and investment I. This is diagrammed in figure 16.1; it is clearly a nonrecursive model. In terms of equations,

$$\begin{aligned} C &= a + bY, \\ Y &= C + I. \end{aligned} \tag{16.1}$$

It is only reasonable, however, to add random components to these equations. Economists prefer to think of these random components as "disturbances" that produce error in the equations. Also, economists are very explicit about the distinction between *endogenous* and *exogenous* variables (see also section 12.2.3). In our example, Y and C are endogenous, since they depend on other variables in the system. In the other hand, I is exogenous, since it is not dependent on other variables in the system. Disturbances also are exogenous.

Equations such as (16.1) are called *structural equations.* They specify relations between variables according to some theoretical model that is assumed to be valid. In a general form, a structural equation could be written as

$$\mathbf{Yc} + \mathbf{Xb} + \mathbf{u} = \mathbf{0}, \tag{16.2}$$

which is a matrix equation. Matrix $\mathbf{Y}$ is an $n \times m$ matrix, and stands for the n observations on each of the m endogenous variables. $\mathbf{X}$ is a $n \times k$ matrix for the n observations of each of k exogenous variables. Finally, $\mathbf{u}$ is an $n \times 1$ vector for the exogenous disturbance values. The vectors $\mathbf{c}$ (of order $m \times 1$) and $\mathbf{b}$ (of order $k \times 1$) are parameter vectors. The linear equation corresponding to (16.2) could then be written as

$$y'c + x'b + u = 0, \tag{16.3}$$

which is a homogeneous equation in the x's, the y's, and u.

A specific economic theory might require more than one structural equation. However, in general it is unwise to specify more than m equations if there are m endogenous variables, since we would then have more equations than variables we want to explain. Therefore, the most general case is a collection of m equations. The general form for them, corresponding to (16.2), could be written as

$$\mathbf{YC} + \mathbf{XB} + \mathbf{U} = \mathbf{0}, \tag{16.4}$$

which differs from (16.2) in that $\mathbf{C}$ is an $m \times m$ matrix of parameters, $\mathbf{B}$ is a $k \times m$ matrix of parameters, and $\mathbf{U}$ is an $n \times m$ matrix of disturbances.

In what follows it will be assumed that the disturbances are independent of the other exogenous variables, so that we have

$$\mathbf{X'U} = \mathbf{0}. \tag{16.5}$$

However, in general there is no need to assume that disturbances are uncorrelated (which would imply $\mathbf{U'U} = \mathbf{0}$).

Given a set of structural equations like (16.4), we can derive other equations in which $\mathbf{Y}$ is expressed as a linear function of exogenous variables alone. They are found, of course, by solving (16.4) for $\mathbf{Y}$:

$$\mathbf{Y} = -\mathbf{XBC}^{-1} - \mathbf{UC}^{-1}, \tag{16.6}$$

where it is assumed that the determinant of $\mathbf{C}$ is different from zero; otherwise there would be linear dependence among the c's, and some equations therefore would be redundant (we assume here that such redundant equations are omitted, if they occur). A set of equations as given in (16.6) is called the *reduced form* of the system. An alternative notation might be

$$\mathbf{Y} = \mathbf{XP} + \mathbf{V}, \tag{16.7}$$

where **P** and **V** are substituted for the corresponding forms in (16.6); i.e., $\mathbf{P} = -\mathbf{BC}^{-1}$, and $\mathbf{V} = -\mathbf{UC}^{-1}$. It follows from (16.5) that

$$\mathbf{X}'\mathbf{V} = \mathbf{0}. \tag{16.8}$$

Throughout the discussion it will be assumed that all variables are measured as deviations from their mean, so that we do not have constants in the equations. This assumption implies no loss of generality.

16.2. Formal Treatment

Given sets of data in matrices **X** and **Y**, we can calculate **P** from

$$\mathbf{P} = (\mathbf{X}'\mathbf{X})^{-1}\mathbf{X}'\mathbf{Y}, \tag{16.9}$$

which is a least-squares solution of (16.7), as explained in section 6.2.3; it guarantees minimum variance in the v's. Strictly speaking, the **P** we calculate from (16.9) is only an estimate of the real population parameters that are meant in equation (16.7); however, since sampling considerations fall outside the scope of this book, we will not make the distinction.

At this moment we encounter the first problem, which is that a solution of (16.7) does not indicate a solution of (16.4). Although the reduced form is uniquely determined if the set of structural equations is given, the reverse is not true. This is easily seen by applying a transformation matrix **T** to the structural equations (16.4), which then become

$$\mathbf{YCT} + \mathbf{XBT} + \mathbf{UT} = \mathbf{0}, \tag{16.10}$$

and this defines a different set of structural equations. However, the reduced form of (16.10) is identical to the reduced form of (16.4), since from (16.10) we have

$$\mathbf{Y} = -\mathbf{XBTT}^{-1}\mathbf{C}^{-1} - \mathbf{UTT}^{-1}\mathbf{C}^{-1} = -\mathbf{XBC}^{-1} - \mathbf{UC}^{-1}. \tag{16.11}$$

It follows that we have an identifiability problem: on the basis of (16.7) we can find infinitely many solutions for (16.4), all equally good. However, in most applications the theory will specify structural equations in which not all variables are included. This means that it is assumed that some of the coefficients in the structural equations are equal to zero. Obviously, this imposes restrictions on the family of solutions for (16.4) that can be derived from (16.7), and if there are enough restrictions, the solution does in fact become identifiable. With too many restrictions, the solution may even become overidentified, so that it becomes impossible to find an expression for (16.4) that is consistent with a given form of (16.7).

To see what this implies, let us take one specific structural equation:

$$\mathbf{Y}_1\mathbf{c}_1 + \mathbf{X}_1\mathbf{b}_1 + \mathbf{u} = \mathbf{0}, \tag{16.12}$$

where we write $\mathbf{X}_1$ and $\mathbf{Y}_1$ to indicate that these matrices do not include all variables x and y but only a selection of them. In other words, $\mathbf{Y}_1$ is a partition of $\mathbf{Y}$, and $\mathbf{X}_1$ a partition of $\mathbf{X}$. The columns that are excluded might be indicated as $\mathbf{Y}_2$ and $\mathbf{X}_2$; we would have $\mathbf{Y} = (\mathbf{Y}_1 \quad \mathbf{Y}_2)$, and $\mathbf{X} = (\mathbf{X}_1 \quad \mathbf{X}_2)$. Equation (16.12) can now be completed as

$$(\mathbf{Y}_1 \quad \mathbf{Y}_2)\begin{pmatrix}\mathbf{c}_1\\ \mathbf{0}\end{pmatrix}(+ \mathbf{X}_1 \quad \mathbf{X}_2)\begin{pmatrix}\mathbf{b}_1\\ \mathbf{0}\end{pmatrix} + \mathbf{u} = \mathbf{0}. \tag{16.13}$$

Suppose now that $\mathbf{Y}_1$ is of order $n \times m_1$, and $\mathbf{Y}_2$ of order $n \times m_2$, with $m_1 + m_2 = m$. Similarly, let $\mathbf{X}_1$ be of order $n \times k_1$, and $\mathbf{X}_2$ of order $n \times k_2$ with $k_1 + k_2 = k$. The partitions in $\mathbf{b}$ and $\mathbf{c}$ are conformable.

Now write the reduced form with similar partitions:

$$(\mathbf{Y}_1 \quad \mathbf{Y}_2) = (\mathbf{X}_1 \quad \mathbf{X}_2)\begin{pmatrix}\mathbf{P}_{11} & \mathbf{P}_{12}\\ \mathbf{P}_{21} & \mathbf{P}_{2}\end{pmatrix} + (\mathbf{V}_1 \quad \mathbf{V}_2), \tag{16.14}$$

where $\mathbf{P}_{11}$ must be of order $k_1 \times m_1$, $\mathbf{P}_{12}$ of order $k_1 \times m_2$, $\mathbf{P}_{21}$ of order $k_2 \times m_1$, and $\mathbf{P}_{22}$ of order $k_2 \times m_2$. If we postmultiply both sides of (16.14) by the vector $\begin{pmatrix}\mathbf{c}_1\\ \mathbf{0}\end{pmatrix}$, we obtain

$$\begin{aligned}(\mathbf{Y}_1 \quad \mathbf{Y}_2)\begin{pmatrix}\mathbf{c}_1\\ \mathbf{0}\end{pmatrix} &= (\mathbf{X}_1 \quad \mathbf{X}_2)\begin{pmatrix}\mathbf{P}_{11} & \mathbf{P}_{12}\\ \mathbf{P}_{21} & \mathbf{P}_{22}\end{pmatrix}\begin{pmatrix}\mathbf{c}_1\\ \mathbf{0}\end{pmatrix} + (\mathbf{V}_1 \quad \mathbf{V}_2)\begin{pmatrix}\mathbf{c}_1\\ \mathbf{0}\end{pmatrix}\\ &= -(\mathbf{X}_1 \quad \mathbf{X}_2)\begin{pmatrix}\mathbf{b}_1\\ \mathbf{0}\end{pmatrix} - \mathbf{u},\end{aligned} \tag{16.15}$$

where the last equality in (16.15) is derived from (16.13). Identifying corresponding forms, we have

$$-\begin{pmatrix} \mathbf{b}_1 \\ \mathbf{0} \end{pmatrix} = \begin{pmatrix} \mathbf{P}_{11} & \mathbf{P}_{12} \\ \mathbf{P}_{21} & \mathbf{P}_{22} \end{pmatrix} \begin{pmatrix} \mathbf{c}_1 \\ \mathbf{0} \end{pmatrix}, \tag{16.16}$$

or

$$\begin{aligned} \mathbf{P}_{11}\mathbf{c}_1 &= -\mathbf{b}_1, \\ \mathbf{P}_{21}\mathbf{c}_1 &= \mathbf{0}. \end{aligned} \tag{16.17}$$

The second equation in (16.17) gives k_2 homogeneous equations in m_1 unknowns. This could be solved if the rank of $\mathbf{P}_{21}$ is equal to $m_1 - 1$ (the solution up to an arbitrary proportionality constant). Then, given a solution for $\mathbf{c}_1$ in this way, we could solve $\mathbf{b}_1$ from the first equation in (16.17), and so equation (16.13) would be identified. In fact, since there is just this one solution if $\mathbf{P}_{21}$ has rank $m_1 - 1$, equations (16.13) is said to be just identified. It can be shown that the solution gives a consistent estimate of the true parameters, in statistical terms.

However, if the rank of $\mathbf{P}_{21}$ is less than $m_1 - 1$ (which happens if there are fewer equations than unknowns), no solution can be identified. The only step then is to reconsider the theory and attempt to specify additional restrictions. The other possibility is that $\mathbf{P}_{21}$ has rank m_1 (because there are m_1 or more equations). This latter alternative requires special measures, and in econometric theory two standard solutions are described (Goldberger 1964) that we shall discuss here. The first one is called the *two-stage least-squares solution* (2SLS, for short); the second is the *limited-information least-generalized residual-variance solution* (LI/LGRV).

16.3. 2SLS Solution

If we look back at equation (16.12), we can easily see that the equation is defined up to an arbitrary multiplier. Therefore we may always select one element out of $\mathbf{c}$ and set it equal to minus one. Suppose now that this is the first element in $\mathbf{c}$, which we can agree on without loss of generality, since it is arbitrary which variable is placed first. Then we may rewrite (16.17) as

$$-\mathbf{y}_1 + \mathbf{Y}_{.1}\mathbf{c}_{.1} + \mathbf{X}_1\mathbf{b}_1 + \mathbf{u} = \mathbf{0}, \tag{16.18}$$

where $\mathbf{Y}_{.1}$ denotes $\mathbf{Y}_1$ with the first column deleted, so that $\mathbf{Y}_{.1}$ is of order $n \times (m_1 - 1)$. Similarly, $\mathbf{c}_{.1}$ is a vector with $m_1 - 1$ elements, obtained from $\mathbf{c}_1$ by omitting the first element. Equation (16.14) can be rearranged in a similar way:

$$(\mathbf{y}_1 \quad \mathbf{Y}_{.1} \quad \mathbf{Y}_2) = (\mathbf{X}_1 \quad \mathbf{X}_2)\begin{pmatrix} \mathbf{p}_{11} & \mathbf{P}_{.11} & \mathbf{P}_{21} \\ \mathbf{p}_{12} & \mathbf{P}_{.12} & \mathbf{P}_{22} \end{pmatrix} + (\mathbf{v}_1 \quad \mathbf{V}_{.1} \quad \mathbf{V}_2)$$
$$= (\mathbf{X}_1 \quad \mathbf{X}_2)(\mathbf{p}_1 \quad \mathbf{P}_{.1} \quad \mathbf{P}_2) + (\mathbf{v}_1 \quad \mathbf{V}_{.1} \quad \mathbf{V}_2), \quad (16.19)$$

where $\mathbf{p}_{11}$ is the first column of $\mathbf{P}_{11}$, and $\mathbf{P}_{.11}$ the remaining matrix of order $k_1 \times (m_1 - 1)$; similarly for $\mathbf{P}_{12}$. Also, $\mathbf{V}_1$ has been partitioned by splitting off its first column. It follows that we may write

$$\mathbf{Y}_{.1} = \mathbf{X}_1\mathbf{P}_{.11} + \mathbf{X}_2\mathbf{P}_{.12} + \mathbf{V}_{.1} = \mathbf{X}\mathbf{P}_{.1} + \mathbf{V}_{.1}. \quad (16.20)$$

The structural equation (16.18) can be written as

$$\mathbf{y}_1 = \mathbf{Y}_{.1}\mathbf{c}_{.1} + \mathbf{X}_1\mathbf{b}_1 + \mathbf{u} = \mathbf{0}. \quad (16.21)$$

Now substitute (16.20) in (16.21) to obtain

$$\mathbf{y}_1 = \mathbf{X}\mathbf{P}_{.1}\mathbf{c}_{.1} + \mathbf{X}_1\mathbf{b}_1 + (\mathbf{u} + \mathbf{V}_{.1}\mathbf{c}_{.1}), \quad (16.22)$$

and we have a reduced-form type of equation for $\mathbf{y}_1$. Strictly speaking, equation (16.22) should be used to estimate *all* parameters on the right side, including $\mathbf{P}_{.1}\mathbf{c}_{.1}$. However, we can assume that values for $\mathbf{P}_{.1}$ that were estimated from the reduced-form equation (16.9) may be inserted in (16.22), so that $\mathbf{P}_{.1}$ in (16.22) is known. Then write $\hat{\mathbf{Y}}_{.1}$ for $\mathbf{X}\mathbf{P}_{.1}$; look back at equation (16.7) to see that $\hat{\mathbf{Y}}_{.1}$ is the systematic part of $\mathbf{Y}_{.1}$, which is dependent on $\mathbf{X}$ and not on the disturbances $\mathbf{V}$). Equation (16.22) then becomes

$$\mathbf{y}_1 = \hat{\mathbf{Y}}_{.1}\mathbf{c}_{.1} + \mathbf{X}_1\mathbf{b}_1 + (\mathbf{u} + \mathbf{V}_{.1}\mathbf{c}_{.1}). \quad (16.23)$$

Now remember that $\mathbf{X}'\mathbf{U} = \mathbf{0}$ and $\mathbf{X}'\mathbf{V} = \mathbf{0}$, so that also $\hat{\mathbf{Y}}_{.1}'\mathbf{U} = \mathbf{0}$ and $\hat{\mathbf{Y}}_{.1}'\mathbf{V} = \mathbf{0}$. Premultiplication of both sides of (16.23), first by $\hat{\mathbf{Y}}_{.1}'$, and subsequently by $\mathbf{X}_1'$, gives

$$\begin{pmatrix} \hat{\mathbf{Y}}_{.1}'\mathbf{y}_1 \\ \mathbf{X}_1'\mathbf{y}_1 \end{pmatrix} = \begin{pmatrix} \hat{\mathbf{Y}}_{.1}'\mathbf{Y}_{.1} & \hat{\mathbf{Y}}_{.1}'\mathbf{X}_1 \\ \mathbf{X}_1'\hat{\mathbf{Y}}_{.1} & \mathbf{X}_1'\mathbf{X}_1 \end{pmatrix} \cdot \begin{pmatrix} \mathbf{c}_{.1} \\ \mathbf{b}_1 \end{pmatrix}. \quad (16.24)$$

However, from (16.9) we have

$$\mathbf{P}_{.1} = (\mathbf{X}'\mathbf{X})^{-1}\mathbf{X}'\mathbf{Y}_{.1}, \quad (16.25)$$

and therefore

$$\hat{\mathbf{Y}}_{.1} = \mathbf{X}\mathbf{P}_{.1} = \mathbf{X}(\mathbf{X}'\mathbf{X})^{-1}\mathbf{X}'\mathbf{Y}_{.1}. \quad (16.26)$$

Equation (16.26) can be substituted in (16.24), and this gives

$$\begin{pmatrix} \mathbf{Y}_{.1}'\mathbf{X}(\mathbf{X}'\mathbf{X})^{-1}\mathbf{X}'\mathbf{y}_1 \\ \mathbf{X}_1'\mathbf{y}_1 \end{pmatrix} = \begin{pmatrix} \mathbf{Y}_{.1}'\mathbf{X}(\mathbf{X}'\mathbf{X})^{-1}\mathbf{X}'\mathbf{Y}_{.1} & \mathbf{Y}_{.1}'\mathbf{X}_1 \\ \mathbf{X}_1'\mathbf{Y}_{.1} & \mathbf{X}_1'\mathbf{X}_1 \end{pmatrix} \cdot \begin{pmatrix} \mathbf{c}_{.1} \\ \mathbf{b}_1 \end{pmatrix}. \quad (16.27)$$

Equation (16.27) gives $m_1 - 1 + k_1$ nonhomogeneous equations in as many unknowns, and therefore we can solve for $\mathbf{c}_{.1}$ and $\mathbf{b}_1$. This defines the 2SLS solution. Its two stages are, in fact, the use of equation (16.9) for the first stage, and of equation (16.27) for the second.

16.4. Numerical Illustration

For an illustration we go back to the example in section 14.3. We there gave a canonical solution to relate background variables and aspiration indicators, with canonical variables η_1 and η_2. As was suggested in section 16.1, we will now repeat the mirror image of the canonical analysis, in order to obtain "factors" with y_3 and y_4 as indicators. First, to simplify notation, we shall agree that $\mathbf{Y}_1 = (\mathbf{y}_1 \mid \mathbf{y}_2)$, and $\mathbf{Y}_2 = (\mathbf{y}_3 \mid \mathbf{y}_4)$. The canonical variables χ_1 and χ_2 in section 14.3 will now be indicated with χ_{11} and χ_{12}, to distinguish them from the canonical variables for the best friend, to be called χ_{21} and χ_{22}. Similarly, the variables η_1 and η_2 now become η_{11} and η_{12}, corresponding to η_{21} and η_{22} for the best friend. For convenience we shall use $\boldsymbol{\chi}_1$ from now on as the joint matrix $(\boldsymbol{\chi}_{11} \mid \boldsymbol{\chi}_{12})$, also $\boldsymbol{\chi}_2 = (\boldsymbol{\chi}_{21} \mid \boldsymbol{\chi}_{22})$, and similarly for $\boldsymbol{\eta}_1$ and $\boldsymbol{\eta}_2$.

Structural equations will be presented later on. First we give the results of the double canonical analysis. In part, this is just a repetition of the results in section 14.3.2.

The variables χ_1 are defined by $\boldsymbol{\chi}_1 = \mathbf{X}\mathbf{C}_1$, with corresponding $\boldsymbol{\eta}_1 = \mathbf{Y}_1\mathbf{D}_1$. Canonical correlations are $\boldsymbol{\chi}_1'\boldsymbol{\eta}_1/n = \mathbf{P}_1$. The set is duplicated as $\boldsymbol{\chi}_2 = \mathbf{X}\mathbf{C}_2$, $\boldsymbol{\eta}_2 = \mathbf{Y}_2\mathbf{D}_2$, and $\boldsymbol{\chi}_2'\boldsymbol{\eta}_2 = \mathbf{P}_2$. The results are:

$$\mathbf{D}_1 = \begin{pmatrix} .4642 & 1.1934 \\ .6419 & -1.1080 \end{pmatrix}; \qquad \mathbf{D}_2 = \begin{pmatrix} .6773 & -1.1119 \\ .4203 & 1.2322 \end{pmatrix};$$

$$\mathbf{C}_1 = \begin{pmatrix} .3258 & -.4107 \\ .4808 & .3401 \\ .4559 & -.7181 \\ .2024 & .6893 \\ .1838 & .1247 \\ -.0266 & .1737 \end{pmatrix}; \qquad \mathbf{C}_2 = \begin{pmatrix} .0534 & -.1705 \\ .1066 & -.1085 \\ .1872 & .0832 \\ .4094 & -.3851 \\ .5764 & .7474 \\ .2680 & -.8434 \end{pmatrix};$$

$$\mathbf{P}_1 = \begin{pmatrix} .6092 & \\ & .1431 \end{pmatrix}; \qquad \mathbf{P}_2 = \begin{pmatrix} .6720 & \\ & .0931 \end{pmatrix}.$$

Note the general symmetry in the results, in that $\rho_{11} \approx \rho_{21}$, and in that $\mathbf{c}_{11}$ is almost the mirror image of $\mathbf{c}_{21}$, just as $\mathbf{d}_{11}$ is the mirror image of $\mathbf{d}_{21}$.

If we want to draw paths directly from the background variables to the aspirations factors η, corresponding path coefficients are found as $\mathbf{Q}_1 = \mathbf{C}_1\mathbf{P}_1$, and $\mathbf{Q}_2 = \mathbf{C}_2\mathbf{P}_2$:

$$\mathbf{Q}_1 = \begin{pmatrix} .1984 & -.0588 \\ .2929 & .0487 \\ .2777 & -.1028 \\ .1233 & .0986 \\ .1120 & .0178 \\ -.0162 & .0249 \end{pmatrix}; \quad \mathbf{Q}_2 = \begin{pmatrix} .0359 & -.0159 \\ .0716 & -.0101 \\ .1258 & .0077 \\ .2751 & -.0359 \\ .3873 & .0696 \\ .1801 & -.0785 \end{pmatrix}.$$

Also, we shall need correlations between original variables and canonical variables. They are given by $\mathbf{X}'\boldsymbol{\eta}_1/n = \mathbf{X}'\mathbf{Y}_1\mathbf{D}_1/n = \mathbf{X}'\mathbf{X}\mathbf{C}_1\mathbf{P}_1/n = \mathbf{R}_{xx}\mathbf{C}_1\mathbf{P}_1$, and similarly for $\mathbf{X}'\boldsymbol{\eta}_2/n$:

$$\mathbf{X}'\boldsymbol{\eta}_1/n = \begin{pmatrix} .2752 & -.0488 \\ .4501 & .0419 \\ .4102 & -.0617 \\ .2905 & .0830 \\ .3228 & .0402 \\ .0803 & .0129 \end{pmatrix}; \quad \mathbf{X}'\boldsymbol{\eta}_2/n = \begin{pmatrix} .1114 & -.0216 \\ .3058 & -.0027 \\ .3239 & .0037 \\ .4296 & -.0120 \\ .5620 & .0398 \\ .2721 & -.0646 \end{pmatrix}.$$

Then, in order to write $\mathbf{Y}_1$ as a function of $\boldsymbol{\eta}_1$, and since $\boldsymbol{\eta}_1 = \mathbf{Y}_1\mathbf{D}_1$, we have $\mathbf{Y}_1 = \boldsymbol{\eta}_1\mathbf{D}_1^{-1}$; similarly $\mathbf{Y}_2 = \boldsymbol{\eta}_2\mathbf{D}_2^{-1}$:

$$\mathbf{D}_1^{-1} = \begin{pmatrix} .8653 & .9321 \\ .5031 & -.3625 \end{pmatrix}; \quad \mathbf{D}_2^{-1} = \begin{pmatrix} .9465 & .8541 \\ -.3228 & .5202 \end{pmatrix}.$$

Correlations between $\mathbf{Y}_1$ and $\boldsymbol{\eta}_1$ are given by $\mathbf{Y}_1'\boldsymbol{\eta}_1/n = \mathbf{D}_1'^{-1}$, and similarly for correlations between $\mathbf{Y}_2$ and $\boldsymbol{\eta}_2$. Further, we want correlations between $\mathbf{Y}_1$ and $\boldsymbol{\eta}_2$, and between $\mathbf{Y}_2$ and $\boldsymbol{\eta}_1$. They are found from $\mathbf{Y}_1'\mathbf{Y}_2\mathbf{D}_2/n$, and $\mathbf{Y}_2'\mathbf{Y}_1\mathbf{D}_1/n$:

$$\mathbf{Y}_1'\boldsymbol{\eta}_2/n = \begin{pmatrix} .3968 & .1560 \\ .3861 & -.0044 \end{pmatrix}; \quad \mathbf{Y}_2'\boldsymbol{\eta}_1/n = \begin{pmatrix} .3873 & -.0164 \\ .4059 & .1403 \end{pmatrix}.$$

Finally, we have $\boldsymbol{\eta}_1'\boldsymbol{\eta}_2/n = \mathbf{D}_1'\mathbf{Y}_1'\mathbf{Y}_2\mathbf{d}_{27}n = \begin{pmatrix} .4321 & .0693 \\ .0478 & .1911 \end{pmatrix}$. This much

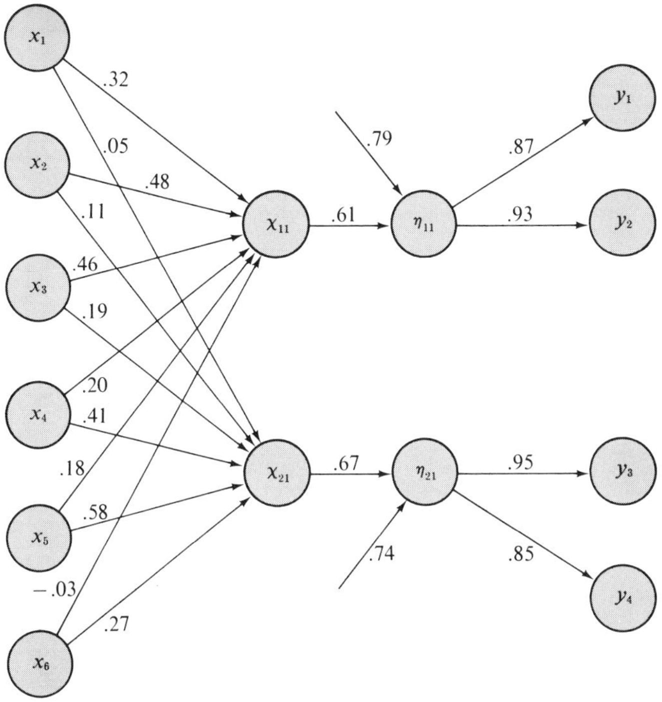

Figure 16.2 Double canonical analysis. Second canonical variables have been omitted. An alternative diagram is to draw paths immediately from the x-variables to η_{11} and η_{21}; path coefficients are then obtained by multiplying the coefficients, as indicated at the left, by their corresponding canonical correlations.

summarizes the canonical analysis. Figure 16.2 gives a path diagram of the results.

We now proceed to the 2SLS analysis. Suppose we agree on the following two structural equations:

$$\eta_{11} = b_{11}x_1 + b_{12}x_2 + b_{13}x_3 + b_{14}x_4 + a_1\eta_{21};$$

$$\eta_{21} = b_{23}x_3 + b_{24}x_4 + b_{25}x_5 + b_{26}x_6 + a_2\eta_{11}.$$

Note that in the first equation the background variables x_5 and x_6 have been omitted, and in the second equation, true to the mirror concept, the variables x_1 and x_2. Also, the equations state that the aspiration factors η_{11} and η_{21} determine each other, which makes the model nonrecursive. (These two structural equations are chosen only as an illustration; there is no special theory behind them.) In words, the equations say that the respondent's aspiration depends on the aspiration of his parents, on his intelligence level, on his family's socio-economic status, the socio-economic status of his best friend's family, and on his best friend's aspiration. The second equation says the same, but where we have merely interchanged respondent and best friend.

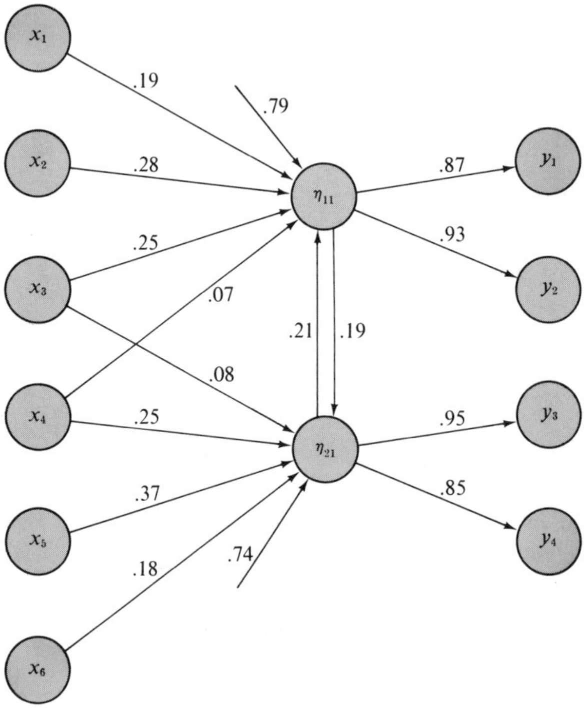

Figure 16.3 Path diagram for a nonrecursive model.

Equation (16.27) can now be identified for the first structural equation as

$$\begin{pmatrix} \boldsymbol{\eta}_{21}'\mathbf{X}(\mathbf{X}'\mathbf{X})^{-1}\mathbf{X}'\boldsymbol{\eta}_{11} \\ \mathbf{X}_1'\boldsymbol{\eta}_{11} \end{pmatrix} = \begin{pmatrix} \boldsymbol{\eta}_{21}'\mathbf{X}(\mathbf{X}'\mathbf{X})^{-1}\mathbf{X}'\boldsymbol{\eta}_{21} & \boldsymbol{\eta}_{21}'\mathbf{X}_1 \\ \mathbf{X}_1'\boldsymbol{\eta}_{21} & \mathbf{X}_1'\mathbf{X}_1 \end{pmatrix} \cdot \begin{pmatrix} a_1 \\ \mathbf{b}_1 \end{pmatrix},$$

with solution $a_1 = .2117$, and $\mathbf{b}_1' = (.1862 \quad .2815 \quad .2493 \quad .0761)$. We could set up the form of (16.27) for the second structural equation in the same way; its solution is $a_2 = .1946$, and

$$\mathbf{b}_2' = (.0737 \quad .2520 \quad .3695 \quad .1835).$$

The result is pictured as a path diagram in figure 16.3.

16.5. LI/LGRV Solution

Suppose again that the structural equations contain the endogenous variables $\mathbf{Y}_1$. From the reduced-form equation (16.7), we have

$$\mathbf{Y}_1 = \mathbf{X}\mathbf{P}_1 + \mathbf{V}_1 \tag{16.28}$$

and we know from (16.9) that $\mathbf{P}_1$ can be estimated as

$$\mathbf{P}_1 = (\mathbf{X}'\mathbf{X})^{-1}\mathbf{X}'\mathbf{Y}_1. \tag{16.29}$$

What we are interested in, for the moment, is the residual variance $\mathbf{W}_1 = \mathbf{V}_1{}'\mathbf{V}_1/n$. From (16.7) we see that $\mathbf{V}_1 = \mathbf{Y}_1 - \mathbf{X}\mathbf{P}_1$, and therefore

$$\begin{aligned}
\mathbf{V}_1{}'\mathbf{V}_1/n &= (\mathbf{Y}_1 - \mathbf{X}\mathbf{P}_1)'(\mathbf{Y}_1 - \mathbf{X}\mathbf{P}_1)/n \\
&= (\mathbf{Y}_1{}'\mathbf{Y}_1/n) + (\mathbf{P}_1{}'\mathbf{X}'\mathbf{X}\mathbf{P}_1/n) - (\mathbf{Y}_1{}'\mathbf{X}\mathbf{P}_1/n) - (\mathbf{P}_1{}'\mathbf{X}'\mathbf{Y}_1/n) \\
&= (\mathbf{Y}_1{}'\mathbf{Y}_1/n) - [\mathbf{Y}_1{}'\mathbf{X}(\mathbf{X}'\mathbf{X})^{-1}\mathbf{X}'\mathbf{Y}_1/n] \\
&= \mathbf{Y}_1{}'[\mathbf{I} - \mathbf{X}(\mathbf{X}'\mathbf{X})^{-1}\mathbf{X}']\mathbf{Y}_1/n.
\end{aligned} \tag{16.30}$$

Suppose now the structural equation is

$$\mathbf{Y}_1\mathbf{c} + \mathbf{X}_1\mathbf{b}_1 + \mathbf{u} = \mathbf{0}. \tag{16.31}$$

Using (16.7), we could predict $\mathbf{Y}_1$ from all x's, and we would have

$$\mathbf{Y}_1\mathbf{c} = \mathbf{X}\mathbf{P}_1\mathbf{c} + \mathbf{V}_1\mathbf{c}, \tag{16.32}$$

with residual variance equal to $\mathbf{c}'\mathbf{V}_1{}'\mathbf{V}_1\mathbf{c}/n = \mathbf{c}'\mathbf{W}_1\mathbf{c}$. Equation (16.32) is, as it were, a completed version of the structural equation, with all predictors in it instead of the set $\mathbf{X}_1$ only. If, however, we use the structural equation (16.31), the residual variance is different; and since this structural equation makes use of less predictors, the residual variance can be expected to be larger than in the complete equation. Let us calculate what the residual variance in (16.32) is. First, we write

$$\mathbf{Y}_1\mathbf{c} = -\mathbf{X}_1\mathbf{b}_1 - \mathbf{u}, \tag{16.33}$$

from which we derive

$$\mathbf{b}_1 = -(\mathbf{X}_1{}'\mathbf{X}_1)^{-1}\mathbf{X}_1{}'\mathbf{Y}_1\mathbf{c}. \tag{16.34}$$

The residual variance is $\mathbf{u}'\mathbf{u}/n$, where $\mathbf{u} = -\mathbf{Y}_1\mathbf{c} - \mathbf{X}_1\mathbf{b}_1$. Therefore

$$\begin{aligned}\mathbf{u}'\mathbf{u} &= (-\mathbf{Y}_1\mathbf{c} - \mathbf{X}_1\mathbf{b})'(-\mathbf{Y}_1\mathbf{c} - \mathbf{X}_1\mathbf{b})\\ &= \mathbf{c}'\mathbf{Y}_1'\mathbf{Y}_1\mathbf{c} + \mathbf{b}_1'\mathbf{X}_1'\mathbf{X}_1\mathbf{b}_1 + \mathbf{b}_1'\mathbf{X}_1'\mathbf{Y}_1\mathbf{c} + \mathbf{c}'\mathbf{Y}_1'\mathbf{X}_1\mathbf{b}_1\\ &= \mathbf{c}'\mathbf{Y}_1'\mathbf{Y}_1\mathbf{c} - \mathbf{c}'\mathbf{Y}_1'\mathbf{X}_1(\mathbf{X}_1'\mathbf{X}_1)^{-1}\mathbf{X}_1'\mathbf{Y}_1\mathbf{c}\\ &= \mathbf{c}'\mathbf{Y}_1'[\mathbf{I} - \mathbf{X}_1(\mathbf{X}_1'\mathbf{X}_1)^{-1}\mathbf{X}_1']\mathbf{Y}_1\mathbf{c}.\end{aligned}$$

If we divide by n, we obtain the residual variance, for which we shall write $\mathbf{u}'\mathbf{u}/n = \mathbf{c}'\mathbf{W}_2\mathbf{c}$, with

$$\mathbf{W}_2 = \mathbf{Y}_1'[\mathbf{I} - \mathbf{X}_1(\mathbf{X}_1'\mathbf{X}_1)^{-1}\mathbf{X}_1']\mathbf{Y}_1/n. \tag{16.35}$$

Note that $\mathbf{W}_1$ and $\mathbf{W}_2$ have very similar expressions, the only difference being that $\mathbf{X}$ in (16.30) is replaced by $\mathbf{X}_1$ in (16.35).

As we remarked earlier, $\mathbf{c}'\mathbf{W}_2\mathbf{c}$ will be larger than $\mathbf{c}'\mathbf{W}_1\mathbf{c}$; the larger their discrepancy, the more we lose by omitting variables as predictors in the structural equation. However, since the parameters c have not yet been defined, we can select them to make the discrepancy minimal. In fact, let us write

$$\lambda = \frac{\mathbf{c}'\mathbf{W}_2\mathbf{c}}{\mathbf{c}'\mathbf{W}_1\mathbf{c}}; \tag{16.36}$$

λ is a measure for relative discrepancy, and we may so select $\mathbf{c}$ that λ has a minimum. It is easier to find the solution if we take logarithms:

$$\log \lambda = \log (\mathbf{c}'\mathbf{W}_2\mathbf{c}) - \log (\mathbf{c}'\mathbf{W}_1\mathbf{c}). \tag{16.37}$$

Setting partial derivatives equal to zero, we have

$$\partial (\log \lambda)/\partial \mathbf{c} = (2\mathbf{W}_2\mathbf{c}/\mathbf{c}'\mathbf{W}_2\mathbf{c}) - (2\mathbf{W}_1\mathbf{c}/\mathbf{c}'\mathbf{W}_1\mathbf{c}) = \mathbf{0}. \tag{16.38}$$

It follows from (16.38) that

$$\mathbf{W}_1\mathbf{c} = \frac{\mathbf{c}'\mathbf{W}_1\mathbf{c}}{\mathbf{c}'\mathbf{W}_2\mathbf{c}} \cdot \mathbf{W}_2\mathbf{c} = \frac{1}{\lambda}\mathbf{W}_2\mathbf{c}. \tag{16.39}$$

To solve, we use the trick to set $\mathbf{W}_2^{\frac{1}{2}}\mathbf{c} = \mathbf{k}$, in order to transform (16.39) into an ordinary eigenvector equation:

$$\mathbf{W}_2^{-\frac{1}{2}}\mathbf{W}_1\mathbf{W}_2^{-\frac{1}{2}}\mathbf{k} = \frac{1}{\lambda}\mathbf{k}, \tag{16.40}$$

where we have to take the eigenvector with the largest eigenvalue $(1/\lambda)$ in order to have a minimum for λ itself.

16.6. Numerical Illustration

For the data described in section 16.4, and for the first structural equation, it can be found that

$$\mathbf{W}_1 = \begin{pmatrix} .6289 & .1190 \\ .1190 & .5485 \end{pmatrix} \quad \text{and} \quad \mathbf{W}_2 = \begin{pmatrix} .6388 & .1542 \\ .1542 & .7278 \end{pmatrix}.$$

The solution of equation (16.39) gives

$$\mathbf{c}' = (1.2872 \quad -.2523),$$

with $\lambda = 1.0048$. The latter shows that very little is lost by omitting variables x_5 and x_6 in the structural equation. Weights for the remaining x-variables can be found from equation (16.34):

$$\mathbf{b}_1' = (-.2409 \quad -.3660 \quad -.3247 \quad -.1049).$$

The structural equation is thus identified as

$$1.2872\eta_{11} - .2523\eta_{21} - .2409x_1 - .3660x_2 - .3247x_3 - .1049x_4 + \mathrm{u} = 0.$$

For comparison with the 2SLS solution in section 16.4, the equation can be rearranged and divided by 1.2872:

$$\eta_{11} = .1960\eta_{21} + .1872x_1 + .2843x_2 + .2523x_3 + .0815x_4 + \mathrm{u} = 0;$$

we see that this equation is very similar to the one obtained by the 2SLS method.

In a similar way, the second structural equation can be identified. For this equation, $\mathbf{W}_1$ is the same, but $\mathbf{W}_2$ is different, since we now omit the background variables x_1 and x_2. We can find

$$\mathbf{W}_2 = \begin{pmatrix} .7590 & .1478 \\ .1478 & .5550 \end{pmatrix},$$

and the solution of (16.3a) appears to be

$$\mathbf{c}' = (-.3049 \quad 1.3777),$$

with $\lambda = 1.0000$; the latter shows that the result could not be better. For

$\mathbf{b}_2$ we find

$$\mathbf{b}_2' = -(.0897 \quad .3420 \quad .5014 \quad .2524).$$

Again, dividing by 1.3777, and rearranging terms, we find for the structural equation

$$\eta_{21} = .2213\eta_{11} + .0651x_3 + .2482x_4 + .3639x_5 + .1832x_6,$$

which also agrees well with the 2SLS result.

16.7. Structural Equations in Terms of Factors

We conclude this chapter with a special application of the nonrecursive model, namely, where the structural equations are specified in terms of latent variables. That is, the procedures sketched in the preceding sections are applied to factor scores, which simplifies computations greatly. If, in fact, the exogenous variables are independent factors, then it follows that $\mathbf{X}'\mathbf{X}/n = \mathbf{I}$, and therefore $(\mathbf{X}'\mathbf{X}/n)^{-1} = \mathbf{I}$. Therefore, in formulas like (16.30) and (16.35), the expressions $(\mathbf{X}'\mathbf{X})^{-1}$ can be omitted, which omission simplifies them considerably.

We shall illustrate the procedure for the example used earlier in this chapter. First, we want to replace the six background variables by factors. For this purpose a principal-components analysis was applied to the matrix of correlations between the six variables, and three factors have been retained. They were rotated in order to obtain more meaningful interpretation; the rotation was found by visual inspection. The results are given in tables 16.1, 16.2, and 16.3. An interpretation of the three

Table 16.1. First three principal components for x-variables in correlation-matrix table 14.3.

	I	II	III
x_1	.2996	.5865	.6616
x_2	.6664	.1177	.2424
x_3	.6003	−.2602	−.0943
x_4	.5725	−.5384	.1020
x_5	.7258	−.0017	−.2215
x_6	.3315	.6380	−.6058

Table 16.2. Transformation matrix applied to table 16.1 to obtain rotated factor matrix in table 16.3.

.3162	.3162	.8944
.6325	.6325	−.4472
.7071	−.7071	—

rotated factors is that the first refers to the respondent's parents' aspiration, the second to the best friend's parents' aspiration, and the third to both intelligence and family socioeconomic status for respondent and best friend jointly. We shall call this factor "intelligence," for convenience.

Similarly, a factor analysis was applied to the y-variables, the four aspiration measures. Two factors from the principal-components analysis were retained, and they were obliquely rotated, resulting in an aspiration factor for the respondent, and an aspiration factor for the best friend, with a correlation of .4450 between them. Results are given in tables 16.4, 16.5, and 16.6.

Let us call the rotated factors for the x-variables χ, and those for the y-variables η. We then postulate a nonrecursive model as sketched in figure 16.4. The model implies two structural equations:

$$a_{11}\boldsymbol{\eta}_1 + a_{12}\boldsymbol{\eta}_2 + b_{11}\boldsymbol{\chi}_1 + b_{13}\boldsymbol{\chi}_3 + \mathbf{u}_1 = \mathbf{0};$$
$$a_{22}\boldsymbol{\eta}_2 + a_{21}\boldsymbol{\eta}_1 + b_{22}\boldsymbol{\chi}_2 + b_{23}\boldsymbol{\chi}_3 + \mathbf{u}_2 = \mathbf{0}.$$

In words, we assume that the respondent's aspiration factor depends on his intelligence, on the aspiration of his parents, and on the aspiration of his best friend. The aspiration of his best friend's parents has no influence on his own aspiration, but might affect him indirectly via his best friend's aspiration. The second equation mirrors the first.

Table 16.3. Rotated factor matrix obtained from table 16.1 after rotation.

x_1	.9335	−.0021	.0057
x_2	.4566	.1138	.5434
x_3	−.0414	.0919	.6533
x_4	−.0874	.0569	.7528
x_5	.0718	.3850	.6499
x_6	.0800	.9369	.0112

Table 16.4. First two principal components for y-variables in correlation matrix table 14.3.

	I	II
y_1	.7738	.4584
y_2	.7520	.5009
y_3	.7603	−.4959
y_4	.7822	−.4530

To solve, we need the matrix of correlations between factors. It can be found from

$$\chi = \mathbf{X}\mathbf{F}_x\Lambda_x^{-1}\mathbf{T}_x, \tag{16.41}$$

where $\mathbf{F}_x$ is the principal-components factor matrix for the x-variables, Λ_x is the matrix of eigenvalues, and $\mathbf{T}_x$ is the rotation matrix that was

Table 16.5. Rotation matrix applied to table 16.4 to obtain table 16.6.

.85	.85
.5268	−.5268

applied to $\mathbf{F}_x$. To verify (16.41), let $\mathbf{A}$ be the matrix of weights that is applied to the observed scores to obtain factor scores, so that $\chi = \mathbf{XA}$. To solve for $\mathbf{A}$, premultiply both sides by $\mathbf{X}'$, which gives $\mathbf{X}'\chi = \mathbf{X}'\mathbf{XA}$, and therefore

$$\mathbf{A} = (\mathbf{X}'\mathbf{X})^{-1}\mathbf{X}'\chi.$$

But $\mathbf{X}'\chi/n$ is the matrix of correlations between observed variables and

Table 16.6. Rotated factor matrix for y-variables obtained from table 16.4 by applying the oblique rotation matrix in table 16.5.

y_1	.8992	.4162
y_2	.9031	.3753
y_3	.3850	.9075
y_4	.4262	.9035

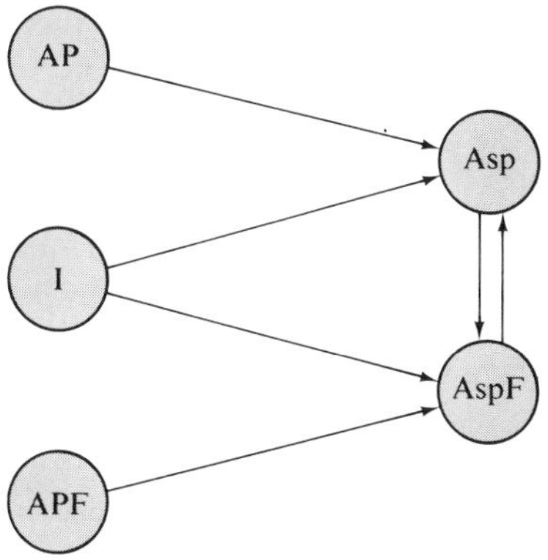

Figure 16.4 Hypothetical nonrecursive model for relations between background factors Aspiration of Parents (AP), Intelligence (I), and Aspiration of Parents of Friend (APF), and intermediating factors Aspiration of Respondent (Asp) and Aspiration of Friend (AspF).

factors, which is the factor matrix $\mathbf{F}_x\mathbf{T}_x$. Also, $(\mathbf{X}'\mathbf{X})^{-1} = n\mathbf{R}^{-1}$. Therefore,

$$\mathbf{A} = \mathbf{R}^{-1}\mathbf{F}_x\mathbf{T}_x = \mathbf{F}_x\Lambda_x^{-1}\mathbf{T}_x,$$

since $\mathbf{F}_x$ gives also eigenvectors for $\mathbf{R}^{-1}$.

Similarly, we have

$$\boldsymbol{\eta} = \mathbf{Y}\mathbf{F}_y\Lambda_y^{-1}\mathbf{T}_y. \qquad (16.42)$$

It follows that

$$\boldsymbol{\chi}'\boldsymbol{\chi}/n = \mathbf{T}_x'\Lambda_x^{-1}\mathbf{F}_x'\mathbf{R}_{xx}\mathbf{F}_x\Lambda_x^{-1}\mathbf{T}_x = \mathbf{T}_x'\Lambda_x^{-1}\mathbf{F}_x'\mathbf{F}_x\mathbf{F}_x'\mathbf{F}_x\Lambda_x^{-1}\mathbf{T}_x$$

$$= \mathbf{T}_x'\mathbf{T}_x = \mathbf{I}.$$

since $\mathbf{T}_x$ is an orthogonal rotation matrix.

In the same way,

$$\boldsymbol{\eta}'\boldsymbol{\eta}/n = \mathbf{T}_y'\mathbf{T}_y = \begin{pmatrix} 1 & .4450 \\ .4450 & 1 \end{pmatrix}.$$

Also

$$\boldsymbol{\chi}'\boldsymbol{\eta}/n = \mathbf{T}_x'\Lambda_x^{-1}\mathbf{F}_x'\mathbf{R}_{xy}\mathbf{F}_y\Lambda_y^{-1}\mathbf{T}_y' = \begin{pmatrix} .2995 & .0754 \\ .0589 & .2684 \\ .5111 & .5800 \end{pmatrix}.$$

But this matrix has the same role as $\mathbf{P}$ in equation (16.9).

To estimate coefficients in the structural equations, we use the LI/LGRV method. Note that $\mathbf{W}_1$ in formula (16.30) simplifies to $\mathbf{W}_1 = (\boldsymbol{\eta}'\boldsymbol{\eta}/n) - \mathbf{P}'\mathbf{P}$, or

$$\mathbf{W}_1 = \begin{pmatrix} 1.0000 & .4450 \\ .4450 & 1.0000 \end{pmatrix} - \begin{pmatrix} .3544 & .3348 \\ .3348 & .4141 \end{pmatrix} = \begin{pmatrix} .6456 & .1102 \\ .1102 & .5859 \end{pmatrix}.$$

Similarly, $\mathbf{W}_2$ can be found if we omit from $\mathbf{P}$ the second row, since in the first structural equation the second variable is omitted. We have

$$\mathbf{W}_2 = \begin{pmatrix} 1.0000 & .4450 \\ .4450 & 1.0000 \end{pmatrix} - \begin{pmatrix} .3509 & .3190 \\ .3190 & .3420 \end{pmatrix} = \begin{pmatrix} .6491 & .1260 \\ .1260 & .6579 \end{pmatrix}.$$

We then compute

$$\mathbf{W}_2^{-\frac{1}{2}} = \begin{pmatrix} 1.2589 & -.1221 \\ -.1211 & 1.2503 \end{pmatrix},$$

and determine the eigenvector of

$$\mathbf{W}_2^{-\frac{1}{2}}\mathbf{W}_1\mathbf{W}_2^{-\frac{1}{2}} = \begin{pmatrix} .9980 & -.0136 \\ -.0136 & .8919 \end{pmatrix}$$

that has the largest eigenvalue. This eigenvector is $\begin{pmatrix} .9921 \\ -.1249 \end{pmatrix}$ and the eigenvalue is .9997. For the coefficients in the structural equation, we still have to multiply the eigenvector by $\mathbf{W}^{-\frac{1}{2}}$, and we find that the coefficients are $\mathbf{a} = \begin{pmatrix} 1.2642 \\ -.2773 \end{pmatrix}$. Finally, the coefficients b are found from $\mathbf{b} = -\mathbf{P}_{13}\mathbf{a}$, where $\mathbf{P}_{13}$ is the matrix obtained from $\mathbf{P}$ by omitting the second row. The result is $\mathbf{b} = -\begin{pmatrix} .3577 \\ .4853 \end{pmatrix}$. The first structural equation is then identified as

$$1.2642\eta_1 = .2773\eta_2 + .3577\chi_1 + .4853\chi_3 + \mathbf{u},$$

or, after division by 1.2642,

$$\eta_1 = .2193\eta_2 + .2829\chi_1 + .3839\chi_3 + \mathbf{u},$$

where it can be shown that the variance of the systematic part on the right is equal to .3825, so that the variance of $\mathbf{u}$ must be equal to $1 - .3825 = .6175$, and its path coefficient equal to the square root of this, .7858.

For the second structural equation, we have to omit the first row of **P**. We find

$$\mathbf{W}_2 = \begin{pmatrix} 1.0000 & .4450 \\ .4450 & 1.0000 \end{pmatrix} - \begin{pmatrix} .2647 & .3122 \\ .3122 & .4084 \end{pmatrix} = \begin{pmatrix} .7353 & .1328 \\ .1328 & .5916 \end{pmatrix}.$$

It appears that

$$\mathbf{W}_2^{-\frac{1}{2}} = \begin{pmatrix} 1.1837 & -.1270 \\ -.1270 & 1.3211 \end{pmatrix}$$

and

$$\mathbf{W}_2^{-\frac{1}{2}}\mathbf{W}_1\mathbf{W}_2^{-\frac{1}{2}} = \begin{pmatrix} .8809 & -.0212 \\ -.0212 & .9960 \end{pmatrix}$$

with eigenvector $\begin{pmatrix} -.1758 \\ .9844 \end{pmatrix}$ and eigenvalue .9998. For the coefficients of the structural equation we obtain $\mathbf{a} = \begin{pmatrix} -.3331 \\ 1.3228 \end{pmatrix}$ and $\mathbf{b} = -\begin{pmatrix} .3354 \\ .5970 \end{pmatrix}$. The equation then becomes

$$1.3228\boldsymbol{\eta}_2 = .3331\boldsymbol{\eta}_1 + .3354\boldsymbol{\chi}_2 + .5970\boldsymbol{\chi}_3 + \mathbf{u},$$

or, after division by 1.3228,

$$\boldsymbol{\eta}_2 = .2518\boldsymbol{\eta}_1 + .2536\boldsymbol{\chi}_2 + .4513\boldsymbol{\chi}_3 + \mathbf{u},$$

where the variance of the error term is .5449, and its path coefficient equal to .7382.

In the path diagram we might wish to show not only these two structural equations, but also the way in which factors are related to the observed variables. This can be done on the basis of (16.41) and (16.42). For (16.41), we have

$$\mathbf{F}_x\Lambda_x^{-1}\mathbf{T}_x = \begin{pmatrix} .8833 & -.1205 & -.0895 \\ .3636 & -.0042 & .2738 \\ -.1161 & .0270 & .3925 \\ -.1285 & -.2833 & .4899 \\ -.0455 & .2905 & .3499 \\ -.0438 & .8754 & -.0946 \end{pmatrix}.$$

This equation determines how the χ-factors depend on the background variables themselves. On the other hand, for the η-factors, we prefer to think of them as determiners of the observed aspiration indicators. To find coefficients, we derive from (16.42) that

$$\mathbf{Y} = \boldsymbol{\eta}\mathbf{T}_y^{-1}\mathbf{F}_y'$$

and

$$\mathbf{T}_y^{-1}\mathbf{F}_y' = \begin{pmatrix} .8902 & .9177 & -.0235 & .0301 \\ .0201 & -.0331 & .9179 & .8900 \end{pmatrix},$$

which gives path coefficients leading from the aspiration factors to the aspiration indicators.

To complete the picture, it should be remarked that the aspiration indicators are not completely accounted for by the two aspiration factors, since we neglected the last two principal components for the y-variables. In fact, the first two aspiration factors explain the correlations between the

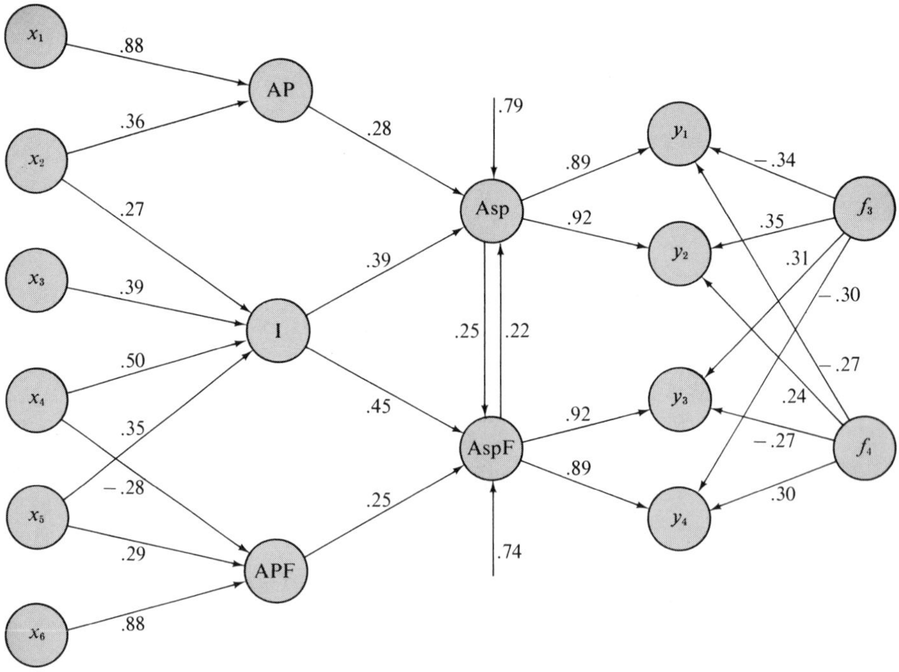

Figure 16.5 Nonrecursive model with intermediary background factors and aspiration factors. Coefficients smaller than .20 have been omitted.

y-variables only for the portion $\mathbf{F}_y\mathbf{F}_y'$. The two neglected factors are added to the path diagram in figure 16.5, and they might be interpreted as random disturbance. Also, the background factors are not the only determiners of the aspiration factors, as we saw from the structural equations; so here, too, we have to add exogenous random components. On the other hand, the background factors are completely determined by the observed background variables, since the factor analysis was applied to **R** and not to **R*** (this suggests another variant). Finally, since we have neglected three principal components in the analysis of the background variables, the diagram does not account for their possible relation with aspiration factors or aspiration indicators.

We want to stress that the illustration here is given *only* as a demonstration of computational techniques. Nevertheless, the result looks simple and plausible, and may help convince the reader that sociological theory might profit from the introduction of latent, synthetic variables by means of which the structure behind observed variables can be explained.

17

Factors and Structural Equations

17.1. Weighted Regression Analysis

Weighted regression analysis, in the form described here, was developed by Koopmans as far back as 1936 (Koopmans 1936). The purpose of the analysis is to find a structural equation with minimum disturbance in it. Such an equation is the best equation to reveal the interdependency between a set of variables, and can be used to see what effect a change in one variable has on the others. The technique has been used mainly in economics, in which the variables might be prices of competing commodities, such as beef and pork, and prices of related producer's goods, wages, etc. The coefficients in such an equation are often called *elasticity coefficients;* they indicate to what extent the value of one variable can be expected to go up or down as a function of changes in other variables.

In general, economists are not interested in finding out the general pattern of interdependencies; this pattern will be a given fact, since only those variables that are known to be closely interrelated are inserted into the equation. Therefore, the general form of the equation will never be surprising, and what one is interested in is the precise quantitative estimation of its parameters.

As usual, it is assumed that the observed variables are subject to error, so that, we may write

$$x_i = m_i + e_i \tag{17.1}$$

with m_i for the systematic component, and e_i for the error component. For convenience of the exposition, we shall also assume that x_i is standardized, although this assumption is not essential and does not affect the final results. It follows that we cannot assume that m and e are standardized.

The total system can be written as

$$\mathbf{X} = \mathbf{M} + \mathbf{E} \tag{17.2}$$

in matrix notation, with $\mathbf{X}$, $\mathbf{M}$, and $\mathbf{E}$ being $n \times k$ matrices (n observations on k variables). We assume, further, that $\mathbf{M}'\mathbf{E} = \mathbf{0}$ (error components are uncorrelated with systematic components), and that $\mathbf{E}'\mathbf{E}/n = \mathbf{U}^2$ is diagonal (error components are independent of each other). We then have

$$\mathbf{R} = \mathbf{X}'\mathbf{X}/n = \mathbf{M}'\mathbf{M}/n + \mathbf{E}'\mathbf{E}/n = \mathbf{R}^* + \mathbf{U}^2, \tag{17.3}$$

where $\mathbf{R}^*$, as usual, is the reduced correlation matrix.

The equation we are looking for will have the general form

$$b_1m_1 + b_2m_2 + b_3m_3 + \cdots + b_km_k = 0. \tag{17.4}$$

This is a homogeneous equation, defined up to an arbitrary multiplier. It might be written more compactly as $\mathbf{b}'\mathbf{m} = 0$. Note that the equation refers to systematic components, only, which therefore are free from error. However, one should not expect real data to obey the equation perfectly, since, even if the terms are errorless, disturbances in economic environment may cause the equation itself to be in error. Therefore, we shall be satisfied with a result

$$\mathbf{b}'\mathbf{m} = v, \tag{17.5}$$

where v indicates the error in the equation, and where v should be small.

With n observations, the equation corresponding to (17.5) is

$$\mathbf{Mb} = \mathbf{v}, \tag{17.6}$$

and the requirement is that $\mathbf{v}'\mathbf{v}/n$ (the error variance) be small. In fact, $\mathbf{v}'\mathbf{v}/n = \mathbf{b}'\mathbf{R}^*\mathbf{b}$, which is the term we want to minimize. But this minimization is meaningful only if we further restrict $\mathbf{b}$ to exclude the degenerate solution $\mathbf{b} = \mathbf{0}$. This restriction is found from realizing that $\mathbf{b}'\mathbf{R}^*\mathbf{b}$ is part of the variance of the transformed variable $\mathbf{Xb}$, which has variance

$$\mathbf{b}'\mathbf{X}'\mathbf{Xb}/n = \mathbf{b}'\mathbf{Rb} = \mathbf{b}'\mathbf{R}^*\mathbf{b} + \mathbf{b}'\mathbf{U}^2\mathbf{b}. \tag{17.7}$$

Therefore it will be a good procedure to minimize $\mathbf{b}'\mathbf{R}^*\mathbf{b}$ relative to $\mathbf{b}'\mathbf{U}^2\mathbf{b}$. Call the ratio of these two components λ, so that

$$\lambda = \frac{\mathbf{b}'\mathbf{R}^*\mathbf{b}}{\mathbf{b}'\mathbf{Rb}}. \tag{17.8}$$

A minimum for λ will coincide with a minimum for $\log \lambda$:

$$\log \lambda = \log (\mathbf{b}'\mathbf{R}^*\mathbf{b}) - \log (\mathbf{b}'\mathbf{U}^2\mathbf{b}). \tag{17.9}$$

Setting partial derivatives equal to zero, we obtain

$$\partial(\log \lambda)/\partial\mathbf{b}' = \frac{2\mathbf{R}^*\mathbf{b}}{\mathbf{b}'\mathbf{R}^*\mathbf{b}} - \frac{2\mathbf{U}^2\mathbf{b}}{\mathbf{b}'\mathbf{U}^2\mathbf{b}} = \mathbf{0}, \tag{17.10}$$

from which it follows that

$$\mathbf{R}^*\mathbf{b} = \frac{\mathbf{b}'\mathbf{R}^*\mathbf{b}}{\mathbf{b}'\mathbf{U}^2\mathbf{b}} \cdot \mathbf{U}^2\mathbf{b} = \lambda\mathbf{U}^2\mathbf{b}. \tag{17.11}$$

To solve, we use the auxiliary variable $\mathbf{q} = \mathbf{Ub}$. Then (17.11) can be rewritten as

$$\mathbf{R}^*\mathbf{U}^{-1}\mathbf{q} = \lambda\mathbf{Uq}$$

or

$$\mathbf{U}^{-1}\mathbf{R}^*\mathbf{U}^{-1}\mathbf{q} = \mathbf{q}\lambda. \tag{17.12}$$

The generalized form of (17.12) becomes

$$\mathbf{U}^{-1}\mathbf{R}^*\mathbf{U}^{-1}\mathbf{Q} = \mathbf{Q}\Lambda. \tag{17.13}$$

Equations (17.12) and (17.13) show that the solution is identical to that for canonical factor analysis, since we have the same equations as (15.12) and (15.14). The analogy can be made even closer if we choose the arbitrary proportionality factor for **b** to make $\mathbf{b}'\mathbf{U}^2\mathbf{b} = 1$, which implies that we have the spherical model for the error component, and that $\mathbf{v}'\mathbf{v}/n = \mathbf{b}'\mathbf{R}^*\mathbf{b} = \lambda$. To identify the structural equation (17.6), we should therefore take the solution for **b** that has the smallest value for λ associated with it.

At first sight this result may seem peculiar, and even shocking to psychologists who habitually neglect factors that contribute little to the total variance. In the economic application, one is interested in exactly the result that psychologists neglect, not in the results that psychologists would emphasize. What should we make of this difference?

It appears that there is a good reason for the difference. Take our economic example, where prices of competing commodities are related to prices of producer goods, wages, etc. First, note that in such a study the observations are collected during successive time periods. For instance, the n observations might refer to n successive years, one observation for each year. All the variables are measured in terms of a monetary unit. It follows that covariances between variables depend on the common underlying variable of monetary value. If, for instance, a gradual inflation causes all prices and wages to rise during the period of observation, then any variable that is measured in monetary units will, by this fact alone, be correlated with *any* other variable that is also measured in monetary units. In fact, this is the classical example of spurious correlation: incomes of Boston reverends are correlated with prices of Jamaican rum. It follows directly that the common factor of monetary value is completely irrelevant for the structural equation. Remember that in the economic study the variables are carefully selected as being intrinsically related: the prices in it are prices of competing goods, like pork and beef, where we can be sure that the price of one directly affects the price of the other. But to the extent that prices depend on monetary value, we might have taken any price to "predict" the price of pork or beef, and the structural equation might as well include earnings of Boston reverends or the price of Jamaican rum. The common factor of monetary values misses the essential point completely, for it defines the over-all economic "environment" in which a specially selected set of variables is embedded. The purpose of the structural equation is rather to specify the intrinsic relations between the variables within the selected set, apart from their joint dependence on the total economic environment. This explains why large common factors are irrelevant and not interesting.

17.2. Numerical Illustration

For an illustration we refer to the example we used for canonical factor analysis in section 15.2.3. We saw there that the matrix of correlations could be explained by four factors, although the analysis also gave a fifth factor with zero contribution; compare table 15.3, last column, which has eigenvalue equal to one for the analysis of $\mathbf{U}^{-1}\mathbf{R}\mathbf{U}^{-1}$, and therefore zero eigenvalue for the analysis of $\mathbf{U}^{-1}\mathbf{R}^{*}\mathbf{U}^{-1}$. This column gives the solution for the elasticity coefficients:

$$.003x_1 - .119x_2 - .528x_3 - .457x_4 + 1.323x_5 = 0, \qquad (17.14)$$

where we may write 0 at the right side instead of v, since the disturbance has zero variance in this special case.

It can be shown that this is the same result as we would have obtained from path analysis, if path analysis were applied to the reduced correlation matrix given in table 15.1. Path analysis would result in a final equation for x_5 like

$$x_5 = b_{51}x_1 + b_{52}x_2 + b_{53}x_3 + b_{54}x_4 + b_{5e}e_5,$$

and the reader might verify that for table 15.1 this equation becomes

$$x_5 = .090x_2 + .399x_3 + .345x_4 \qquad (17.15)$$

with zero error term. Equation (17.15) is identical with (17.14), which is seen if (17.14) is divided by 1.323. The reason why the results are identical is that the error contribution in (17.14) and (17.15) happens to be zero. Equation (17.14), which relates the five variables with minimum disturbance, therefore is also the best equation to predict x_5 (or any other variable in it, for that matter) from the remaining variables. The reader may remember that the data in table 15.1 were somewhat modified from the empirical correlations obtained by Blau and Duncan. I can now confess that the reason for this slight falsification was to demonstrate the present point: canonical factor analysis and path analysis (and square-root factor analysis, too, for that matter) converge to the same final, structural equation that relates all variables, if the reduced correlation matrix $\mathbf{R}^*$ has rank smaller than m.

To see the point more clearly, it might be helpful to look at a case with two variables only. Imagine that systematic components are mapped as a point in a V_2, with m_1 and m_2 as coordinates. The points will form a kind of elliptic cloud. In weighted regression analysis, we are looking for a vector in V_2 with the smallest variance. This vector will coincide with the shortest axis of the ellipse. In regression analysis, however, we want to predict one variable from the other, and the error variance will correspond to the width of the ellipse as measured along one of the coordinate axes. In general, the latter variance will be larger than the variance along the shortest axis of the ellipse, the only exceptions being when the variables are uncorrelated (so that the shortest axis of the ellipse is parallel to one of the coordinate axes), or when the ellipse degenerates to a straight line (so that both variances are zero). In our example, we have the latter alternative, or what is analogous to it for an ellipsoid in a V_5.

17.3. Alternative Structural Equations

In the example in section 17.2, we found one zero eigenvalue for $\mathbf{U}^{-1}\mathbf{R}^*\mathbf{U}^{-1}$, and therefore we could identify a structural equation with zero disturbance. We may ask whether we could not use, as an alternative solution for the structural equation, the second-best solution with which second-smallest eigenvalue is associated. This would be a particularly reasonable question if the two smallest eigenvalues are both small. In our example, this would lead to an equation from the fourth column in table 15.3, with eigenvalue 2.24. The equation becomes

$$1.98x_1 - .93x_2 - 1.24x_3 + .84x_4 + .02x_5 = v. \qquad (17.16)$$

One reason why we might be interested in this second-best equation, in our example, is that it contains x_1, whereas equation (17.14) does not—more precisely, in equation (17.14) the coefficient of x_1 approximates the rounding error. Equation (17.14) therefore cannot be used to make statements about how x_1 is affected by the other variables. In an economic example, however, this situation could hardly be expected to turn up, because the special selection of intrinsically related variables seems to preclude a solution with one variable missing. Apart from that, it would not be considered good practice, in economic applications, to consider alternative structural equations, again because one knows pretty well what the structural equation should look like; if the empirical result deviates from the theoretical expectation, the conclusion should be that the model has failed. This somewhat rigorous point of view is reinforced by the consideration that, once we allow for a second structural equation, we run into problems of identifiability, for, if both (17.14) and (17.16) are acceptable equations, then any linear combination of the two also must have disturbance within acceptable limits. But this fact means that there are not just two alternatives, but infinitely many.

Still, we might have some good hypothesis about what the structural equation should be, and we might then try to find a linear combination of the two best structural equations that is as much in agreement with theory as we can make it. To illustrate this on our example, we shall formulate a most fanciful theory that should by no means be taken seriously; it merely serves the demonstration. Suppose, then, that the son's first and final jobs will depend upon the son's education and the father's occupational status. However, fathers with high occupational status are relatively older and will have received their educations in earlier times, when higher education was still for the very few. Then fathers with higher status will tend to have less education, with the consequence that, in the structural equation, the son's first and final jobs will be dependent on the father's education but with a negative sign. This negative sign would then reveal our fanciful assumption about a society that is very rapidly developing in educational matters.

Suppose, further, that the structural equation is something like

$$x_1 - .5x_2 - .6x_3 + .5x_4 + .5x_5 = v$$

or, rearranged,

$$.5x_4 + .5x_5 = -x_1 + .5x_2 + .6x_3. \qquad (17.17)$$

If this idea is acceptable, then there must be some combination of equations (17.14) and (17.16) that is most similar to the ideal equation (17.17). To solve, we might collect the coefficients of (17.14) and (17.16)—i.e., the two final columns of table 15.3—in a 5×2 matrix $\mathbf{B}$, say, and find weights $\mathbf{x}$ for which $\mathbf{Bx} = \mathbf{y}$, so that $\mathbf{y}$ agrees with the coefficients in (17.17), which we shall call $\mathbf{c}$. One criterion is to stipulate that $\mathbf{y'y}$ must be equal to $\mathbf{c'c}$, and that we maximize $\mathbf{y'c}$ under this condition. The solution is to take $\mathbf{x}$ proportional to $(\mathbf{B'B})^{-1}\mathbf{B'c}$, and to so select the proportionality constant that $\mathbf{y'y} = \mathbf{c'c}$. The solution appears to be

$$1.01x_1 - .51x_2 - .78x_3 + .30x_4 + .37x_5 = v,$$

which is perhaps not too much different from the ideal.

In fact, all we are doing is rotating factors to agree with some criterion. As we have seen, there is an essential identity between structural equations and factors: factor loadings are directly related to coefficients in structural equations, and rotating factors results in synthetic structural equations. Our point here is not so much to recommend the procedure as to show that there are such basic relations between factors and structural equations, and to show that arguments against the combination of structural equations will also be arguments against factor rotation. It is up to the researcher's own insights, and it will depend on what exactly he is studying, whether he decides to allow for rotation or not.

17.4. Nonlinear Structural Equations

17.4.1. Statement of the Problem. Sometimes one will be interested in nonlinear compounds of the observed variables. The idea is best illustrated for the simplest case, where we have points in a plane. Usually we take it for granted that the points are scattered around their mean according to a unimodal bivariate random distribution. However, it might appear that the points are located in a more systematic way, in that they follow a certain curved trend in the plane. Figure 17.1 gives an example. One may argue that, in such a case, the underlying structure is essentially one-dimensional, and that the relevant dimension is given by the order of the observations along the curve. Why this dimension appears as a curve poses a quite different problem in interpretation.

Formally, the problem is to find some expression for the curve. The simplest class of functions that can be used for this purpose is given by the

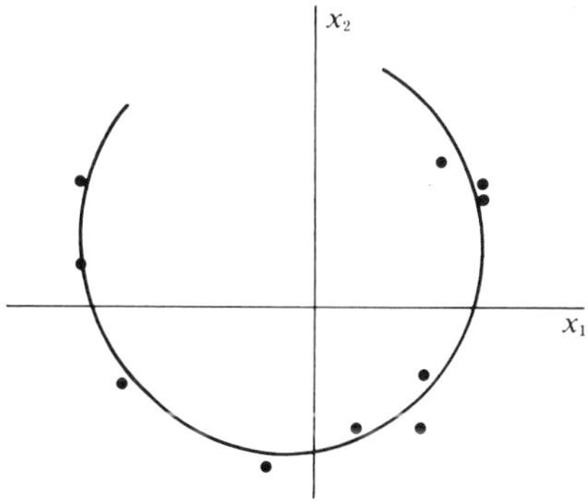

Figure 17.1 Graph of the points given in table 17.1 with best-fitting quadratic equation.

quadratic equation

$$c_1x_1^2 + c_2x_1x_2 + c_3x_2^2 + c_4x_1 + c_5x_2 + c_6 = 0, \qquad (17.18)$$

which equation covers the conic sections. A given point with coordinates (x_{1i}, x_{2i}) is located on the curve if the coordinates satisfy the equation. For points not on the curve, the equation is not satisfied unless we allow the term at the right to be different from zero. If the difference is small, the point will be located near the curve. Our problem can therefore be phrased as follows: for a given set of points, identify equation (17.18) in such a way that the terms at the right are as small as possible. Note, however, that the term at the right does not represent the distance between a point and the curve.

17.4.2. Solution. A solution can be obtained easily if we transform the problem to the familiar linear case. To see this, think of x_1^2, x_1x_2, and x_2^2 as additional coordinates that, together with x_1 and x_2, define an S_5. It follows that, if all points in the original plane are located on a quadratic curve, the points in the S_5 are restricted to a V_4 hyperplane, since equation (17.18) now means a linear restriction in the S_5. In other words, there is a vector in the S_5 on which the points in the S_5 have identical projections, equal to the constant c_6 in (17.18), provided that the sum of the squares of the other coefficients c_1 to c_5 is set equal to unity, for here these coefficients represent direction cosines in the S_5. It also follows that if we center the coordinates in the S_5 (i.e., take deviations from the mean for all five coordinates), there must be a vector with zero projections. The obvious approach to a best fit, therefore, is to find the coefficients c_1 to c_5 as direction cosines of the vector in the S_5 that has the smallest sum of squared projections on it.

To formalize this, we introduce new coordinates as follows. Let $t_1 = x_1^2 - \overline{x_1^2}$, $t_2 = x_1x_2 - \overline{x_1x_2}$, and $t_3 = x_2^2 - \overline{x_2^2}$, where it is assumed that x_1 and x_2 are already taken as deviations from the mean. For symmetry, write $t_4 = x_1$, and $t_5 = x_2$. Then, given an $n \times 2$ matrix for the original coordinates of n points in a plane, we can easily compute the corresponding $n \times 5$ matrix of centered coordinates $\mathbf{T}$ in which $\mathbf{t}_1$ to $\mathbf{t}_5$ are columns. Take a linear compound $\mathbf{Tc} = \mathbf{y}$, where $\mathbf{c}$ is a 5×1 vector of coefficients, and $\mathbf{y}$ the $n \times 1$ vector of linear compounds. The objective is to minimize $\mathbf{y}'\mathbf{y}$ under the condition that $\mathbf{c}'\mathbf{c} = 1$. The solution then is found easily. For the auxiliary function

$$F = \mathbf{y}'\mathbf{y} - \mu\mathbf{c}'\mathbf{c} = \mathbf{c}'\mathbf{T}'\mathbf{Tc} - \mu\mathbf{c}'\mathbf{c}, \qquad (17.19)$$

we obtain the partial derivatives

$$\partial F/\partial \mathbf{c} = 2\mathbf{T}'\mathbf{T}\mathbf{c} - 2\mu\mathbf{c} \tag{17.20}$$

and set them equal to zero. The resulting equation in **c** is

$$\mathbf{T}'\mathbf{T}\mathbf{c} = \mu\mathbf{c}, \tag{17.21}$$

which shows that **c** is an eigenvector of **T′T**. The eigenvalue μ represents the sum of the squared y's, as is seen immediately if we premultiply (17.21) by **c′**. It follows that, for the best-fitting quadratic equation, we should take the smallest eigenvalue solution.

17.4.3. Numerical Example. Table 17.1 gives coordinates of ten points

Table 17.1. Coordinates of ten points in a plane that follow a curved trend.

−1.116	.613
−1.124	.220
−.932	−.322
−.218	−.750
.200	−.555
.504	−.320
.468	−.598
.799	.535
.801	.496
.618	.680

in two coordinates centered around the mean.* The solution of (17.21) produces an eigenvalue of .0517 with eigenvector

$$\mathbf{c}' = (.6506 \quad .0353 \quad .6417 \quad .2201 \quad -.3397).$$

The best-fitting curve therefore is an ellipse that we can identify as

$$.65x_1^2 + .04x_1x_2 + .64x_2^2 + .22x_1 - .34x_2 - .55 = 0,$$

* The points represent Dutch political parties, and were obtained from a study in which subjects judged similarity between parties presented in pairs. The spatial map was obtained by a Kruskal nonmetric multidimensional analysis. The order of the ten points in table 17.1 corresponds roughly to a dimension that could be interpreted as political "left" to "right."

where the constant term (−.55) is obtained by calculating the value of

$$-(c_1\overline{x_1{}^2} + c_2\overline{x_1 x_2} + c_3\overline{x_2{}^2}).$$

Figure 17.1 pictures the result.

17.4.4. Generalization. A generalization of the procedure is straightforward. For instance, if we want to fit a quadratic equation to points given in an S_3, the equivalent of matrix **T** in section 17.4.2 becomes an $n \times 9$ matrix, since, apart from x_1, x_2, and x_3, we have to introduce coordinates $x_1{}^2$, x_1x_2, x_1x_3, $x_2{}^2$, x_2x_3, and $x_3{}^2$. The function represents an ellipsoid or hyperboloid, as the case may be. As a further generalization, for an algebraic equation of degree p fitted to original points in an S_m, the number of columns of the auxiliary matrix **T** will equal $\binom{m+p}{m} - 1$. Of course, we always can find a perfectly fitting equation if the original number of points is $n = \binom{m+p}{m}$; such an equation is trivial.

17.4.5. Nonlinear Factor Analysis. The procedure I have described is sometimes called "nonlinear factor analysis." The term is perhaps a bit misleading, since in factor analysis we generally are interested in compounds with large variance, whereas here we are looking for a (nonlinear) compound with small variance. I therefore prefer to interpret the result as a structural equation rather than as a factor. On the other hand, the nonlinear procedure can very well be used as a refinement of factor analysis. Linear factor analysis could then be applied in order to reduce the dimensionality of the original data space; subsequently the remaining factor space could be inspected for nonlinear structures. If there is an acceptable one, the dimensionality of the space can be further reduced. For instance, in Figure 17.1, if we project the ten points on the ellipse, we obtain a one-dimensional solution. Similarly, if a quadratic equation fits points in an S_3, these points are located on the surface of an ellipsoid (if it is an ellipsoid), and this surface is two-dimensional.

18

Discriminant Analysis

Discriminant analysis is concerned with the following problem. We are given a matrix **X** of order $n \times m$; it contains n observations on each of m variables. Further, **X** is partitioned by rows into subgroups of observations. An example is a matrix that gives scores of n persons in m tests, where the persons are classified into subgroups such as: n_1, psychotic patients; n_2, neurotic patients; and n_3, normal persons $(n_1 + n_2 + n_3 = n)$. Discriminant analysis enables one to find out whether there is a compound score of the variables that differentiates optimally between the subgroups, to specify this compound score, and to find out how far it can be used to decide which subgroup an individual probably belongs to.

18.1. Distance Method

A traditional psychometric approach to such problems is to work with test *profiles*. An individual profile is just a row of **X**, assuming that **X** gives deviations from the mean for each variable. An individual profile can be compared with average profiles, which are obtained by averaging the variables for individuals that belong to the same subgroup. If an individual profile resembles the average neurotic profile more than it does the average normal profile, one may reasonably guess that the individual is more likely to be neurotic than normal.

A common criterion for similarity between profiles is the distance between them. In the geometric picture of **X**, the distance between profiles $\mathbf{x}_i'$ and $\mathbf{x}_j'$ is the distance between the points representing these two profiles in an S_m. Average profiles can also be mapped as points in the S_m. The distance method then implies the following decision rule: if an individual point x_i is closer to the average point for subgroup j than to any other average point, then assign the individual represented by x_i to subgroup j.

One should generally standardize **X** before the distances between profiles are calculated, since otherwise the decision rule will be biased in favor of the variables with larger variance. This would introduce some arbitrariness into the decision, especially if the variance depends on an arbitrary choice of the unit of measurement.

Distances in an S_m can be computed on the basis of the generalized Pythagorean theorem. If $\mathbf{x}_i'$ and $\mathbf{x}_j'$ are two rows of **X**, then the squared distance between the points representing $\mathbf{x}_i$ and $\mathbf{x}_j$ in the S_m is found by taking the difference vector $(\mathbf{x}_i - \mathbf{x}_j)$ and summing the squared elements of this vector:

$$d_{ij}^2 = (\mathbf{x}_i - \mathbf{x}_j)'(\mathbf{x}_i - \mathbf{x}_j).$$

To illustrate the procedure, take the matrix **X** of table 18.1, which is a

Table 18.1. Hypothetical scores in three tests for three neurotic and three normal persons.

x_1	x_2	x_3
4	6	4
3	7	6
4	4	5
6	3	8
2	1	8
5	3	5

6×3 matrix that gives raw scores for six persons in three tests, the persons being divided into two subgroups, a group of three neurotics and a group of three normals. First, we standardize scores over the whole matrix, obtaining table 18.2. Then we compute averages for the two subgroups. They are, for the normals, $(-.26 \quad .83 \quad -.65)$, and for the neurotics $(.26 \quad -.83 \quad .65)$; since the two subgroups are the same size $(n_1 = n_2)$, the first vector is the opposite of the second. We then have that the first individual's squared distance from the average of the normals is

$$d_{11}^2 = (.00 + .26)^2 + (1.00 - .83)^2 + (-1.31 + .65)^2 = .53,$$

Table 18.2. Standardized version of table 18.1.

x_1	x_2	x_3
.00	1.00	−1.31
−.77	1.50	.00
.00	.00	−.65
1.55	−.50	1.31
−1.55	−1.50	1.31
.77	−.50	−.65

so that $d_{11} = .73$. This same individual's distance from the average of the neurotics is $d_{12} = 2.69$. So we may conclude that he is more like a normal person than like a neurotic, which, of course, is not surprising in this example, where the subgroups are so small.

A formal notation for the decision rule is the following. Let $\bar{\mathbf{x}}_1$ and $\bar{\mathbf{x}}_2$ be the average profiles for the two subgroups. Then an individual with profile $\mathbf{x}_i$ will be assigned to the first subgroup if the distance in the S_m between the points $\mathbf{x}_i$ and $\bar{\mathbf{x}}_1$ is smaller than the distance between the points $\mathbf{x}_i$ and $\bar{\mathbf{x}}_2$. In a formula,

$$(\mathbf{x}_i - \bar{\mathbf{x}}_1)'(\mathbf{x}_i - \bar{\mathbf{x}}_1) < (\mathbf{x}_i - \bar{\mathbf{x}}_2)'(\mathbf{x}_i - \bar{\mathbf{x}}_2). \tag{18.1}$$

If we write $\bar{\mathbf{x}}_1 - \bar{\mathbf{x}}_2 = \mathbf{d}$, rule (18.1) can be rearranged to read

$$\mathbf{x}_i'\mathbf{d} > \tfrac{1}{2}(\bar{\mathbf{x}}_1'\bar{\mathbf{x}}_1 - \bar{\mathbf{x}}_2'\bar{\mathbf{x}}_2). \tag{18.2}$$

The expression at the right becomes zero if the two subgroups are the same size, since then $\bar{\mathbf{x}}_1 = -\bar{\mathbf{x}}_2$, and $\bar{\mathbf{x}}_1'\bar{\mathbf{x}}_1 = \bar{\mathbf{x}}_2'\bar{\mathbf{x}}_2$ (assuming there are only two subgroups). There are various alternative notations for (18.2), such as

$$\mathbf{x}_i'\mathbf{d} > \tfrac{1}{2}\mathbf{d}'(\bar{\mathbf{x}}_1 + \bar{\mathbf{x}}_2).$$

Rule (18.2) can be given a somewhat different interpretation. To see this, imagine a line connecting the two averages $\bar{\mathbf{x}}_1$ and $\bar{\mathbf{x}}_2$, and project the individual points $\mathbf{x}_i$ onto this line. Then it is clear that if $\mathbf{x}_i$ is nearer to $\bar{\mathbf{x}}_1$ than to $\bar{\mathbf{x}}_2$, the projection of $\mathbf{x}_i$ will also be nearer to $\bar{\mathbf{x}}_1$. Somewhat more generally, if we project the points on some line parallel to the line connecting the two averages, then the projection of $\mathbf{x}_i$ on this line will be nearer to the projection of $\bar{\mathbf{x}}_1$.

Such a line has direction cosines proportional to $\mathbf{d}$. The projection of $\mathbf{x}_i$ will therefore be proportional to $\mathbf{x}_i'\mathbf{d}$, and the projections of $\bar{\mathbf{x}}_1$ and $\bar{\mathbf{x}}_2$ are proportional to $\bar{\mathbf{x}}_1'\mathbf{d}$ and $\bar{\mathbf{x}}_2'\mathbf{d}$, respectively. Therefore, $\mathbf{x}_i$ is nearer to $\bar{\mathbf{x}}_1$ if

$$(\mathbf{x}_i'\mathbf{d} - \bar{\mathbf{x}}_1'\mathbf{d})^2 < (\mathbf{x}_1'\mathbf{d} - \bar{\mathbf{x}}_2'\mathbf{d})^2.$$

But this rule is equivalent to (18.1) and (18.2), as can be seen after some simple algebraic operations.

18.2. Discriminant Analysis for Two Subgroups

The objective of discriminant analysis can be easily developed from the decision rules valid for the distance model. Let us go back, therefore, to the final paragraph of the preceding section, where the decision rule was given in terms of projections of points on a line parallel to the line connecting the two averages. We have two distributions of projected points on this line: a distribution for projected points in the first subgroup, and a distribution for projected points in the second subgroup. The projections of the average points are also the means of these two distributions.

The success of the decision procedure will clearly depend on the amount of overlap between the two distributions. If the overlap is large, many wrong decisions will be made; if the distributions are far apart, only a few decisions will be in error. A measure for the amount of overlap is the ratio of the distance between the means to the spread within the distributions (we shall assume here that both distributions have equal spread): The larger this ratio is, the better the two groups can be discriminated.

Now it is clear that the numerator of this ratio is largest if we project points on a line connecting the averages or on a line parallel to such a connecting line. Here the distance between the projected averages is equal to the distance between the average points themselves, and for any other vector in the S_m the projected distance will be smaller. However, there might well be a different vector in the S_m that makes the spread within distributions become smaller, and the ratio between projected distance of the means and within-groups spread larger. We would then have a vector that discriminates better between the subgroups than does the vector through the two averages. Discriminant analysis is a technique for finding such an optimal vector.

18.2.1. Formal Treatment. We begin with the matrix $\mathbf{X}$ of order $n \times m$, which gives n observations on m variables, and assume that the observations are given as deviations from the mean. The matrix $\mathbf{X}$ can be partitioned into two submatrices: $\mathbf{X}_1$ of order $n_1 \times m$, and $\mathbf{X}_2$ of order $n_2 \times m$, with $n_1 + n_2 = n$. Subgroup averages are given in vectors $\bar{\mathbf{x}}_1$ and $\bar{\mathbf{x}}_2$, and we shall write $\bar{\mathbf{x}}_1 - \bar{\mathbf{x}}_2 = \mathbf{d}$ for the difference vector.

The first step is to transform the submatrix $\mathbf{X}_1$ into a matrix of deviations from its own subgroup means. This can be written as

$$\mathbf{X}_1 - \mathbf{u}\bar{\mathbf{x}}_1', \tag{18.3}$$

where $\mathbf{u}$ is a unit vector. For the matrix of sums of squares and cross-products, we therefore obtain

$$(\mathbf{X}_1 - \mathbf{u}\bar{\mathbf{x}}_1')'(\mathbf{X}_1 - \mathbf{u}\bar{\mathbf{x}}_1') = \mathbf{X}_1'\mathbf{X}_1 - n_1\bar{\mathbf{x}}_1\bar{\mathbf{x}}_1'. \tag{18.4}$$

Similarly for the second subgroup, we obtain the matrix

$$\mathbf{X}_2'\mathbf{X}_2 - n_2\bar{\mathbf{x}}_2\bar{\mathbf{x}}_2'. \tag{18.5}$$

We assume that the subgroups are sampled from populations with identical variance-covariance matrices. This implies that the best estimate of the within-groups matrix is obtained if we pool the matrices of sums of squares and cross-products for both subgroups. This pooled matrix must then be divided by the appropriate number of degrees of freedom to obtain the within-groups variance-covariance matrix.

The pooled matrix of sums of squares and cross-products is simply the sum of the expressions (18.4) and (18.5):

$$\mathbf{W} = \mathbf{X}'\mathbf{X} - n_1\bar{\mathbf{x}}_1\bar{\mathbf{x}}_1' - n_2\bar{\mathbf{x}}_2\bar{\mathbf{x}}_2', \tag{18.6}$$

where we have used $\mathbf{X}_1'\mathbf{X}_1 + \mathbf{X}_2'\mathbf{X}_2 = \mathbf{X}'\mathbf{X}$. The number of degrees of freedom for the matrix in (18.6) equals $n_1 - 1 + n_2 - 1 = n - 2$.

An alternative expression for 18.6 can be found by using the two equations

$$\begin{aligned} \bar{\mathbf{x}}_1 - \bar{\mathbf{x}}_2 &= \mathbf{d}, \\ n_1\bar{\mathbf{x}}_1 + n_2\bar{\mathbf{x}}_2 &= 0. \end{aligned} \tag{18.7}$$

The second equation follows directly from the assumption that $\mathbf{X}$ was given in terms of deviations from the means. The two equations (18.7) can be solved for $\bar{\mathbf{x}}_1$ and $\bar{\mathbf{x}}_2$, with the result

$$\begin{aligned} \bar{\mathbf{x}}_1 &= n_2\mathbf{d}/n, \\ \bar{\mathbf{x}}_2 &= -n_1\mathbf{d}/n. \end{aligned} \tag{18.8}$$

Substituting (18.8) in (18.6), we obtain the alternative form

$$\mathbf{W} = \mathbf{X}'\mathbf{X} - \mathbf{dd}'(n_1 n_2/n). \tag{18.9}$$

The next step is to find expressions for the projections on a line with direction cosines $\mathbf{k}$. First, the sum of the squared within-subgroups deviations from the mean is simply found from (18.9) after pre- and postmultiplication by $\mathbf{k}$. The result is

$$\mathbf{k}'\mathbf{X}'\mathbf{X}\mathbf{k} - \mathbf{k}'\mathbf{dd}'\mathbf{k}(n_1 n_2/n). \tag{18.10}$$

Again, if we divide by the number of degrees of freedom, $n - 2$, we obtain the within-subgroups pooled mean-square estimate.

We also need an expression for the projected distance between the means. Since the projections of the averages are $\bar{\mathbf{x}}_1'\mathbf{k}$ and $\bar{\mathbf{x}}_2'\mathbf{k}$, respectively, the distance between them equals

$$\bar{\mathbf{x}}_1'\mathbf{k} - \bar{\mathbf{x}}_2'\mathbf{k} = \mathbf{d}'\mathbf{k}. \tag{18.11}$$

Here we meet the difficulty that $\mathbf{d}'\mathbf{k}$ may take a negative value, whereas thus far our use of the concept of distance has implied that it has a positive value. We want to maximize the ratio of the distance between the projected means to the within-subgroups spread, but cannot do so if we take the expression $\mathbf{d}'\mathbf{k}$ for the distance between the means. The simplest way out is to square the ratio. We then maximize the squared ratio of distance between means to within-subgroup spread, and such maximization is no longer ambiguous.

Another amendment may be added. We could as well maximize the squared distance with respect to the within-subgroups sum of squared deviations rather than to the within-subgroups variance, since the division by the number of degrees of freedom—a fixed number—in the denominator of the ratio does not affect the maximum. The resulting ratio will be called α, and is defined as

$$\alpha = \frac{\mathbf{k'dd'k}}{\mathbf{k'X'Xk} - \mathbf{k'dd'k}(n_1n_2/n)}, \tag{18.12}$$

where the numerator is the square of (18.11)—note that since $\mathbf{d'k}$ is a single number, we may write $(\mathbf{d'k})^2 = \mathbf{k'dd'k}$—and the denominator is expression (18.10).

A closer look at (18.12) reveals that the numerator reappears in the second term of the denominator. This suggests that we might as well determine a maximum for the expression

$$\beta = \frac{\mathbf{k'dd'k}}{\mathbf{k'X'Xk}}. \tag{18.13}$$

A formal proof follows from the identity

$$\beta = \frac{\alpha}{1 + \alpha n_1n_2/n}. \tag{18.14}$$

We are interested only in the positive range of α and β (both are ratios of expressions that contain squares only, and are therefore positive); β is a monotonic increasing function of α, and will reach a maximum when α reaches a maximum. We may therefore concentrate on the more simple function β.

The easiest way to find the $\mathbf{k}$ for which β is maximum is to take logarithms. We then have

$$\log \beta = 2 \log (\mathbf{k'd}) - \log (\mathbf{k'X'Xk}). \tag{18.15}$$

Again, in the positive range β is a monotonic increasing function of $\log \beta$, and β and $\log \beta$ therefore will reach a maximum for the same $\mathbf{k}$. The derivative of (18.15) with respect to $\mathbf{k'}$ is

$$\frac{\partial(\log \beta)}{\partial \beta} = \frac{2\mathbf{d}}{\mathbf{k'd}} - \frac{2\mathbf{X'Xk}}{\mathbf{k'X'Xk}}, \tag{18.16}$$

and must be set equal to zero to identify $\mathbf{k}$. The result is, somewhat rearranged,

$$\mathbf{d}/(\mathbf{k}'\mathbf{d}) = \mathbf{X}'\mathbf{X}\mathbf{k}/(\mathbf{k}'\mathbf{X}'\mathbf{X}\mathbf{k}). \tag{18.17}$$

If we multiply both sides by $(\mathbf{k}'\mathbf{d})^2$, (18.17) becomes

$$(\mathbf{k}'\mathbf{d})\mathbf{d} = \frac{(\mathbf{k}'\mathbf{d})^2}{\mathbf{k}'\mathbf{X}'\mathbf{X}\mathbf{k}}\mathbf{X}'\mathbf{X}\mathbf{k} = \beta\mathbf{X}'\mathbf{X}\mathbf{k}. \tag{18.18}$$

Since β and $\mathbf{k}'\mathbf{d}$ are constants, it follows that

$$\mathbf{d} \propto \mathbf{X}'\mathbf{X}\mathbf{k}, \tag{18.19}$$

and therefore

$$\mathbf{k} \propto (\mathbf{X}'\mathbf{X})^{-1}\mathbf{d}. \tag{18.20}$$

Let us stop and reflect a little on this result. First note that if $\mathbf{X}'\mathbf{X}\mathbf{k}$ is proportional to $\mathbf{d}$, then $\mathbf{X}'\mathbf{X}\mathbf{k}/(n-1)$ is proportional to $\mathbf{d}$. Since $\mathbf{X}'\mathbf{X}/(n-1)$ is the total variance-covariance matrix $\mathbf{V}_t$, we can write

$$\mathbf{V}_t\mathbf{k} \propto \mathbf{d}, \quad \text{and} \quad \mathbf{k} \propto \mathbf{V}_t^{-1}\mathbf{d}.$$

It will also be true that

$$\mathbf{d} \propto \mathbf{X}'\mathbf{X}\mathbf{k} - \mathbf{d}(\mathbf{d}'\mathbf{k} \cdot n_1 n_2/n), \tag{18.21}$$

since $\mathbf{d}$ is proportional to the first term according to (18.19), and the second term is just $\mathbf{d}$ itself multiplied by a constant. It follows that

$$\mathbf{d} \propto (\mathbf{X}'\mathbf{X} - \mathbf{d}\mathbf{d}' \cdot n_1 n_2/n)\mathbf{k} = \mathbf{W}\mathbf{k}, \tag{18.22}$$

and the expression within parentheses is the matrix of the within-subgroups sum of squares and cross-products, $\mathbf{W}$. Therefore we also could have calculated $\mathbf{k}$ from

$$\mathbf{k} \propto \mathbf{W}^{-1}\mathbf{d}. \tag{18.23}$$

All this implies that we can calculate $\mathbf{k}$ in several ways, and that it does not matter whether we base this calculation on the total variance-covariance matrix or on the within-subgroups variance-covariance matrix, or on matrices of sums of squares and cross-products.

A second reflection refers to the choice of a particular normalization of $\mathbf{k}$. Thus far $\mathbf{k}$ is determined up to an arbitrary scalar. The maximum for α

or β does not depend on this choice, since both ratios are unaffected by the scalar. Therefore we are free to define **k** in various ways.

1. We may define $\mathbf{k}'\mathbf{k} = 1$. We then have a vector of direction cosines for the vector in the S_m that produces maximum discrimination between the two subgroups. This is how we have thought of **k** thus far.

2. We may define $\mathbf{k}'\mathbf{V}_w\mathbf{k} = 1$, where $\mathbf{V}_w$ is the within-subgroups variance-covariance matrix $\mathbf{W}/(n - 2)$. This normalization therefore implies that for the linear compound $\mathbf{x}'\mathbf{k}$ the pooled within-subgroups variance is unity, and that under this condition we maximize the squared distance between the subgroup means. That is, we might also have developed discriminant analysis from recognizing that a disadvantage of the classical distance method is that it does not account for the within-subgroups spread. We then might have followed the following line of reasoning. Let us first transform the space S_m in such a way that the within-subgroups variance becomes equal in all directions. In other words, the within-subgroups distribution of the points is made spherical. After this transformation, the best discrimination between the subgroups is given by the vector with maximum distance between the subgroup means. We see now that this philosophy would have produced the same solution for **k**. (Compare section 18.2.5.)

3. We may define $\mathbf{k}'\mathbf{V}_t\mathbf{k} = 1$. We then normalize **k** to ensure that the linear compound $\mathbf{x}'\mathbf{k}$ is standardized over the total group.

18.2.2. Numerical Example. For a simple example, we may refer to the data in table 18.3. These data are artificial, but we may imagine that they refer to scores in two tests taken by two groups of psychiatric patients, the first a group of 12 patients with organic illness (brain damage, say), and the second a group of 12 patients with functional illness (no traceable organic cause of the symptoms). The scores in table 18.3 are given as deviations from the means for the total group. The matrix could be mapped in the (x_1, x_2)-plane as two sets of 12 points, as in figure 18.1, details of which will be explained as we proceed. The objective is to find a vector in the plane on which the projections of the 24 points make the two subgroups optimally discriminable.

According to formula (18.20), these direction cosines can be found from the matrix of sums of squares and cross-products for the total set. We have

$$\mathbf{X}'\mathbf{X} = \begin{pmatrix} 184 & -1 \\ -1 & 140 \end{pmatrix},$$

Table 18.3. Hypothetical scores for 12 patients with organic illness (left side of table) and 12 patients with functional illness (right side).

X_1		X_2	
x_1	x_2	x_1	x_2
−5	1	−3	0
−4	−1	−2	−2
−4	3	−1	2
−3	−2	−1	4
−3	2	1	−3
−1	−4	1	−1
−1	−2	1	3
−1	0	2	1
1	−3	3	4
2	−4	4	−1
2	−1	4	2
5	−1	3	3
−12	−12	12	12

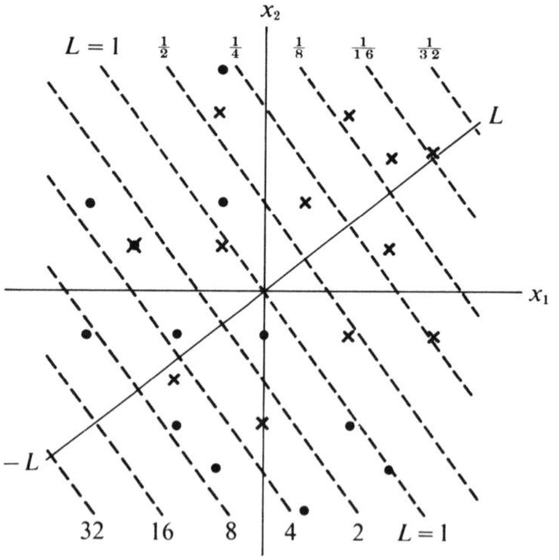

Figure 18.1 Graph of the data given in table 18.3, with discriminant direction (L to $-L$), and various levels of L.

$\bar{\mathbf{x}}_1' = (-1 \quad -1)$, and $\bar{\mathbf{x}}_2' = (1 \quad 1)$; so $\mathbf{d}' = (-2 \quad -2)$. It follows that $\mathbf{k}$ must be proportional to

$$\begin{pmatrix} 184 & -1 \\ -1 & 140 \end{pmatrix}^{-1} \begin{pmatrix} -2 \\ -2 \end{pmatrix} = \frac{1}{25{,}759} \begin{pmatrix} 140 & 1 \\ 1 & 184 \end{pmatrix} \begin{pmatrix} -2 \\ -2 \end{pmatrix}$$

$$= \begin{pmatrix} -282 \\ -370 \end{pmatrix} \Big/ 25{,}759.$$

The division by 25,759 is inessential, since we want to determine $\mathbf{k}$ up to an arbitrary scalar. If we require $\mathbf{k}'\mathbf{k} = 1$, the solution becomes

$$\mathbf{k} = \begin{pmatrix} -.606 \\ -.795 \end{pmatrix}.$$

The reader might ponder the meaning of the minus signs in $\mathbf{k}$. They are there because our procedure implied (by the definition of $\mathbf{d} = \bar{\mathbf{x}}_1 - \bar{\mathbf{x}}_2$) that the compound variable $\mathbf{Xk} = \mathbf{y}$ has positive "distance" for $\bar{\mathbf{y}}_1 - \bar{\mathbf{y}}_2$. That is, we want a high positive score in the compound to correspond to "organic illness"; so the sign of the projections must be reversed to conform to this prescription.

A second approach would be to use the within-subgroups matrix of sums of squares and cross-products. We know from (18.9) that this matrix becomes

$$\begin{pmatrix} 184 & -1 \\ -1 & 140 \end{pmatrix} - \frac{144}{24} \begin{pmatrix} -2 \\ -2 \end{pmatrix} (-2 \quad -2) = \begin{pmatrix} 160 & -25 \\ -25 & 116 \end{pmatrix}.$$

The vector $\mathbf{k}$ is proportional to $\mathbf{W}^{-1}\mathbf{d}$:

$$\begin{pmatrix} 160 & -25 \\ -25 & 116 \end{pmatrix}^{-1} \begin{pmatrix} -2 \\ -2 \end{pmatrix} = \frac{1}{17{,}935} \begin{pmatrix} 116 & 25 \\ 25 & 160 \end{pmatrix} \begin{pmatrix} -2 \\ -2 \end{pmatrix}$$

$$= \begin{pmatrix} -282 \\ -370 \end{pmatrix} \Big/ 17{,}935,$$

which is the same result we obtained before.

The ratio of the squared distance to the within-subgroups variance is 1.70; so the distance between the projected means is 1.27 times the standard deviation within subgroups.*

---

* It is interesting to note a striking analogy between this measure and the measure d' used in signal-detection theory (Green and Swets 1966). In fact, the analogy is so close that discriminant analysis indicates how signal-detection theory can be generalized to the multidimensional case. An example is given in a study by De Klerk (1968).

A unique solution for **k** can also be identified by setting the within-subgroups variance of the linear compound equal to unity. This solution should obey $\mathbf{k}'\mathbf{V}_w\mathbf{k} = 1$, and becomes

$$\mathbf{k} = \begin{pmatrix} -.274 \\ -.359 \end{pmatrix}.$$

The proof is left to the reader as an exercise. One might also calculate compound scores $\mathbf{y} = \mathbf{Xk}$, and verify directly that the pooled within-groups variance is unity. In the course of this process, it will be noticed that the two within-subgroups matrices of sums of squares and cross-products are dissimilar; so one cannot merely assume that both sets of data are sampled from populations with identical variance-covariance matrices. This problem is taken up in section 18.4.

18.2.3. Discriminant Analysis Based on the Correlation Matrix. If the data matrix is standardized first, the procedure is completely equivalent to that developed in section 18.2.1. If $\mathbf{X}_s$ is the standardized matrix, $\mathbf{X}_s'\mathbf{X}_s/n$ is the correlation matrix. It follows from equation (18.20) that we can calculate a vector of weights $\mathbf{k}_s$ proportional to $\mathbf{R}^{-1}\mathbf{d}_s$, where **R** is the correlation matrix and $\mathbf{d}_s$ the vector of differences between the subgroup averages (for the standardized variables).

The relation between $\mathbf{d}_s$ and **d** is

$$\mathbf{d}_s = \mathbf{S}^{-1}\mathbf{d}, \tag{18.24}$$

where **S** is the diagonal matrix of standard deviations. In fact, we can also write $\mathbf{R} = \mathbf{S}^{-1}\mathbf{X}'\mathbf{X}\mathbf{S}^{-1}/n$. According to equation (18.20), the vector **k** is proportional to $(\mathbf{X}'\mathbf{X})^{-1}\mathbf{d}$. For $\mathbf{k}_s$ we have

$$\mathbf{k}_s \propto (\mathbf{S}^{-1}\mathbf{X}'\mathbf{X}\mathbf{S}^{-1})^{-1}\mathbf{d}_s = \mathbf{S}(\mathbf{X}'\mathbf{X})^{-1}\mathbf{S}\mathbf{d}_s = \mathbf{S}(\mathbf{X}'\mathbf{X})^{-1}\mathbf{S}\mathbf{S}^{-1}\mathbf{d} = \mathbf{S}(\mathbf{X}'\mathbf{X})^{-1}\mathbf{d}$$

and therefore

$$\mathbf{k}_s \propto \mathbf{Sk}. \tag{18.25}$$

This means simply that $\mathbf{k}_s$ gives essentially the same weights as **k**, but they are applied to standard scores; so each weight in $\mathbf{k}_s$ has to be multiplied by the corresponding standard deviation to obtain the weight in **k**. Another way to see this is to calculate the compound scores

$$\mathbf{y}_s = \mathbf{X}_s\mathbf{k}_s = \mathbf{X}\mathbf{S}^{-1}\mathbf{S}\mathbf{k} = \mathbf{Xk}:$$

the compound scores $\mathbf{y}_s$ are identical to $\mathbf{y} = \mathbf{Xk}$.

18.2.4. Numerical Illustration. To illustrate, we take the data of table 17.1. The corresponding correlation matrix is

$$\mathbf{R} = \begin{pmatrix} 1.00 & .00 & -.08 \\ .00 & 1.00 & -.60 \\ -.08 & -.60 & 1.00 \end{pmatrix}.$$

For the vector $\mathbf{d}_s'$, we find $\mathbf{d}_s' = (-.52 \quad 1.67 \quad -1.31)$. It follows that $\mathbf{k}_s$ must be proportional to $\mathbf{R}^{-1}\mathbf{d}_s$:

$$\mathbf{k}_s \propto \frac{1}{.6336}\begin{pmatrix} .64 & .048 & .08 \\ .048 & .9936 & .60 \\ .08 & .60 & 1.00 \end{pmatrix}\begin{pmatrix} -.52 \\ 1.67 \\ -1.31 \end{pmatrix}$$

$$= \begin{pmatrix} -.357 \\ .848 \\ -.350 \end{pmatrix} \Big/ (.6336).$$

If we normalize $\mathbf{k}_s$ so that $\mathbf{k}_s'\mathbf{R}\mathbf{k}_s = 1$ (i.e., the compound score $\mathbf{y} = \mathbf{X}_s\mathbf{k}_s$ is standardized), the result becomes

$$\mathbf{k}_s' = (-.312 \quad .742 \quad -.306).$$

To compare this result with the one that would have been obtained from **X** before it was standardized, we first calculate the matrix of squares and cross-products:

$$\mathbf{X}'\mathbf{X} = \begin{pmatrix} 10 & 0 & -1 \\ 0 & 24 & -11 \\ -1 & -11 & 14 \end{pmatrix}.$$

For the difference vector **d**, we find: $\mathbf{d}' = (-2 \quad 10 \quad -6)/3$. The vector **k** therefore must be proportional to

$$(\mathbf{X}'\mathbf{X})^{-1}\mathbf{d} = \frac{1}{2{,}126}\begin{pmatrix} 215 & 11 & 24 \\ 11 & 139 & 110 \\ 24 & 110 & 240 \end{pmatrix}\begin{pmatrix} -2 \\ 10 \\ -6 \end{pmatrix} \Big/ 3$$

$$= \begin{pmatrix} -464 \\ 708 \\ -382 \end{pmatrix} \Big/ 6{,}378.$$

Here, too, we normalize $\mathbf{k}$ so that $\mathbf{k}'\mathbf{X}'\mathbf{X}\mathbf{k}/n = 1$, to standardize compound $\mathbf{y} = \mathbf{X}\mathbf{k}$. The result is

$$\mathbf{k} = (-.243 \quad .371 \quad -.200).$$

As a check, one may divide the elements of $\mathbf{k}_s$ by their respective standard deviations, which procedure should result in $\mathbf{k}$. The standard deviations are $s_1 = 1.29$, $s_2 = 2.00$, and $s_3 = 1.53$.

18.2.5. Mahalanobis Distance. In section 18.2, discriminant analysis was introduced as a kind of correction to the naive distance method. The correction becomes necessary if the variables are correlated. If they are uncorrelated, the distance method and discriminant analysis lead to the same result, provided that the distance method is applied to standardized variables (which is a good idea anyway, for most problems).

First, let us show that both methods do lead to the same result for uncorrelated variables. The proof is simple: $\mathbf{k}_s$ (as defined in section 18.2.3) is proportional to $\mathbf{R}^{-1}\mathbf{d}_s$, but when $\mathbf{R} = \mathbf{I}$, $\mathbf{k}_s$ is proportional to $\mathbf{d}_s$. If we normalize $\mathbf{k}_s$ to unity, in order to obtain a vector of direction cosines, the proportionality between $\mathbf{k}_s$ and $\mathbf{d}_s$ just means that we project on the line connecting the two subgroup averages, which is essentially what the distance method does.

In all other situations, discriminant analysis and the distance method diverge. But one can reestablish the distance method by introducing a more general concept of distance. This distance is named for Mahalanobis, who proposed the concept (Mahalanobis 1936). A formal definition is as follows. Let $\mathbf{x}_i$ and $\mathbf{x}_j$ be vectors of coordinates for two points in an S_m, and let the coordinates for the whole set of points have a variance-covariance matrix $\mathbf{V}$. Then the squared Mahalanobis distance is defined as

$$D_{ij}^2 = (\mathbf{x}_i - \mathbf{x}_j)'\mathbf{V}^{-1}(\mathbf{x}_i - \mathbf{x}_j). \tag{18.26}$$

Note that for $\mathbf{V} = \mathbf{I}$ this reduces to the classical Pythagorean theorem.

To visualize what a Mahalanobis distance means, it is easiest to take the simplest situation, with only two correlated variables. Observations on these variables can be pictured in the (x_1, x_2)-plane. If the variables are correlated, these points will be located in an elliptical region. Suppose now that the coordinates are rotated to coincide with the two axes of symmetry of the ellipse. These rotated coordinates are no longer correlated. Then we can apply the traditional distance method to the points in the plane after the new coordinates are standardized. By such standardization, the plane is shortened along the longest axis of the ellipse and expanded along the shortest axis until both coordinates have variance equal to unity. In other words, the ellipse is transformed into a circle.

In an S_m, the situation is essentially the same. Here we have points located in a hyperellipsoid region, which has m orthogonal axes of symmetry. To find Mahalanobis distances, the coordinates are rotated to coincide with these axes, and the axes are made equal in length, so that the cloud of points becomes a hypersphere with equal spread in all directions. D^2 then agrees with the classical distance measure applied to the new, transformed space.

The reader may recall that this procedure was applied in factor analysis. A brief formal proof would run as follows. $\mathbf{X}'\mathbf{X}/n = \mathbf{V}$ is the variance-covariance matrix. Let $\mathbf{VK} = \mathbf{K}\Lambda$, where $\mathbf{K}$ is the matrix of eigenvectors normalized to unity, and Λ the diagonal matrix of eigenvalues ordered in decreasing order along the diagonal. Then the transformation $\mathbf{Y} = \mathbf{XK}\Lambda^{-\frac{1}{2}}$ defines the transformation of the space we have suggested above. To prove this, we would have to show that $\mathbf{Y}'\mathbf{Y}/n = \mathbf{I}$. This is easily done:

$$\begin{aligned}\mathbf{Y}'\mathbf{Y}/n &= \Lambda^{-\frac{1}{2}}\mathbf{K}'\mathbf{X}'\mathbf{XK}\Lambda^{-\frac{1}{2}}/n = \Lambda^{-\frac{1}{2}}\mathbf{K}'\mathbf{VK}\Lambda^{-\frac{1}{2}} = \Lambda^{-\frac{1}{2}}\mathbf{K}'\mathbf{K}\Lambda\Lambda^{-\frac{1}{2}} \\ &= \Lambda^{-\frac{1}{2}}\Lambda\Lambda^{-\frac{1}{2}} = \mathbf{I}.\end{aligned}$$

The classical squared distance between two points y_i and y_j equals

$$\begin{aligned}(\mathbf{y}_i - \mathbf{y}_j)'(\mathbf{y}_i - \mathbf{y}_j) &= (\mathbf{x}_i - \mathbf{x}_j)'\mathbf{K}\Lambda^{-\frac{1}{2}}\Lambda^{-\frac{1}{2}}\mathbf{K}'(\mathbf{x}_i - \mathbf{x}_j) \\ &= (\mathbf{x}_i - \mathbf{x}_j)'\mathbf{K}\Lambda^{-1}\mathbf{K}'(\mathbf{x}_i - \mathbf{x}_j) \\ &= (\mathbf{x}_i - \mathbf{x}_j)'\mathbf{V}^{-1}(\mathbf{x}_i - \mathbf{x}_j),\end{aligned}$$

and the latter expression is the squared Mahalanobis distance, D_{ij}^2.

To show the similarity between use of the Mahalanobis distance and discriminant analysis, we should apply the Mahalanobis version of formula (18.1). This rule says that an individual point representing the vector $\mathbf{x}_i$ will be assigned to the first subgroup if its squared Mahalanobis distance to the first subgroup average is smaller than its squared Mahalanobis distance to the second subgroup average. However, we first have to agree on which matrix we should use as the variance-covariance matrix that enters the Mahalanobis formula. Should we use the total variance-covariance matrix $\mathbf{V}_t$, or the within-subgroups matrix $\mathbf{V}_w$? We can decide in favor of the latter, for reasons which will be discussed in section 18.3.1.

The statement of the rule can now be expressed as

$$(\mathbf{x}_i - \bar{\mathbf{x}}_1)'\mathbf{V}_w^{-1}(\mathbf{x}_i - \bar{\mathbf{x}}_1) < (\mathbf{x}_i - \bar{\mathbf{x}}_2)'\mathbf{V}_w^{-1}(\mathbf{x}_i - \bar{\mathbf{x}}_2). \quad (18.27)$$

Expanding terms and rearranging, we find, as an equivalent expression,

$$\mathbf{x}_i'\mathbf{V}_w^{-1}\mathbf{d} > \tfrac{1}{2}(\bar{\mathbf{x}}_1'\mathbf{V}_w^{-1}\bar{\mathbf{x}}_1 - \mathbf{x}_2'\mathbf{V}_w^{-1}\bar{\mathbf{x}}_2). \quad (18.28)$$

The term at the left is proportional to $\mathbf{x}_i'\mathbf{k}$, since $\mathbf{k}$ is proportional to $\mathbf{V}_w^{-1}\mathbf{d}$, and is therefore proportional to the linear compound score used in discriminant analysis; the expression as a whole says that x_i will be assigned to the first subgroup if the compound score exceeds some constant defined on the right. The precise meaning of this constant is not yet very clear, but it will be discussed in section 18.3.1; for the moment, it is sufficient to see that the Mahalanobis distance method is in fact fully comparable to the procedure used in discriminant analysis.

18.3. A Bayesian Interpretation of Discriminant Analysis

In 1763 a posthumous paper was published from the hand of the Reverend Thomas Bayes, "An Essay towards Solving a Problem in the Doctrine of Chances" (Bayes 1763). In this paper, Bayes derives a theorem that indicates how a priori odds in favor of some hypothesis can be corrected on the basis of empirical data. The reasoning, modernized here, is as follows. Suppose $H_1, H_2, \ldots, H_r$ is a collection of r mutually exclusive hypotheses. E.g., H_1 says that individual i is neurotic, H_2 that i is a psychotic patient, H_3 that i is normal, etc. It is assumed that each hypothesis has a certain a priori likelihood. In our example, it is perhaps not immediately clear what this might mean. However, imagine a group of men being recruited into military service. On the basis of a priori experience, it might be known what the probabilities are that a randomly selected man from this group is neurotic, psychotic, normal, etc.; these probabilities might serve as a priori probabilities $p(H_1)$, $p(H_2)$, $p(H_3)$, etc. Here $p(H_1)$ gives the probability that H_1 is true for an observation i sampled from a certain population. (Some people would object to this formulation, since H_1 is either true or not true, and there is no probability associated with the truth of H_1; an alternative expression then would be that $p(H_1)$ gives the probability that the statement "H_1 is true for observation i" is correct.)

Suppose now that an empirical datum x is collected for individual i (his score in a perception test, say). According to simple rules from probability calculus, we then have

$$p(H_g, x) = p(H_g) \cdot p(x \mid H_g) = p(x) \cdot p(H_g \mid x), \qquad (18.29)$$

where $p(H_g, x)$ is the probability that both H_g is true and x is observed; $p(H_g \mid x)$ is the probability that H_g is true if x is observed; and $p(x \mid H_g)$ is the conditional probability that x is observed if H_g is true. It follows that

$$p(H_g \mid x) = p(H_g) \cdot p(x \mid H_g)/p(x). \qquad (18.30)$$

If the set H contains all possible alternatives, then one of the hypotheses must be true. This can be written as $\sum_i p(H_i \mid x) = 1$. Then it follows from summation over (18.30) that

$$\sum_i [p(H_i) \cdot p(x \mid H_i)]/p(x) = 1, \qquad (18.31)$$

and therefore

$$p(x) = \sum [p(H_i) \cdot p(x \mid H_i)]. \qquad (18.32)$$

Substitution of (18.32) in (18.30) gives

$$p(H_g \mid x) = \frac{p(H_g) \cdot p(x \mid H_g)}{\sum [p(H_i) \cdot p(x \mid H_i)]}. \qquad (18.33)$$

Similarly, for a different hypothesis H_f, we would have

$$p(H_f \mid x) = \frac{p(H_f) \cdot p(x \mid H_f)}{\sum [p(H_i) \cdot p(x \mid H_i)]}. \qquad (18.34)$$

The ratio of (18.33) and (18.34) becomes

$$\frac{p(H_g \mid x)}{p(H_f \mid x)} = \frac{p(H_g)}{p(H_f)} \cdot \frac{p(x \mid H_g)}{p(x \mid H_f)}. \qquad (18.35)$$

On the left of (18.35), we have the ratio of a posteriori probabilities. On the right we have two factors, first the ratio of a priori probabilities, and then a factor that is called the *likelihood ratio L*. Formula (18.35) tells how a priori odds should be corrected on the basis of observation x to obtain a posteriori odds.

18.3.1. Use of the Likelihood Ratio. Suppose we consider two hypotheses, H_1 and H_2 (in terms of the earlier example, H_1 might be the hypothesis that a given patient has an organic illness, H_2 the hypothesis that his illness is functional). Suppose we collect the observation x_i (a patient's score in some test). We shall assume that we know the distribution of x under hypothesis H_1, indicated as $f_1(x)$, and also the distribution $f_2(x)$ under hypothesis H_2. Then the ratio $f_1(x_i)/f_2(x_i)$ corresponds to the likelihood ratio. We could substitute this ratio in formula (18.35) to see how the a priori odds must be corrected on the basis of observation x_i.

Very often it is much more convenient to use the logarithm of the likelihood ratio, especially if the distributions $f_1(x)$ and $f_2(x)$ are normal. We then have

$$\log [f_1(x)] = -\log \sigma_1 - \log (\sqrt{2\pi}) - (x - \bar{x}_1)^2/2\sigma_1^2, \qquad (18.36)$$

and a similar expression would be obtained for $\log [f_2(x)]$. The logarithm of the likelihood ration $f_1(x)/f_2(x)$ becomes

$$\log L = -\log \sigma_1 + \log \sigma_2 - [(x - \bar{x}_1)^2/2\sigma_1^2] + [(x - \bar{x}_2)^2/2\sigma_2^2]. \qquad (18.37)$$

If variances are equal, so that $\sigma_1 = \sigma_2 = \sigma$, equation (18.37) becomes

$$\log L = -\tfrac{1}{2}[(x - \bar{x}_1)^2 - (x - \bar{x}_2)^2]/\sigma^2. \qquad (18.38)$$

For example, suppose that $f_1(x)$ and $f_2(x)$ are normal distributions with equal variance, $\sigma^2 = 25$. For the means, assume $x_1 = 35$, and $x_2 = 29$. Suppose, further, that an individual has an observed test score $x_i = 31$. Then the log of the likelihood ratio is

$$\log L = -\tfrac{1}{2}[(31 - 35)^2 - (31 - 29)^2]/25 = -.24.$$

Let us assume that the a priori probabilities have ratio 5; in words, the odds are 5 to 1 that a randomly selected individual belongs to the first group. What are the a posteriori odds after observation x_i is made? We have, from (18.35),

$$\log \frac{p(H_1 \mid x)}{p(H_2 \mid x)} = \log \frac{p(H_1)}{p(H_2)} + \log L. \qquad (18.39)$$

Substituting what is known, (18.39) becomes

$$\log \frac{p(H_1 \mid x)}{p(H_2 \mid x)} = \log 5 - .24 = 1.609 - .24 = 1.369,$$

from which it follows that

$$\frac{p(H_1 \mid x)}{p(H_2 \mid x)} = 3.93,$$

so that the a posteriori odds are 3.93 to 1 that the individual with score $x_i = 31$ is sampled from the first population.

We shall now generalize to the multinormal case. Here we have, from equation (8.14),

$$\log [f_1(x)] = -\tfrac{1}{2}m \log (2\Pi) - \tfrac{1}{2} \log |\mathbf{V}_1| - \tfrac{1}{2}(\mathbf{x} - \bar{\mathbf{x}}_1)'\mathbf{V}_1^{-1}(\mathbf{x} - \bar{\mathbf{x}}_1), \tag{18.40}$$

where $\mathbf{V}_1$ is the variance-covariance matrix within the first subgroup. A similar form applies for $\log [f_2(x)]$. The log of the likelihood ratio becomes

$$\log [f_1(x)/f_2(x)] = -\tfrac{1}{2} \log |\mathbf{V}_1| + \tfrac{1}{2} \log |\mathbf{V}_2| - \tfrac{1}{2}(\mathbf{x} - \bar{\mathbf{x}}_1)'\mathbf{V}_1^{-1}(\mathbf{x} - \bar{\mathbf{x}}_1) + \tfrac{1}{2}(\mathbf{x} - \bar{\mathbf{x}}_2)'\mathbf{V}_2^{-1}(\mathbf{x} - \bar{\mathbf{x}}_2). \tag{18.41}$$

If the within-groups variance-covariance matrices are equal, $\mathbf{V}_1 = \mathbf{V}_2 = \mathbf{V}$, equation (18.41) simplifies to

$$\log L = -\tfrac{1}{2}(\mathbf{x} - \bar{\mathbf{x}}_1)'\mathbf{V}^{-1}(\mathbf{x} - \bar{\mathbf{x}}_1) + \tfrac{1}{2}(\mathbf{x} - \bar{\mathbf{x}}_2)'\mathbf{V}^{-1}(\mathbf{x} - \bar{\mathbf{x}}_2). \tag{18.42}$$

If we expand the terms on the right, and rearrange, a further simplification follows:

$$\log L = \mathbf{x}'\mathbf{V}^{-1}\mathbf{d} - \tfrac{1}{2}(\bar{\mathbf{x}}_1'\mathbf{V}^{-1}\bar{\mathbf{x}}_1 - \bar{\mathbf{x}}_2'\mathbf{V}^{-1}\bar{\mathbf{x}}_2). \tag{18.43}$$

It is instructive to look back at this moment to equation (18.28) and to note that it is the same as (18.43) when $\log L = 0$, or $L = 1$. That is, the Mahalanobis distance criterion is exactly the same as a criterion with a likelihood ratio equal to unity, which implies $f_1(x) = f_2(x)$. In such a situation the observation x gives no information, so to speak. The implications of this identity for the procedure under discussion in this section are that a priori odds and a posteriori odds are the same, and, for the Mahalanobis distance procedure, that no decision can be made about whether the observation is sampled from the first subgroup or from the second.

To illustrate the procedure, we go back to the example of section 18.2.2. We there found, for the within-groups variance-covariance matrix,

$$\mathbf{V}_w = \begin{pmatrix} 160 & -25 \\ -25 & 116 \end{pmatrix} \Big/ 22.$$

Also, we have $\bar{\mathbf{x}}_1' = (-1 \quad -1)$, and $\bar{\mathbf{x}}_2' = (1 \quad 1)$, so that $\mathbf{d}' = (-2 \quad -2)$. For $\mathbf{V}_w^{-1}\mathbf{d}$, we find

$$\mathbf{V}_w^{-1}\mathbf{d} = \frac{22}{17{,}935}\begin{pmatrix} 116 & 25 \\ 25 & 160 \end{pmatrix}\begin{pmatrix} -2 \\ -2 \end{pmatrix} = -\frac{22}{17{,}935}\begin{pmatrix} 282 \\ 370 \end{pmatrix}.$$

Further, since $\bar{\mathbf{x}}_1 = -\bar{\mathbf{x}}_2$, the constant term on the right of (18.43) is zero. It follows that we can identify (18.43) as

$$\log L = \mathbf{x}'\mathbf{V}_w^{-1}\mathbf{d} = -(.346 \quad .454)\mathbf{x}.$$

To illustrate further, let us take an individual with test scores $\mathbf{x}_i' = (-1 \quad -2)$. The corresponding log of the likelihood ratio is

$$\log L = -(-1 \quad -2)\begin{pmatrix} .346 \\ .454 \end{pmatrix} = 1.254,$$

from which we can find $L = 3.5$. Therefore, if the a priori odds were equal, the a posteriori odds are 3.5 to 1 that this individual is sampled from the first group.

In figure 18.1, lines are drawn for some constant values of L. The line $L = 1$, defined by $\mathbf{x}'\mathbf{V}_w^{-1}\mathbf{d} = 0$, divides the plane into two regions, one region where it is more likely that points represent patients with organic illnesses, and the other where points are more likely to represent patients with functional illnesses; this is on the assumption that a priori probabilities are equal.

If a priori probabilities are not equal, we should use the general formula (18.39) to find a posteriori odds. The latter then are no longer equal to the likelihood ratio, but to this ratio multiplied by the a priori odds. This implies that the plane should be divided into two regions according to the line

$$L \cdot p(H_1)/p(H_2) = 1, \quad \text{or} \quad L = p(H_2)/p(H_1).$$

Sometimes it may be reasonable to identify a priori odds as the ratio n_1/n_2. For instance, if in a study as in section 18.2.2 patients were drawn at random from a population of patients (in practice we may not be able to define such a population), then the ratio of the frequencies with which the two kinds of patients are sampled is a consistent estimate of their frequencies in the population, and therefore of the a priori probability ratio.

The procedure, explained so far, also assumes that errors are equally weighted. In fact, the procedure could have been developed from the intuitive idea that we want to divide the space S_m into two regions (in our example, this is done by the line $L = 1$) in such a way that the total number of misclassifications is minimized. It is one type of misclassification to decide that a patient has a functional illness whereas in fact he has an organic illness, another type to assume he has an organic illness whereas his illness is in fact functional. Using the line $L = 1$ (in a higher-dimensional space S_m, the expression $L = 1$ specifies a hyperplane V_{m-1}), we assume that the two types of misclassification are equally serious, and that we should minimize their total frequency. This assumption might not be realistic. Suppose, for instance, that patients with an organic illness have a chance to be cured by a somewhat hazardous brain operation, but, since there are limited facilities for brain surgery, not all these patients can be subjected to the operation anyway. Then one could consider that operating on a patient with a functional illness is a much more serious error than not operating on a patient with an organic illness. In such cases it is possible to correct formula (18.39). The correction is to add, on the right, a term $\log g$, where g is a number that expresses the ratio of the degree of seriousness of deciding that H_2 is true whereas in fact H_1 is true, to the degree of seriousness of the other type of misclassification.

18.4. Discriminant Analysis for Two Subgroups with Different Within-Group Variance-Covariance Matrices

In section 18.2 we assumed throughout that we could pool within-groups sums of squares and cross-products. Also, in section 18.3 we made the same assumption in order to derive expressions for likelihoods. We shall now see what should be done if the assumption cannot be made.

Obviously, we then have to accept equation (18.41) without simplification. To see what this implies, let us suppose for the moment that $m = 2$, so that (18.41) refers to a bivariate distribution. For a constant value of the log of the likelihood ratio on the left, equation (18.41) becomes a quadratic form in x_1 and x_2, and therefore defines a conic section in an S_2. With $m > 2$, the equation remains a quadratic form of higher order, so that it specifies a ellipsoid or other generalized quadratic surface in an S_m.

To illustrate, take the example in section 18.2.2. The two within-group variance-covariance matrices $\mathbf{V}_1$ and $\mathbf{V}_2$, which we have pooled earlier, are in fact rather different. We have

$$\mathbf{V}_1 = \begin{pmatrix} 112 & -25 \\ -25 & 66 \end{pmatrix} \Big/ 11, \quad \text{and} \quad \mathbf{V}_2 = \begin{pmatrix} 72 & 24 \\ 24 & 74 \end{pmatrix} \Big/ 11.$$

In order to identify equation (18.30), we calculate

$$\mathbf{V}_1^{-1} = \frac{11}{6{,}767}\begin{pmatrix} 66 & 25 \\ 25 & 112 \end{pmatrix} \quad \text{and} \quad \mathbf{V}_2^{-1} = \frac{11}{4{,}752}\begin{pmatrix} 74 & -24 \\ -24 & 72 \end{pmatrix}.$$

It follows that the determinants are

$$|\mathbf{V}_1| = 6{,}767/121, \quad \text{and} \quad |\mathbf{V}_2| = 4{,}752/121.$$

What we shall need are the logarithms of these values:

$$\log|\mathbf{V}_1| = 4.0240, \quad \text{and} \quad \log|\mathbf{V}_2| = 3.6705.$$

The constant

$$-\tfrac{1}{2}\log|\mathbf{V}_1| + \tfrac{1}{2}\log|\mathbf{V}_2|$$

then becomes $-.1767$.

We proceed by expanding $(\mathbf{x} - \bar{\mathbf{x}}_1)'\mathbf{V}_1^{-1}(\mathbf{x} - \bar{\mathbf{x}}_1)$. It becomes

$$\frac{11}{6{,}767}(66x_1^2 + 50x_1x_2 + 112x_2^2 + 182x_1 + 274x_2 + 228),$$

whereas for $(\mathbf{x} - \bar{\mathbf{x}}_2)'\mathbf{V}_2^{-1}(\mathbf{x} - \bar{\mathbf{x}}_2)$ we obtain

$$\frac{11}{4{,}752}(74x_1^2 - 48x_1x_2 + 72x_2^2 - 100x_1 - 96x_2 + 98).$$

It can be verified that equation (18.42) now can be identified as

$$\log L = .0320x_1^2 - .0962x_1x_2 - .0077x_2^2 - .2637x_1 - .3338x_2 - .2486.$$

For $L =$ constant, this is the equation of a hyperbola. In particular, the hyperbola $\log L = 0$ divides the plane into two regions, which correspond to the decision that is more likely (assuming equal a priori probabilities and equal seriousness of the two types of misclassification). Figure 18.2 gives a picture that can be compared with figure 18.1.

In section 8.1 we showed that $f_1(x)$ = constant can be represented as an equiprobability contour. For $\log L = 0$ we have $f_1(x) = f_2(x)$, and it follows that the likelihood curve $\log L = 0$ passes through the intersections of corresponding equiprobability contours. I.e., if $f_1(x) = f_2(x) = c$, with c some constant, then we have two equiprobability contours,

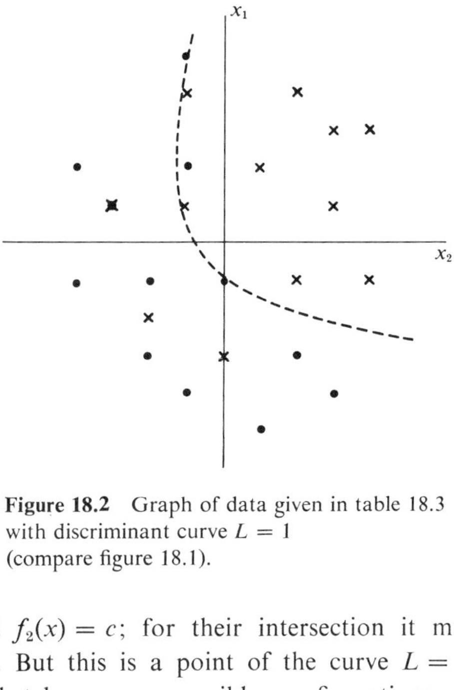

Figure 18.2 Graph of data given in table 18.3 with discriminant curve $L = 1$ (compare figure 18.1).

$f_1(x) = c$, and $f_2(x) = c$; for their intersection it must be true that $f_1(x)/f_2(x) = 1$. But this is a point of the curve $L = 1$, or $\log L = 0$. Figure 18.3 sketches some possible configurations. If the variance-covariance matrices $\mathbf{V}_1$ and $\mathbf{V}_2$ are known, it is not too difficult to make a rough sketch of the situation in the spirit of figure 18.3. This will give a first idea of how the likelihood curve will look like.

18.5. Relation Between Discriminant Analysis and Multiple Correlation Theory

The only difference between discriminant analysis and multiple correlation analysis is that in multiple correlation analysis a continuous criterion variable is given, whereas in discriminant analysis the criterion is dichotomic if we have two subgroups. In fact, we might distinguish between the two subgroups by introducing a dummy variable that has, say, a value of one for the first subgroup, and of zero for the second. Discriminant analysis can then be reduced to multiple correlation analysis by using this

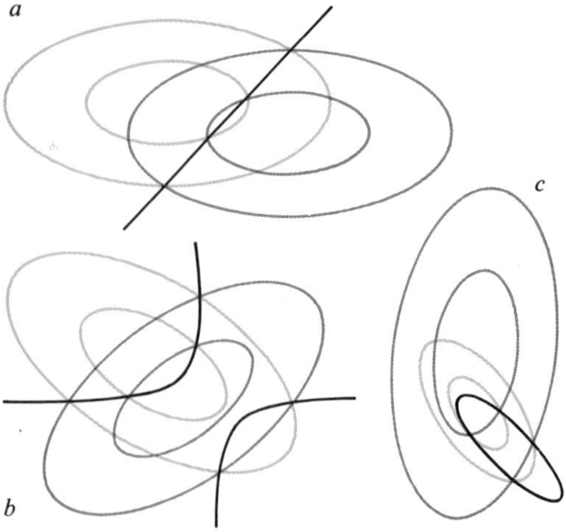

Figure 18.3 Discriminant curves $L = 0$ as intersections of corresponding equiprobability contours. In (a) the variances and the covariance are identical for both distributions, and the discriminant curve becomes a straight line. Example (b) illustrates a hyperbolic curve, and (c) an ellipse.

dummy variable as the criterion variable, and by calculating correlations between the observed variables and the dummy. The correlations should be point biserial correlations if we have reason to believe that the dummy variable is really dichotomous, or biserial correlations if we assume that the distinction in the two subgroups is an arbitrary dichotomy applied to an underlying continuous variable. Whatever the case, the result would be that such correlations between the dummy variable and an observed variable are proportional to the difference between the means for the subgroups in the observed variable. It follows that the weights (we called them **b**) in the multiple correlation analysis will be proportional to the weights **k** obtained from discriminant analysis. In fact, note that **b** was defined as proportional to $\mathbf{R}^{-1}\mathbf{r}$—see equation (11.15)—and **k** as proportional to $\mathbf{R}^{-1}\mathbf{d}$ (see section 18.2.3); and the two expressions come to the same thing.

18.6. Discriminant Analysis for More Than Two Subgroups

If the matrix **X** of observed measurements can be partitioned into more than two subgroups (such as neurotics, psychotics, normals), the treatment can be just a generalization of the two-groups case. All we do is calculate likelihood curves for pairwise sets of groups. We shall illustrate the idea for three subgroups; the generalization to a larger number is straightforward.

With three subgroups, we can form three pairs of two groups and can therefore define three likelihood curves. We shall assume that within-groups variance-covariance matrices can be pooled, so that these curves are straight lines. We then have

$$\begin{aligned} \log L_{12} &= \log [f_1(x)] - \log [f_2(x)], \\ \log L_{23} &= \log [f_2(x)] - \log [f_3(x)], \\ \log L_{31} &= \log [f_3(x)] - \log [f_1(x)], \end{aligned} \tag{18.44}$$

and the space S_m could be divided into regions by using the hyperplanes $\log L_{12} = 0$, $\log L_{23} = 0$, and $\log L_{31} = 0$. Note that the last equation of (18.44) is linearly dependent on the first two (since the sum of the right sides is zero). It follows that the three hyperplanes intersect in a V_{m-2}. In particular, if there are two observed variables only, so that (18.44) gives three lines in an S_2, then these three lines will intersect in one point. They divide the plane into $3! = 6$ different regions, each of which has a certain rank order for the a posteriori probabilities of the three hypotheses.

To illustrate, we use the data in table 18.4. They are artificial data, and might refer to scores in two tests taken by three groups of four persons

Table 18.4. Hypothetical scores for three groups of four individuals in two tests.

I		II		III	
7	2	5	1	2	2
6	3	3	2	1	3
2	−5	−3	7	−9	−2
5	−2	−9	−4	−10	−7
5	−.5	−1	1.5	−4	−1

(normals, neurotic patients, psychotic patients, say). The table gives the scores as deviations from the means.

We have, from table 18.4,

$$\mathbf{d}_{12}' = (6 \quad -2),$$
$$\mathbf{d}_{23}' = (3 \quad 2.5),$$
$$\mathbf{d}_{31}' = (-9 \quad -.5),$$

where the reader might verify that $\mathbf{d}_{12}' + \mathbf{d}_{23}' + \mathbf{d}_{31}' = \mathbf{0}$. The pooled within-groups variance-covariance matrix $\mathbf{V}_w$ is

$$\mathbf{V}_w = \begin{pmatrix} 256 & 133 \\ 133 & 164 \end{pmatrix} \Big/ 9,$$

where we divide by the number of degrees of freedom. It can be found that

$$|\mathbf{V}_w| = 24{,}295/81,$$

and

$$\mathbf{V}_w^{-1} = 9\begin{pmatrix} 164 & -133 \\ -133 & 256 \end{pmatrix} \Big/ 24{,}295 = \begin{pmatrix} .0608 & -.0493 \\ -.0493 & .0948 \end{pmatrix}.$$

We then calculate

$$\mathbf{V}_w^{-1}\mathbf{d}_{12} = \begin{pmatrix} .463 \\ -.485 \end{pmatrix}, \qquad \mathbf{V}_w^{-1}\mathbf{d}_{23} = \begin{pmatrix} .059 \\ .089 \end{pmatrix}, \qquad \mathbf{V}_w^{-1}\mathbf{d}_{31} = \begin{pmatrix} -.523 \\ .396 \end{pmatrix},$$

where again it can be verified that these three vectors add up to zero.

For the constants in the expressions for log L—equation (18.43), used thrice for the set (18.44)—we calculate

$$\bar{\mathbf{x}}_1'\mathbf{V}_w^{-1}\bar{\mathbf{x}}_1 = 1.790, \qquad \bar{\mathbf{x}}_2'\mathbf{V}_w^{-1}\bar{\mathbf{x}}_2 = .422, \qquad \bar{\mathbf{x}}_3'\mathbf{V}_w^{-1}\bar{\mathbf{x}}_3 = .673.$$

The set of equations (18.44) then can be identified as

$$\log L_{12} = (.463 \quad -.485)\mathbf{x} - .684,$$
$$\log L_{23} = (.059 \quad .089)\mathbf{x} + .126,$$
$$\log L_{31} = (-.523 \quad .396)\mathbf{x} + .558,$$

where it may be checked that the sum of the expressions on the right is zero. Another check is that, if in the equation for L_{12} we insert $\mathbf{x} = \frac{1}{2}(\bar{\mathbf{x}}_1 + \bar{\mathbf{x}}_2)$, which represents a point midway between the two means, the result must be $\log L_{12} = 0$, and similarly for the other two equations.

Figure 18.4 gives a picture of the results. The twelve observed values are mapped in it, and lines log $L = 0$ are drawn. On the assumption that a priori probabilities are equal, and that all possible misclassifications would be equally serious, these three lines divide the plane into the six regions that we can distinguish. In the figure three sectors are indicated,

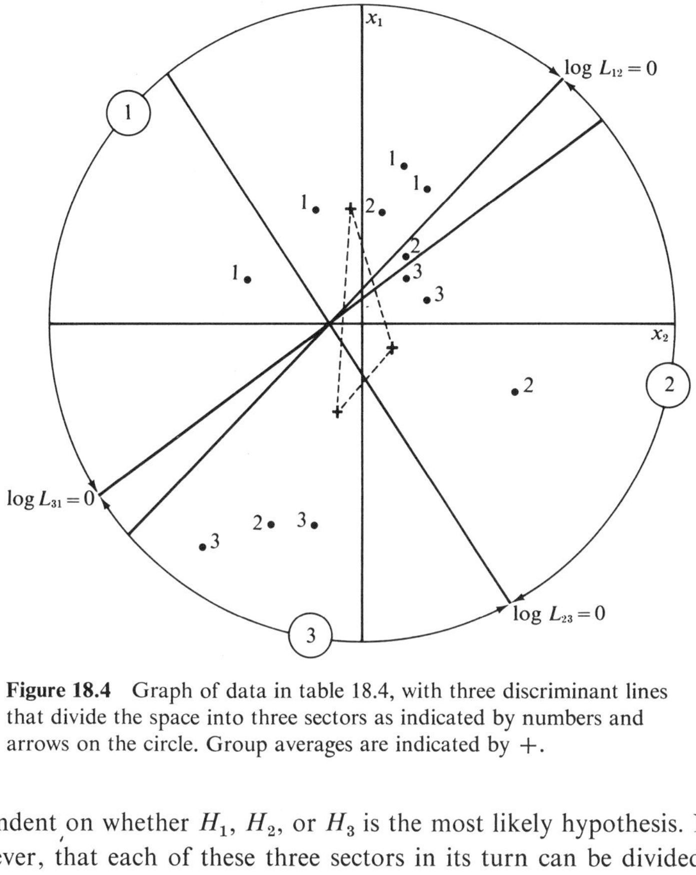

Figure 18.4 Graph of data in table 18.4, with three discriminant lines that divide the space into three sectors as indicated by numbers and arrows on the circle. Group averages are indicated by +.

dependent on whether H_1, H_2, or H_3 is the most likely hypothesis. Note, however, that each of these three sectors in its turn can be divided into two regions. For instance, the H_1 sector contains a region where H_2 is the second-best hypothesis and H_3 the least likely one, and another region where H_3 is second-best and H_2 least likely.

If we do not assume equal a priori probabilities, boundaries between these regions must be translated; they would still have one point as their intersection, though. The same is true if we give different weights to the seriousness of misclassifications.

An obvious refinement of the procedure is to take into account the fact that within-group variance-covariance matrices might be different. The regions then would be bounded by conic sections.

18.6.1. Canonical Discriminant Factor Analysis. Another approach to the situation with multiple categories would be to use canonical discriminant factor analysis. As was shown in section 15.2.4, this technique specifies vectors in an S_m in such a way that projected subgroup means are differentiated maximally relative to the within-groups variance of the projections.

In general, with k subgroups, we would obtain $k - 1$ such vectors, since the k averages of the subgroups are located in a V_{k-1}. With three subgroups, for instance, we have three points to indicate subgroup means, and these three points will always be located in a plane. It follows that there are $m - k + 1$ vectors in an S_m on which the k averages have equal projections; these vectors do not discriminate between the groups. Then, within the V_{k-1}, canonical discriminant factor analysis specifies directions that successively optimize between-means variance relative to within-groups variance.

Since the technique was discussed extensively as a variety of factor analysis, we need not repeat details here. We will point out only that there is a relationship to the approach just discussed in 18.6, in that constant likelihood-ratio hyperplanes, defined by $\log L =$ constant (boundaries between regions in an S_m if the constant is zero), will be parallel to the $m - k + 1$ vectors that do not differentiate between the means. For instance, suppose there are three subgroups, and also three observed variables: $k = 3$, $m = 3$. Then the three averages are contained in a V_2, a plane, and there is one vector in an S_3 that does not differentiate between the means (the vector perpendicular to the plane on which the means are located). Boundaries between regions, defined by $\log L = 0$, now will also be planes, orthogonal to the plane of the means, and therefore parallel to the vector perpendicular to this plane.

18.7. Multivariate Analysis of Variance

To conclude this chapter, we may mention that there is a very close analogy between canonical discriminant analysis and techniques for multivariate analysis of variance (MANOVA). We shall assume that the reader is familiar with simple designs for univariate analysis of variance. In the simplest form, data on one dependent variable are collected under several different treatments. Then the variance between treatment means is tested against the variance within treatment subgroups. Usually this is done by calculating sums of squares, dividing them by the appropriate number of degrees of freedom to obtain mean-square estimates, and then computing the ratio, F, of these mean-square estimates. However, the ratio of the sums of squares themselves is an equivalent statistic, given that, for tables with percentage points of the statistic, we have to take degrees of freedom for numerator and denominator into account anyway.

Suppose now that we have p dependent variables instead of one. An example is that the "treatments" are defined in terms of different groups of psychiatric patients who take p tests. These tests then are the p dependent variables. Generally, such tests are correlated. Therefore it would be unwise to do a separate analysis of variance for each test, for, if treatments have a significant effect for test x_1, and variable x_1 is highly correlated with variable x_2, then effects for variable x_2 will also tend to be significant. The latter is then not a new result; it follows more or less (depending on the amount of correlation) from the first. With separate analyses of variance for each variable, we would never know how much the results are duplicating each other.

The approach taken in multivariate analysis of variance is easily visualized. Suppose we have two dependent variables, so that the observations can be represented in the (x_1, x_2)-plane. If there are k different treatments, the points will fall into k different subgroups. Imagine that we also draw points for the k subgroup averages. Then if the k subgroups are sampled from one identical population (this is the null hypothesis), spread among subgroup averages should be within limits that are consistent with the spread of points within subgroups. To be somewhat more precise: on the basis of spread within subgroups, we should like to set up a confidence region for the subgroup means, so that if the means scatter outside this region the null hypothesis can be rejected at the chosen level of significance.

However, to reject the null hypothesis, we need only show that there is at least one vector in the plane such that the variance between the means projected on it relative to variance within subgroups exceeds the critical ratio. In multivariate analysis of variance, the appropriate statistic therefore is to take the vector for which this ratio of variance between means to variance within groups is maximum. Obviously, if even this largest ratio does not reach significance, there will be no reason at all to reject the null hypothesis. On the other hand, if the ratio is large enough to allow us to reject the null hypothesis, we might search for another vector orthogonal to the first and apply a similar test (of course, for our two-dimensional illustration, there is only one other direction left, but if there are m dependent variables, so that points are mapped in an S_m, the search process becomes more pertinent).

As mentioned above, instead of testing a ratio of variance estimates, we may also test the ratio of sums of squares. The solution then becomes identical to canonical discriminant analysis. For example, let **W** be the matrix of sums of squares and cross-products within groups, and **B** the matrix of sums of squares and cross-products between groups. Then $\mathbf{t}'\mathbf{B}\mathbf{t}$ is the sum of the squared projections between groups for a vector **t**, and $\mathbf{t}'\mathbf{W}\mathbf{t}$ is the pooled within-groups sum of squared projections. Let α be the ratio $\mathbf{t}'\mathbf{B}\mathbf{t}/\mathbf{t}'\mathbf{W}\mathbf{t}$. We want a maximum for α. The solution, as shown earlier in formula (15.48), is

$$\mathbf{B}\mathbf{t} = \alpha\mathbf{W}\mathbf{t}, \quad \text{or} \quad \mathbf{W}^{-1}\mathbf{B}\mathbf{t} = \alpha\mathbf{t},$$

so that **t** is an eigenvector of the nonsymmetric matrix $\mathbf{W}^{-1}\mathbf{B}$, and α the corresponding eigenvalue. The appropriate statistic obviously is the largest eigenvalue. Charts of percentage points of the largest eigenvalue have been constructed by Heck (1960; also Morrison 1967) for various parameters such as k, m, n.

The approach in multivariate analysis of variance thus appears to be completely parallel to that in canonical discriminant factor analysis. In addition, the approach can be generalized to more complex designs, such as where we have row effects, column effects, and interaction effects for the treatments. Here, too, we can calculate the $m \times m$ matrices for sums of squares and cross-products for the effect we want to test, and a similar matrix for the residual effects, and test the largest eigenvalue of the "ratio" of these matrices.

However, we will not enter into details, but will settle for these general indications, the point of which is to emphasize again the very intricate similarity that may exist between multivariate techniques that go under different names.

An Iterative Solution for Eigenvalues and Eigenvectors

Let **A** be a given matrix of order $n \times n$, and let $\mathbf{b}_{11}$ be an arbitrary $n \times 1$ vector. Calculate $\mathbf{Ab}_{11} = \mathbf{b}_{12}$; $\mathbf{b}_{12}$ is also an $n \times 1$ vector. Calculate $\mathbf{Ab}_{12} = \mathbf{b}_{13}$. If we continue this process, it will ultimately lead to a vector $\mathbf{b}_{1p}$ that is equal to the first eigenvector of **A**, so that $\mathbf{Ab}_{1p}$ is proportional to $\mathbf{b}_{1p}$.

A good practical procedure is to start with a trial vector $\mathbf{b}_{11}$ in which all elements are equal to one. Then $\mathbf{b}_{12}$ will give row totals of **A**. Then divide $\mathbf{b}_{12}$ by its largest element, so that in the resulting vector the largest element is equal to one. Continue in this way until $\mathbf{b}_{1,\,p+1}$ is proportional to $\mathbf{b}_{1p}$; i.e., if we divide $\mathbf{Ab}_{1p}$ by its largest element, we get $\mathbf{b}_{1p}$ back; the divider is then equal to the eigenvalue.

The method can be shown to converge to the eigenvector that has the largest eigenvalue associated with it. For an illustration, we take the matrix in table 12.1. The first trial vector is $\mathbf{b}_{11}$ with unit elements. The output vector is

$$\mathbf{Ab}_{11} = \begin{pmatrix} 2.623 \\ 2.776 \\ 3.025 \\ 2.828 \\ 2.864 \end{pmatrix},$$

which, divided by 3.025, becomes

$$\mathbf{b}_{12} = \begin{pmatrix} .8671 \\ .9176 \\ 1.0000 \\ .9348 \\ .9467 \end{pmatrix}.$$

Then calculate

$$\mathbf{A}\mathbf{b}_{12} = \begin{pmatrix} 2.4088 \\ 2.5764 \\ 2.8619 \\ 2.6556 \\ 2.6994 \end{pmatrix},$$

and divide by 2.8619 to obtain

$$\mathbf{b}_{13} = \begin{pmatrix} .8416 \\ .9002 \\ 1.0000 \\ .9278 \\ .9431 \end{pmatrix}.$$

Continuing in this way, we find ultimately

$$\mathbf{A}\mathbf{b}_{18} = \mathbf{A} \begin{pmatrix} .8336 \\ .8942 \\ 1.0000 \\ .9275 \\ .9439 \end{pmatrix} = \begin{pmatrix} 2.3599 \\ 2.5314 \\ 2.8309 \\ 2.6259 \\ 2.6723 \end{pmatrix};$$

dividing by 2.8309,

$$\mathbf{b}_{19} = \begin{pmatrix} .8336 \\ .8942 \\ 1.0000 \\ .9275 \\ .9439 \end{pmatrix}.$$

We see that $\mathbf{b}_{19} = \mathbf{b}_{18}$, and we have the eigenvector accurate to four decimal points, with eigenvalue 2.8309. If $\mathbf{b}_{19}$ is normalized to the eigenvalue 2.8309, we have the first column of table 13.6.

In order to find subsequent eigenvectors, normalize the eigenvector we have found to the eigenvalue. Let us call the result $\mathbf{f}_1$. Then calculate the residual matrix $\mathbf{A}_1 = \mathbf{A} - \mathbf{f}_1\mathbf{f}_1'$. In the example, the result is

$$\mathbf{A}_1 = \begin{pmatrix} .5366 & .0189 & -.1028 & -.1835 & -.2026 \\ .0189 & .4668 & -.1581 & -.1360 & -.1578 \\ -.1028 & -.1581 & .3332 & -.0804 & -.0333 \\ -.1835 & -.1360 & -.0804 & .4263 & -.0427 \\ -.2026 & -.1578 & -.0333 & -.0427 & .4058 \end{pmatrix}.$$

Then the procedure is repeated for this residual matrix. The result would be

$$\mathbf{A}_1\mathbf{b}_{21} = \begin{pmatrix} .0666 \\ .0338 \\ -.0415 \\ -.0164 \\ -.0307 \end{pmatrix};$$

dividing by .0666,

$$\mathbf{b}_{22} = \begin{pmatrix} 1.0000 \\ .5079 \\ -.6233 \\ -.2468 \\ -.4619 \end{pmatrix}.$$

Continuing, we would find at the 21st trial that

$$\mathbf{b}_{2,21} = \begin{pmatrix} 1.0000 \\ .7042 \\ -.2982 \\ -.5728 \\ -.6713 \end{pmatrix},$$

and that this vector is proportional to $\mathbf{b}_{2,22}$, with proportionality factor equal to .8219, the second eigenvalue.

The procedure in this second stage could be speeded up considerably if we first change the sign of any row and column in $\mathbf{A}_1$ where there are many negative elements. In the example, this might be done for the last three rows and columns, so that the only negative elements remaining would be a_{34}, a_{35}, and a_{45}, and their symmetric counterparts, a_{43}, a_{53}, and a_{54}. The procedure then would converge much more rapidly. We should keep in mind, however, that for the result of the procedure we must change signs again for the three last elements of the vector.

A second residual matrix $\mathbf{A}_2$ is then obtained from $\mathbf{A}_1 - \mathbf{f}_2\mathbf{f}_2'$, and the procedure can be repeated until the final eigenvector is determined.

Another way to speed up computations is to apply the procedure to a power of $\mathbf{A}$, such as $\mathbf{A}^2$ or $\mathbf{A}^3$. This does not change eigenvectors, but it results in proportionality factors that are also powers (of the same degree as for $\mathbf{A}$) of the eigenvalues of $\mathbf{A}$.

We might also warn that enough decimal places must be carried, since otherwise residual matrices can become increasingly in error and later eigenvectors will be inaccurate.

ANSWERS TO SELECTED EXERCISES

Chapter 2

2. Let $\mathbf{C} = \mathbf{AB}$. Then $c_{ij} = a_i{}'b_j$.
   Let $\mathbf{D} = \mathbf{B}'\mathbf{A}'$. Then $d_{ji} = b_j{}'a_i$.
   But $c_{ij} = \mathbf{a}_i{}'\mathbf{b}_j = \mathbf{b}_j{}'\mathbf{a}_i = d_{ji}$; therefore $\mathbf{C}' = \mathbf{D}$.

5.
$$\begin{pmatrix} 4 & 3 & 4 \\ 6 & 6 & 5 \\ 8 & 9 & 6 \\ 16 & 3 & 9 \end{pmatrix}$$

6. **DA**, with **D** diagonal, and diagonal elements equal to d_{ii}.

7. **AD**, with diagonal matrix **D**, for which $d_{22} = d_1$, and $d_{ii} = 1$ $(i \neq 2)$.

8. **AK**, with matrix **K**, for which $k_{12} = 1$, $k_{21} = 1$, $k_{ii} = 1$ $(i \neq 1, 2)$, and all other elements equal to zero.

10. **E** is a diagonal matrix with elements $e_{ii} = 1/d_{ii}$.

12. Take an 8×8 matrix **K** for which k_{15}, k_{21}, k_{36}, k_{48}, k_{53}, k_{64}, k_{72}, and k_{87} are equal to one, all other elements equal to zero. Let the given matrix be **A**, then compute $\mathbf{K}'\mathbf{AK} = \mathbf{B}$. The result is:

			k	*k*	*k*	*k*	*k*	2
			k	*k*	*k*	*k*	*k*	2*p*
					k	*k*	*k*	12
k	*k*				*n*	*n*	*n*	1*p*
k	*k*				*n*	*n*	*n*	1
k	*k*	*k*	*n*	*n*			*w*	3
k	*k*	*k*	*n*	*n*			*w*	3*p*
k	*k*	*k*	*n*	*n*	*w*	*w*		4

The result can be interpreted as: prefix k is used for a combination of 2 with either 1 or 3 or 4; prefix n is used for a combination of 1 and not 2 with either 3 or 4; prefix w is used for a combination of not 1 and not 2 with either 3 or 4. This applies only to filled cells of **B**.

13. (a)

$$\mathbf{C} = \begin{pmatrix} 0 & 0 & 1 & 0 & 1 \\ 1 & 0 & 0 & 0 & 0 \\ 1 & 1 & 0 & 1 & 0 \\ 1 & 1 & 0 & 0 & 1 \\ 1 & 1 & 1 & 0 & 0 \end{pmatrix}.$$

(b) $\mathbf{u}'\mathbf{C} = (4 \quad 3 \quad 2 \quad 1 \quad 2)$ tells how often a person is chosen.
(c) $\mathbf{Cu}$ tells how often a person chooses somebody else.
(d) Element n_{kh} tells how beliefs of k correspond with the way choices of h are actually made. If $n_{kh} > n_{kk}$, this suggests that k perceives other's attitudes toward h are if they were directed toward k himself.
(e) Element n_{hh} tells how good the agreement is between how h believes he is chosen and how he is actually chosen.
(f) The diagonal element of this product matrix tells how good the agreement is between actual choices made by h, and how h is believed to choose.
(g) An element (i, j) of $\mathbf{C}^2$ tells how often i selects somebody who in his turn selects j. Diagonal elements give the number of reciprocal choices.
(h) An element (i, j) of $\mathbf{C}^n$ tells how often j is indirectly chosen by i via $n - 1$ intermediary persons.
(i) An element (i, j) of $\mathbf{C}'\mathbf{C}$ tells whether the pattern of choices given to i is similar to that given to j. Diagonal elements are the measure for popularity. Diagonal elements of $\mathbf{CC}'$ give the measure for generosity; other elements express similarity between patterns of active choices.

14. (b)

$$\mathbf{X} = \begin{pmatrix} 0 & 0 & 1 & 1 & 1 \\ 1 & 0 & 1 & 1 & 1 \\ 0 & 0 & 0 & 0 & 0 \\ 0 & 0 & 1 & 0 & 0 \\ 0 & 0 & 1 & 1 & 0 \end{pmatrix}.$$

(c) $\mathbf{u}'\mathbf{X} = (1 \quad 0 \quad 4 \quad 3 \quad 2)$; this corresponds to the order of preference.
(d) The row and column order must be 2 1 5 4 3.
(e) $\mathbf{u}'\mathbf{X}$ no longer contains numbers from 0 to $m - 1$. The number of circular triads can be computed from this vector. Let $\mathbf{u}'\mathbf{X} = \mathbf{a}'$; then the number of circular triads is equal to $\frac{1}{12}m(m - 1)(2m - 1) - \frac{1}{2}\mathbf{a}'\mathbf{a}$.

15. (a) $\bar{\mathbf{x}}' = \mathbf{u}'\mathbf{X}/n$.
(b) $\mathbf{A} = \mathbf{X} - \mathbf{u}\bar{\mathbf{x}}'$.

(c) $\mathbf{u}'\mathbf{A} = \mathbf{u}'\mathbf{X} - \mathbf{u}'\mathbf{u}\bar{\mathbf{x}}' = n\bar{\mathbf{x}}' - n\bar{\mathbf{x}}' = \mathbf{0}$.

(d)

$$\mathbf{A} = \begin{pmatrix} 0 & 2 \\ -1 & -6 \\ -3 & -4 \\ 1 & 0 \\ 3 & 8 \end{pmatrix}.$$

(e)

$$\mathbf{A}'\mathbf{A}/n = \begin{pmatrix} 20 & 42 \\ 42 & 120 \end{pmatrix} \bigg/ 5 = \begin{pmatrix} 4 & 8.4 \\ 8.4 & 24 \end{pmatrix}.$$

$\mathbf{A}'\mathbf{A}/n$ is the variance-covariance matrix. It could also have been calculated directly from $\mathbf{T}$, by $\mathbf{A}'\mathbf{A} = \mathbf{T}'\mathbf{T} - n\bar{\mathbf{x}}\bar{\mathbf{x}}' =$

$$\begin{pmatrix} 520 & 1042 \\ 1042 & 2120 \end{pmatrix} - 5\begin{pmatrix} 10 \\ 20 \end{pmatrix}(10 \quad 20) = \begin{pmatrix} 520 & 1042 \\ 1042 & 2120 \end{pmatrix} - \begin{pmatrix} 500 & 1000 \\ 1000 & 2000 \end{pmatrix}.$$

(f) $\mathbf{X} = \mathbf{A}\mathbf{S}^{-1}$, with $\mathbf{S}^{-1}$ diagonal, and diagonal elements are $1/\sqrt{4}$, $1/\sqrt{24}$.

(g) $\mathbf{R} = \mathbf{S}^{-1}(\mathbf{A}'\mathbf{A}/n)\mathbf{S}^{-1} = \begin{pmatrix} 1.00 & .86 \\ .86 & 1.00 \end{pmatrix}$. $\mathbf{R}$ is the correlation matrix.

Chapter 3

2. $\mathbf{K}'$ gives direction cosines of the original coordinates with respect to the orthogonal set of vectors specified by $\mathbf{K}$. It follows that $\mathbf{K}\mathbf{K}' = \mathbf{I}$.

3. Let p have coordinates $\mathbf{x}$. The distance to the origin then is equal to the square root of $\mathbf{x}'\mathbf{x}$. Projections on the new vectors are $\mathbf{y}' = \mathbf{x}'\mathbf{K}$. The squared distance to the origin now becomes $\mathbf{y}'\mathbf{y} = \mathbf{x}'\mathbf{K}\mathbf{K}'\mathbf{x} = \mathbf{x}'\mathbf{I}\mathbf{x} = \mathbf{x}'\mathbf{x}$.

   The angle between the two vectors has cosine equal to $\mathbf{x}_1'\mathbf{x}_2$ divided by the product of the distances to the origin. The latter are invariant, and for the new set of directions, we have $\mathbf{y}_1'\mathbf{y}_2 = \mathbf{x}_1'\mathbf{K}\mathbf{K}'\mathbf{x}_2 = \mathbf{x}_1'\mathbf{x}_2$.

4. If $\mathbf{K}'\mathbf{K} \neq \mathbf{I}$, and therefore $\mathbf{K}\mathbf{K}' \neq \mathbf{I}$, the derivations in exercise 3 are no longer valid.

Chapter 4

1. $|\mathbf{A}| = 1$; $|\mathbf{B}| = 12{,}345$; $|\mathbf{C}| = 1{,}000$.

2. Subtract the second column from the third, and the first column from the second; the second and third column then have equal elements, and the value of the determinant is zero.

4. $-1{,}071.$

5.
$$\mathbf{A}^{-1} = \tfrac{1}{2}\begin{pmatrix} 6 & -5 & 1 \\ -6 & 8 & -2 \\ 2 & -3 & 1 \end{pmatrix}.$$

6.
$$\mathbf{B}^{-1} = \tfrac{1}{2}\begin{pmatrix} 12 & -16 & 6 \\ -7 & 12 & -5 \\ 1 & -2 & 1 \end{pmatrix}.$$

7. If **A** is symmetric, then the matrix of cofactors is symmetric.

8. If **A** is triangular with zero elements below the diagonal, then the matrix of cofactors is also triangular with zero elements above the diagonal; therefore the inverse matrix is triangular with zero elements below the diagonal.

9. $\mathbf{T} = p\mathbf{C} + p^2\mathbf{C}^2 + p^3\mathbf{C}^3 + \cdots = p\mathbf{C}(\mathbf{I} + p\mathbf{C} + p^2\mathbf{C}^2 + \cdots)$
$= p\mathbf{C}(\mathbf{I} + \mathbf{T})$. Rearranged: $\mathbf{T} - p\mathbf{CT} = p\mathbf{C}$, or $(\mathbf{I} - p\mathbf{C})\mathbf{T} = p\mathbf{C}$; therefore $\mathbf{T} = p(\mathbf{I} - p\mathbf{C})^{-1}\mathbf{C}$.

Chapter 5

1.
$$\mathbf{x} = \tfrac{1}{2}\begin{pmatrix} 6 & -5 & 1 \\ -6 & 8 & -2 \\ 2 & -3 & 1 \end{pmatrix}\begin{pmatrix} 6 \\ 14 \\ 36 \end{pmatrix} = \begin{pmatrix} 1 \\ 2 \\ 3 \end{pmatrix}.$$

2.
$$\begin{pmatrix} x \\ y \\ z \end{pmatrix} = \tfrac{1}{6}\begin{pmatrix} 1 & 0 & 1 \\ 8 & 6 & -4 \\ -13 & -6 & 5 \end{pmatrix}\begin{pmatrix} 0 \\ 2 \\ 6 \end{pmatrix} = \begin{pmatrix} 1 \\ -2 \\ 3 \end{pmatrix}.$$

3. Let $z = 1$, and solve for the other unknowns. Result: $(1 \quad -1 \quad -1 \quad 0)$. It follows that no solution is obtained if at the beginning w were set equal to -1.

4. The rank of the matrix of coefficients equals 3. The solution is $(4 \quad 3 \quad 2 \quad 1)$.

5. The first column of **K** has elements $1/\sqrt{3}$. The other columns are arbitrary as long as they satisfy $\mathbf{K}'\mathbf{K} = \mathbf{I}$. One solution is to take the elements of the second column equal to $1/\sqrt{2}$, $-1/\sqrt{2}$, and 0, respectively. The third column then is fixed and has elements $1/\sqrt{6}$, $1/\sqrt{6}$, and $-2/\sqrt{6}$.

Chapter 6

1. $x = 1/3$; minimum.

2. Minimum for $x = 2$, maximum for $x = -3$.

3. -14, -18.

4. $x = 1/\sqrt{2}$, $y = 1/\sqrt{2}$, $z = \sqrt{2}$.

5. $x = 4/34$, $y = 14/38$, $z = 86/144$.

6. $F(c) = (\sum x_i^2 - 2c \sum x_i + nc^2)/n$. Set $\partial F/\partial c$ equal to zero. This gives $-2 \sum x_i + 2nc = 0$, and therefore $c = \sum x_i/n$. It follows that the second moment has a minimum if c is equal to the mean of the variable (so that the moment equals the variance).

7. (a) The formal problem is to obtain a minimum of

$$F(a, b) = \sum y_i^2 + na^2 + b^2 \sum x_i^2 - 2a \sum y_i - 2b \sum x_i y_i + 2ab \sum x_i.$$

Take partial derivatives $\partial F/\partial b$ and $\partial F/\partial a$, and set them equal to zero. The result, collected in matrix notation, is

$$\begin{pmatrix} \sum x_i^2 & \sum x_i \\ \sum x_i & n \end{pmatrix} \begin{pmatrix} b \\ a \end{pmatrix} = \begin{pmatrix} \sum x_i y_i \\ \sum y_i \end{pmatrix}.$$

This gives two nonhomogeneous equations in two unknowns, which can be solved.

(b) The solution in terms of t and u becomes:

$$\begin{pmatrix} \sum t_i^2 & 0 \\ 0 & n \end{pmatrix} \begin{pmatrix} b \\ a \end{pmatrix} = \begin{pmatrix} \sum t_i u_i \\ 0 \end{pmatrix},$$

since $\sum t_i = \sum u_i = 0$. The solution is $a = 0$, $b = \sum t_i u_i / \sum t_i^2$. In the example, $b = .8$. The best-fitting line is $u = .8t$, or $y = .8x + 10$.

(c) Take partial derivatives $\partial F/\partial c$, $\partial F/\partial b$, and $\partial F/\partial a$, and set them equal to zero. The result is

$$\begin{pmatrix} \sum x_i^4 & \sum x_i^3 & \sum x_i^2 \\ \sum x_i^3 & \sum x_i^2 & \sum x_i \\ \sum x_i^2 & \sum x_i & n \end{pmatrix} \begin{pmatrix} c \\ b \\ a \end{pmatrix} = \begin{pmatrix} \sum x_i^2 y_i \\ \sum x_i y_i \\ \sum y_i \end{pmatrix}.$$

If we use transformed coordinates t and u, the result is

$$\begin{pmatrix} \sum t_i^4 & 0 & \sum t_i^2 \\ 0 & \sum t_i^2 & 0 \\ \sum t_i^2 & 0 & n \end{pmatrix} \begin{pmatrix} c \\ b \\ a \end{pmatrix} = \begin{pmatrix} \sum t_i^2 u_i \\ \sum t_i u_i \\ 0 \end{pmatrix}.$$

In the example, we obtain

$$\begin{pmatrix} 34 & 0 & 10 \\ 0 & 10 & 0 \\ 10 & 0 & 5 \end{pmatrix} \begin{pmatrix} c \\ b \\ a \end{pmatrix} = \begin{pmatrix} -5.6 \\ 8 \\ 0 \end{pmatrix}.$$

The second equation gives immediately $b = .8$. From the first and third equations, we obtain $a = .8$ and $c = -.4$. The best-fitting quadratic polynomial therefore is

$$u = -.4t^2 + .8t + .8$$

or

$$y = -.08x^2 + 8.8x - 186.$$

(d) Formally we can find a solution from the auxiliary function $G = F - 2\mu g$. The undetermined multiplier is written as -2μ in order to simplify resulting equations. F has the same meaning as in part (c), and g is the condition

$$g(a, b, c) = 100c - 10b + a + 10 = 0.$$

Partial derivatives of G are set equal to zero. This produces three equations in a, b, c, and μ. A fourth equation is provided by the condition g itself. The four equations are:

$$\begin{pmatrix} \sum t_i^4 & 0 & \sum t_i^2 & -100 \\ 0 & \sum t_i^2 & 0 & 10 \\ \sum t_i^2 & 0 & n & -1 \\ 1000 & -10 & 1 & 0 \end{pmatrix} \begin{pmatrix} c \\ b \\ a \\ \mu \end{pmatrix} = \begin{pmatrix} \sum t_i^2 u_i \\ \sum t_i u_i \\ 0 \\ -10 \end{pmatrix}.$$

In the example,

$$\begin{pmatrix} 34 & 0 & 10 & -100 \\ 0 & 10 & 0 & 10 \\ 10 & 0 & 5 & -1 \\ 100 & -10 & 1 & 0 \end{pmatrix} \begin{pmatrix} c \\ b \\ a \\ \mu \end{pmatrix} = \begin{pmatrix} -5.6 \\ 8 \\ 0 \\ -10 \end{pmatrix}.$$

To solve, we may find an expression for μ from the third equation, and substitute in the other three. In the remaining equations, we can eliminate b by adding the second and third equations. This gives

$$\begin{pmatrix} -966 & -490 \\ 200 & 51 \end{pmatrix} \begin{pmatrix} c \\ a \end{pmatrix} = \begin{pmatrix} -5.6 \\ -2 \end{pmatrix},$$

with solution $c = -1{,}265.6/48{,}734$, and $a = 3{,}052/48{,}734$. Then b can be calculated: $b = 36{,}383.2/48{,}734$. The polynomial therefore is

$$48{,}734u = -1{,}265.6t^2 + 36{,}383.2t + 3{,}052$$

or

$$9{,}746.8y = -50.624x^2 + 12{,}339.04x.$$

8. The vector $\mathbf{k}'$ is proportional to $(19 \quad 17)$ and must be normalized to unity to give direction cosines. Therefore $\mathbf{k}' = (19 \quad 17)/\sqrt{650}$. The proportionality constant c equals $\mathbf{m}'\mathbf{y}/\mathbf{m}'\mathbf{m} = \sqrt{6.5}$.

Chapter 7

1. Largest root 1.89, with eigenvector $\mathbf{k}' = (1 \quad .8 \quad .5)$.
2. $\mathbf{k}' = (.87 \quad .49)$.
4. Let $\mathbf{X} = \mathbf{P\Lambda Q}'$. Then $\mathbf{X}'\mathbf{X} = \mathbf{Q\Lambda P}'\mathbf{P\Lambda Q}' = \mathbf{Q\Lambda}^2\mathbf{Q}'$, so that $\mathbf{Q}$ can be found as the matrix of eigenvectors of $\mathbf{X}'\mathbf{X}$. Identify $\mathbf{Y}'$ with $\mathbf{Q}'$ normalized to unity. Result:

$$\mathbf{Y}' = \begin{pmatrix} -1/\sqrt{2} & 0 & 1/\sqrt{2} \\ 1/\sqrt{6} & -2/\sqrt{6} & -1/\sqrt{6} \end{pmatrix}.$$

Identify $\mathbf{A}$ with $\mathbf{P\Lambda}$. $\mathbf{P}$ can be found as the matrix of eigenvectors of $\mathbf{XX}'$ Result:

$$\mathbf{A}' = \begin{pmatrix} 3\sqrt{2} & 6\sqrt{2} & 0 & 6\sqrt{2} & 3\sqrt{2} \\ -2\sqrt{6} & -\sqrt{6} & -\sqrt{6} & 0 & 4\sqrt{6} \end{pmatrix}.$$

Two basic curves are sufficient because $\mathbf{X}$ has rank 2 (since $\mathbf{Xu}' = 0$).

REFERENCES

*Aitken, A. C. 1958. *Determinants and Matrices.* London: Oliver & Boyd.

*Anderson, T. W. 1958. *An Introduction to Multivariate Analysis.* New York: Wiley.

Bayes, T. 1763. "An Essay towards Solving a Problem in the Doctrine of Chances." *Philos. Trans. Roy. Soc.*, **53.** [Reprinted in *Biometrika*, **45** (1958), 293–315.]

Beaton, A. E. 1968. *Statistical Methods for Computers for Social Sciences.* Reading, Pa.: Addison-Wesley.

Blalock, H. M. 1961. *Causal Inferences in Non-experimental Research.* Chapel Hill: Univ. No. Carolina Press.

———. 1963. "Making causal inferences for unmeasured variables from correlations among indicators." *Amer. J. Sociol.*, **69,** 53–62.

Blalock, H. M., and A. B. Blalock, eds. 1968. *Methodology in Social Research.* New York: McGraw-Hill.

Blau, P. M., and O. D. Duncan. 1967. *The American Occupational Structure.* New York: Wiley.

Boudon, R. 1965. "A method of linear causal analysis: dependence analysis." *Amer. Sociol. Rev.*, **30,** 365–73.

———. 1967. *L'Analyse mathématique des faits sociaux.* Paris: Plon.

———. 1968. "A New Look at Correlation Analysis; Dependence Analysis." In Blalock and Blalock 1968, pp. 199–235.

Cattell, R. B. 1957. *Personality and Motivation, Structure and Measurement.* New York: World Book.

De Klerk, L. F. W. 1968. *Probabilistic Concept Learning.* Voorschoten, The Netherlands: Uitgeverij VAM.

---

* General reference textbooks are indicated by the asterisks.

Duncan, O. D. 1966. "Path analysis: sociological examples." *Amer. J. Sociol.*, **72,** 1–16.

Duncan, O. D., A. O. Haller, and A. Portes. 1968. "Peer influences on aspiration: a reinterpretation." *Amer. J. Sociol.*, **74,** 119–37.

*Goldberger, A. S. 1964. *Econometric Theory.* New York: Wiley.

Green, D. M., and J. A. Swets. 1966. *Signal Detection Theory and Psychophysics.* New York: Wiley.

*Harman, H. 1960. *Modern Factor Analysis.* Chicago: Univ. Chicago Press.

Heck, D. L. 1960. "Charts of some upper percentage points of the distribution of the largest characteristic root: *Ann. Math. Statistics*, **31,** 625–42.

*Horst, P. 1965. *Factorial Analysis of Data Matrices.* New York: Holt, Rinehart & Winston.

*Johnston, J. 1963. *Econometric Methods.* New York: McGraw-Hill.

Jorgerson, D. W., and Z. Griliches. 1967. "The explanation of productivity change." *Rev. Econ. Stud.*, **34,** 249–83.

Kaiser, H. F. 1958. "The varimax criterion for analytic rotation in factor analysis." *Psychometrika*, **23,** 187–200.

Kaiser, H. F., and J. Caffrey. 1965. "Alpha factor analysis." *Psychometrika*, **30,** 1–14.

*Kendall, M. G. 1957. *A Course in Multivariate Analysis.* London: Griffin.

*Kendall, M. G., and A. Stuart. 1966. *The Advanced Theory of Statistics.* London: Griffin, 3 vols.

Koopmans, T. C. 1936. *Linear Regression Analysis in Economic Time Series.* Haarlem: Bohn N.V.

———. 1953. "The Identification Problem in Economic Model Construction." In W. C. Hood and T. C. Koopmans, eds., *Studies in Econometric Method* (New York: Wiley), pp. 27–48.

Koopmans, T. C., and A. F. Bausch, 1959. "Selected topics in economics involving mathematical reasoning." *SIAM Rev.*, **1,** 79–148.

Koopmans, T. C., and O. Reiersøl. 1950. "The identification of structural characteristics." *Ann. Math. Stat.*, **21,** 165–81.

Koopmans, T. C., H. Rubin, and R. B. Leipnik. 1950. "Measuring the Equation Systems of Dynamic Economics." In J. Marschak, ed., *Statistical Inference in Dynamic Models*, Cowles Monogr. **10** (New York: Wiley), pp. 53–237.

Lawley, D. N., and A. E. Maxwell, 1963. *Factor Analysis as a Statistical Method.* London:Butterworth.

Lazarsfeld, P. F. 1955. "The Interpretation of Statistical Relations as a Research Operation." In P. F. Lazarsfeld and M. Rosenberg, eds., *The Language of Social Research* (New York: Free Press), pp. 115–25.

Mahalanobis, P. C. 1936. "On the generalized distance in statistics." *Proc. Nat. Inst. Sci. Calcutta*, **12,** 49–55.

*Malinvaud, E. 1966. *Statistical Methods of Econometrics.* Chicago: Rand McNally.

*Morrison, D. F. 1967. *Multivariate Statistical Methods.* New York: McGraw-Hill.

* General reference textbooks are indicated by the asterisks.

Munk, B. 1969. "The welfare costs of content protection: the automotive industry in Latin America." *J. Polit. Econ.*, **77,** 85–98.

Pike, K. L., and B. Erickson. 1964. "Conflated field structures in Potawatomi and in Arabic." *Int. J. Amer. Linguistics*, **30,** 201–12.

Pike, K. L., and I. Lowe. 1968. "Pronomial reference in English conversation and discourse." *Proc. Linguistic Institute, Kiel.*

*Rao, C. R. 1952. *Advanced Statistical Methods in Biometric Research.* New York: Wiley.

Schönemann, P. H., R. D. Bock, and L. H. Tucker. 1965. *Some Notes on a Theorem by Eckart and Young.* Research memorandum no. 25. Chapel Hill: the Psychometric Laboratory, Univ. No. Carolina.

Simon, H. A. 1957. *Models of Man.* New York: Wiley.

*Tintner, G. T. 1952. *Econometrics.* New York: Wiley.

Tukey, J. W. 1954. "Causation, Regression, and Path Analysis." In O. Kempthorne *et al.*, eds., *Statistics and Mathematics in Biology* (Ames: Iowa State Coll. Press), pp. 34–45.

Turner, M. E., and C. D. Stevens. 1959. "The regression analysis of causal paths." *Biometrics*, **15,** 236–58.

*Van de Geer, J. P. 1967. *Inleiding in de Multivariate Analyse.* Arnhem: Van Loghum Slaterus.

———. 1968. *Matching k Sets of Configurations.* Univ. Leiden, Dept. Data Theory, report RN-005-68.

Wright, S. 1934. "The method of path coefficients." *Ann. Math. Stat.*, **5,** 161–215.

———. 1954. "The Interpretation of Multivariate Systems." In O. Kempthorne *et al.*, eds., *Statistics and Mathematics in Biology* (Ames: Iowa State Coll. Press), pp. 11–33.

———. 1960a. "Path coefficients and path regressions: alternative or complementary concepts?" *Biometrics*, **16,** 189–202.

———. 1960b. "The treatment of reciprocal interaction, with or without lag, in path analysis. *Biometrics*, **16,** 423–45.

* General reference textbooks are indicated by the asterisks.

INDEX